ULTIMATE GUIDE
FARM MECHANICS

ULTIMATE GUIDE
TO FARM MECHANICS

A Practical How-To Guide for the Farmer

Fred D. Crawshaw, E. W. Lehmann,
Byron D. Halsted, and James H. Stephenson

Skyhorse Publishing

Farm Mechanics by Fred D. Crawshaw and E.W. Lehmann was first published 1922.
Farm Engines and How to Run Them by James H. Stephenson was first published in 1903.
Farm Conveniences and How to Make Them by Byron D. Halsted was first published in 1884.

Skyhorse Publishing books may be purchased in bulk at special discounts for sales promotion, corporate gifts, fund-raising, or educational purposes. Special editions can also be created to specifications. For details, contact the Special Sales Department, Skyhorse Publishing, 307 West 36th Street, 11th Floor, New York, NY 10018 or info@skyhorsepublishing.com.

Skyhorse® and Skyhorse Publishing® are registered trademarks of Skyhorse Publishing, Inc.®, a Delaware corporation.

Visit our website at www.skyhorsepublishing.com.

10 9 8 7 6 5 4 3 2 1

Library of Congress Cataloging-in-Publication Data is available on file.

Cover design by Jane Sheppard

Print ISBN: 978-1-62914-445-0
Ebook ISBN: 978-1-63220-164-5

Printed in the United States of America

CONTENTS

FARM MECHANICS

by Fred D. Crawshaw and E. W. Lehmann

PREFACE

THIS book has been prepared to meet the increasing need for a textbook on the mechanical processes commonly taught in agricultural high schools and colleges, and in industrial schools. Many teachers of vocational agriculture who find it difficult to organize suitable projects for their students will find that the exercises in this text have been worked out to meet their needs. The book should also be widely useful as a reference and instruction book on the farm.

The types of work covered, while primarily representing the common branches of mechanical activity required under rural conditions, are, in most cases, applicable to the requirements of the industry upon which each type has a bearing.

Each part of the book deals exclusively and comprehensively with one particular type of work, as woodwork, cement work, forging, etc.; a fact which should contribute to its usefulness, both as a text and as a reference book. Thru further divisions into chapters and numbered topics, a greater possibility of locating, at any time, the various details and descriptions is offered.

The treatment thruout the book is thoroly practical. Emphasis is placed upon the proper use of tools and materials in their application to projects. The projects are selected from the standpoint of the practical application to the needs of the student. The gradation of projects within each of the parts has been kept in mind. The plan has been to treat each topic in such detail that the teacher who has a variety of mechanical work going on

3

in his classes at one time may be largely relieved of the burden of class instruction, and can devote his energies to the needs of the individual pupil. Working drawings and specifications for many of the projects have been given. Each of these projects is analyzed into its sequential operations with numerous references to the previous projects for specific details. Many supplementary projects are provided.

The authors are indebted to their many friends who have given freely of their material and advice. They wish particularly to acknowledge the use of material furnished by the University of Illinois, the University of Missouri, the Iowa State College, the Portland Cement Association, and of cuts furnished by several trade journals and taken from state bulletins.

FRED D. CRAWSHAW.
E. W. LEHMANN.

CONTENTS

PART I

WOODWORKING

PART II

CEMENT AND CONCRETE

PART III

BLACKSMITHING

PART IV

SHEET-METALWORK

PART V

FARM MACHINERY REPAIR AND ADJUSTMENT

PART VI

Belts and Belting

PART VII

Farm Home Lighting and Sanitary Equipment

PART VIII

Rope and Harness Work on the Farm

PART I

Woodworking

CHAPTER I

Trees and Lumber

1. Logging. The student is familiar with wood in two forms. One is logs and the other is lumber. It is not only desirable as information that you know the common trees, but it is necessary for practical purposes that you know different kinds of wood when you see them in boards.

Timber is first "spotted" by men who go thru the forest to mark with an ax those trees which are to be cut. It is then felled (chopped or sawed down) and trimmed by having all limbs cut off. The body, or trunk, of the tree and the limbs which are large enough to be sawed into boards are cut to board lengths of twelve, fourteen, or sixteen feet, etc., forming logs. These logs are rolled, hauled or skidded into a clearing to be piled up, measured and later transported to a saw-mill.

While in large piles in the clearing, which is an open space in the woods where the logs are said to be "banked," they are scaled. This is measuring and estimating the number of board feet in each log. Each end

9

of the log is measured and marked with the owner's number.

The banking ground is frequently near a river and on a level above that of the water in the river, so that the logs can easily be rolled down into the stream, where they are allowed to drift to some point down stream, to be collected in a bog, or set-back, near a mill, and then to be sorted and later run into the mill and sawed into lumber. In case it is not possible to transport logs in the natural way, as just described, they must be hauled by team or train to the mill.

This description is very brief and is designed merely to give the outstanding facts in the process of felling trees and conveying them cut up to the mill. The reader is referred to Noyes' *Handwork in Wood*, published by The Manual Arts Press, Peoria, Illinois, for an adequate description of this process and for a bibliography on logging.

2. Milling. The logs are conveyed from the mill pond or yard into the mill by means of an endless chain and the "jack ladder" which is an inclined platform running from the mill into the water of the mill yard. The endless chain which runs over this inclined platform is fitted with studs which engage with the logs as they are directed toward the jack ladder by men with long spiked poles. The logs are carried end to end into the mill and there are inspected for stones which may be lodged in the bark. A flipper, controlled by steam, throws each log to the side when the operator of the machine throws a lever. The log now rolls down an inclined plane to a stop made of heavy iron which is located at the edge of the saw table. When the operator of the saw wants a log, he releases the stop. This operation permits one log to roll onto the saw table, where it is dogged, or clamped, to the table.

The saw table moves backward and forward. With each passage of the table, a large circular, or band, saw cuts off a board. When two or three boards have been removed from the side, the log is turned completely over and a similar operation is performed on the opposite side.

By easily-controlled machinery, the log is revolved or moved into different positions to be sawed into boards. It is sent from the saw to the edger and the cross-cut, or butting, saw on "live" rollers which revolve on a horizontal table and transmit the boards at a rate of 200 to 250 feet per minute from one place to another. Finally,

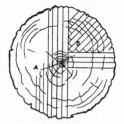

Fig. 1. Methods of sawing lumber. *A*, slash-sawing; *B*, quarter-sawing.

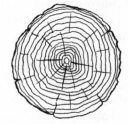

Fig. 2. End of log, showing annual rings and medullary rays.

the boards, now known as lumber, are transferred to a shed, where they are sorted as to size, quality and cut, and then again transferred out of doors to be piled for air-seasoning until sold for construction purposes.

Boards are usually slash-sawed, the term used for parallel sawing (*A*, Fig. 1). However, they are also rift-sawed or quarter-sawed, which means that the saw cut is radial, as shown in *B*, Fig. 1. The advantage of the radially-sawed board is that the edges rather than the sides of the fiber of the wood form the surface of the board and thereby make a more even grain and one which wears better.

3. Tree Growth. When a tree is sawed down, the sawed end will show concentric rings (Fig. 2). Those near the center are more compact than the ones near the outside. The center portion is called heart wood: the outer portion, sap wood, because it conducts the sap which gives vitality to the tree.

Each ring, if observed closely, will be found to be made up of two layers—one denser than the other. These are called annual rings because one pair of rings is formed each year. The dense portion of the ring is the result of winter growth, and the porous part is that formed in the spring and summer when the growth is most rapid.

Upon closer inspection, it will be observed that these rings are crossed by radial lines running from the center to the bark. These are called medullary rays. In a sense, they help to bind the rings together. When cut at a slant, as they may be in radial- or quarter-sawing, these rays, which are very solid, will appear as light spots in the grain of the wood shown on the surface of a board. The beauty of quarter-sawed wood when polished makes certain kinds of it very desirable for interior finish and furniture construction. One of the woods which has this particular feature emphasized is oak. Other grain irregularities, such as wanes and gnarls, make attractive wood surfaces. Curly birch and bird's-eye maple are conspicuous examples.

4. Seasoning. One of the most important parts of the preparation of wood for construction use is its seasoning or drying. A properly-seasoned board is lighter than one not seasoned. It is stronger and is not subject to change of volume which causes checking and warping. Of the several methods of seasoning, the best is natural-air-drying, which takes from two to six years. In this process, boards are piled up with broad surfaces horizontal and separated one from another by thin strips of wood known as sticks. The boards in a particular layer are placed so that edges will not touch; hence, air is permitted to circulate throughout the pile and come in contact with all surfaces. The piles are set up a foot or more from the ground, one end being a few inches higher than the opposite one. They are covered with boards to protect the drying lumber from rain and sun.

In order to produce lumber quickly for construction use, it is artificially seasoned or kiln-dried. This reduces the moisure of the wood to perhaps five per cent, whereas, in the natural process, ten per cent is the approximate minimum. However, kiln-dried lumber will more quickly re-absorb moisture. As most lumber nowadays is seasoned by some artificial means, it is advisable to pile it in shops as for air-seasoning. In case there is a tendency to warp, it is sometimes advisable to clamp a board to a flat surface, concave surface down, or clamp two boards together with the concave surfaces facing each other.

Whenever a board is dressed, it is well to plane both broad surfaces, especially in the case of air-dried lumber, in order to open the pores, as It were, on both sides and thus make the exposure conditions uniform throughout. If the ordinary means of overcoming warping are not sufficient, it is sometimes possible to straighten a board by heating the convex side and, possibly, at the same time moistening the concave side. The heating can be done by laying the board on top of a furnace.

5. Measurements and Calculations. Lumber is measured by the so-called board foot, which is one foot square and one inch thick.

There are two satisfactory methods of calculating the number of board feet in a board or a number of boards:

Rule 1. Multiply thickness in inches by width in inches by length in feet, and divide by 12. Example: $\dfrac{2'' \times 7'' \times 14''}{12} = 16-1/3$ board feet.

Rule 2. Multiply the thickness in inches by width in feet by length in feet. Example: $\dfrac{2 \times 7 \times 14}{1 \times 12 \times 1} = 16-1/3$ board feet.

The possibility of cancellation in the second method makes it shorter and, consequently, preferable.

When purchasing lumber, give the dimensions in the order of thickness, width and length, as: 8 pieces 5″ x 9″ x 12′.

In quantities, lumber should be ordered as follows:

Example 1. 1000′ Norway pine dressed two sides to 7/8″, 9″ and up. This makes the minimum width 9″.

Example 2, 1000′ White Pine S4S 7/8″ x 5″ x 12″. This means all boards are to be surfaced on all four surfaces and the dimensions are to be uniform, viz.; 7/8″ thick by 5″ wide by 12′ long.

6. Trees. Trees are divided into two general classes known as the broadleaf, or hardwoods, and the needle-leaf, or softwoods. In each of these classes, there are many varieties which are of great value in some one or more forms of construction work. Those listed below are only a few of particular significance, either because of their general use, or because of their prevalence in agricultural or industrial communities:

BROADLEAF OR HARDWOODS

NAME	VARIETY	LOCATION	QUALITIES	USES
Oak.	White,	North Central and East U.S.	Durable, easily worked. Does not warp or check easily. Polishes well.	Cabinet work and Interior finishes.
	Red Oak.	North Central and East U.S.	Same,	Same.
	Burr.		Same.	Same.
	Black.	East of long. 96, westward to Mo. and Tex.	Same.	Same. Also outdoor construction.
	Live.	West of Rockies.	Durable, tough.	Implements.

Distinguishing
Tree Features: Heartwood, light brown to red or dark brown.
Sapwood, light brown to yellow.
Height, 75 feet; diameter, 4-1/2 feet. Rough bark.

Ash.	White.	Eastern U.S.	Tough, elastic, straight-grained, brittle.	Cheap interior finish and cabinet work.
	Black.	North and Northeast U.S.	Soft, heavy, tough, not strong.	Same and splints.
	Green	East of Rockies.	Hard heavy, strong, brittle.	Same
	Oregon.	Pacific Coast.	Light, hard, strong.	Furniture, cooperage, carriage frames.

BROADLEAF OR HARDWOODS (*Continued*)

Name	Variety	Location	Qualities	Uses
Ash (*Cont.*)				

Distinguishing

Tree Feature; Heartwood, yellow to brown, or reddish-brown.
Sapwood, light yellow.
Height, 65 feet; diameter, 2-1/2 feet.

Name	Variety	Location	Qualities	Uses
Maple.	Hard.	Northeast and East U.S.	Straights-grained, strong, tough, shrinks.	Furniture, interior finish, implements.
	Silver.	East U. S. Ohio Basin.	Light, brittle, easily worked.	Interior finish, woodenware.
	Red.	East U.S.	Same.	Cabinet work.
	Oregon.	Western Coast.	Light, hard, strong.	Furniture, tool handles.

Distinguishing

Tree Features: Heartwood, light to dark yellow.
Sapwood, white to dark yellow.
Height, 75 feet; diameter, 2 feet.

Name	Variety	Location	Qualities	Uses
Walnut.	Black.	East and Central U. S.	Heavy, hard, strong, firm, easily worked.	Furniture, fixtures, interior finish.
	White (Butternut).	Northeast and Central U.S.	Light, soft, not strong.	Interior finish, cabinet work.

Distinguishing

Tree Features: Heartwood, dark brown to reddish brown.
Sapwood, light brown to dark brown.
Height, 80 feet; diameter, 1 foot and larger.

BROADLEAF OR HARDWOODS (*Continued*)

NAME	VARIETY	LOCATION	QUALITIES	USES
Hickory.	Shagbark.	Eastern U. S.	Very tough, elastic, resilient, heavy.	Carriage and implement work, ax handles.

Distinguishing
Tree Features: Heartwood, light to dark brown.
Sapwood, ivory to cream.
Height, 85 feet; diameter, 2-1/2 feet.

NAME	VARIETY	LOCATION	QUALITIES	USES
Chestnut,		East of Mississippi river except in central portion of this section.	Weak, brittle, durable, easy to work, checks and warps in drying.	Cabinet work and furniture.

Distinguishing
Tree Features: Heartwood, brown.
Sapwood, lighter brown.
Height, 65 feet; diameter, 7-1/2 feet.

NAME	VARIETY	LOCATION	QUALITIES	USES
Beech.		Eastern and Central U.S.	Hard, heavy, strong.	Ship and wagon work, plane stocks.
	Ironwood (Blue Beach).	Same.	Same.	Liners, tool handles.

Distinguishing
Tree Features: Heartwood, light reddish brown.
Sapwood, nearly white.
Height, 55 feet; diameter, 2-1/2 feet. (Dimension of ironwood less.)

BROADLEAF OR HARDWOODS (*Continued*)

NAME	VARIETY	LOCATION	QUALITIES	USES
Birch.	White.	Canada, Atlantic Coast to Delaware.	Soft, light, weak.	Small woodenware, cheap furniture.
	Red.	Massachusetts and Florida.	Light and strong.	Furniture and woodenware.
	Yellow.	Eastern U.S.	Same,	Same.
	Sweet.	Northeastern U.S.	Heavy, hard, strong.	Furniture, ships.

Distinguishing Tree Features: Heartwood, light brown.
Sapwood, white to yellow.
Height, 50 feet; diameter, 2 feet.

NAME	VARIETY	LOCATION	QUALITIES	USES
Whitewood.	Yellow.	Eastern Coast.	Light, soft, difficult to season, durable.	Boxes, cabinet work, interior trim.
				Note: The pine of the hardwoods.
	Poplar.	Scattered. Central U.S.	Same.	Same.
	Basswood.	Eastern U.S. Coasts	Tough, weak, very soft.	Boxes, cheap furniture, carriage bodies.

Distinguishing Tree Features: Heartwood, greenish yellow to brownish yellow.
Sapwood, almost white.
Height, 80 feet; diameter, 5 feet.

NAME	VARIETY	LOCATION	QUALITIES	USES
Mahogany.		Central America, West Indies.	Strong, durable, easily warped, beautiful polish.	Furniture, interior trim.

BROADLEAF OR HARDWOODS (*Continued*)

NAME	VARIETY	LOCATION	QUALITIES	USES
Mahogany (*Cont.*)				
	White.	Mexico and Central America.	Same. More yellow.	Same.
	Spanish.	Mexico, Cuba, West Indies.		Same. Veneers.
	Cedar.			

Distinguishing
Tree Features: Heartwood reddish brown, darkens easily.
Sapwood, light brown to yellow.
Height, 50 feet; diameter, 3 feet.

NEEDLELEAF OR SOFTWOODS

NAME	VARIETY	LOCATION	QUALITIES	USES
Pine.	White.	North Central and Eastern U.S.	Uniform grain, strong, elastic, light, easily worked, weakest of pines.	General carpentry, boxes and crates.
	Georgia ("Hard," "Yellow" or "Longleaf").	South Atlantic and Gulf states	Resinous, strong and heavy. Durable.	Heavy and outside construction flooring.
	Norway (Red).	New England and Lake states.	Light, hard, resinous.	Poles, masts, flooring. *Note:* The oak of the softwoods.

Distinguishing
Tree Features: Heartwood, yellowish to reddish brown.
Sapwood, white to whitish yellow.
Height, 80 feet; diameter, 3 feet.

NEEDLELEAF OR SOFTWOODS *(Continued)*

NAME	VARIETY	LOCATION	QUALITIES	USES
Spruce.	Black.	Eastern U. S.	Soft, light, not durable when exposed.	Structural substitute for white pine.
	White.	Western states.	Close, straight-grain, soft, light.	Lumber, ordinary carpentry.
	Sitka.	Pacific Coast.		Construction, interior finish.

*Distinguishing
Tree Features:*   Heartwood, reddish brown.
Sapwood nearly white.
Height, 75 to 100 feet; diameter, 2-1/2 feet.

Fir.	Great Silver.	Washington, Oregon, Texas and Mexico.	Soft, easily split.	Interior finish boxes.
	Red.	Northwestern U.S.	Light, hard, strong.	House trimmings.

*Distinguishing
Tree Features:*   Heartwood, light red to brownish yellow.
Sapwood, white to yellow.
Height, 200 feet; diameter, 5 feet.

Cedar.	Red.	Atlantic Coast, Southeastern U.S.	Fine-grained, light, soft, weak, durable.	Chests, boxes, pencils.
	White.	Northern states, mountains of North Carolina and Tennessee.	Light, soft, weak, durable.	Poles, fencing, railroad ties.

NEEDLELEAF OR SOFTWOODS *(Continued)*

NAME	VARIETY	LOCATION	QUALITIES	USES
Cedar*(Cont.)*	Incense	Southeastern U.S.	Same	Furniture, interior finish, shipbuildig.

Distinguishing
Tree Features: Heartwood. reddish brown.
Sapwood. nearly white.
Height. 40 feet; diameter. 2-1/2 **feet.**

Name	Variety	Location	Qualities	Uses
Cypress.		Southern Coast.	Soft. very durable.	Cooperage, carpentry.
	Redwood.	Western Coast. California.	Soft, durable, light weight.	Construction, shingles.

Distinguishing
Tree Features: Heartwood, reddish brown.
Sapwood, yellow.
Height, 85 feet; diameter, 3 feet.
Giant, 250 feet; diameter , 25 feet.

The trees above listed are "exogenous," which means that they grow from the inside out. There are a few trees which are "endogenous," or inward-growing. These are the palm, yucca and bamboo, all of which grow in southern countries, principally in the tropical region. They have little value in this country except for novelty furniture and, when shredded into cane, for chair seats, etc.

CHAPTER II
WOODWORKING TOOLS

7. Classification. Practically all woodworking tools are listed below under a classification based on use (Figs. 3, 4, 5, 6 and 7). The particular use of each tool is

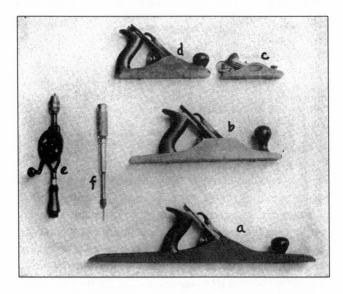

FIG. 3. *a*, jointer plane; *b*, jack plane; *c*, block plane; *d*, smooth plane; *e*, hand drill; *f*, automatic drill.

explained in the instructions given for the several projects. It is believed that one will learn best how to use a tool by actually using it in making something of material value.

Dividing Tools: Planes (jack, smooth, block, jointer, rabbet, moulding, tongue and groove, router), Chisels (firmer, paring, framing, mortise), Saws (rip, crosscut, back, turning, compass, dovetail), Knife,

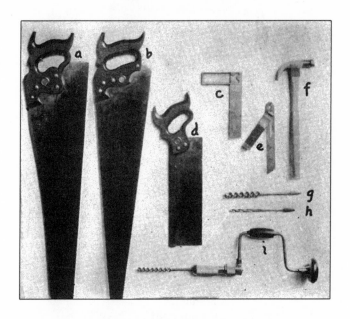

FIG. 4. *a*, rip-saw; *b*, crosscut-saw; *c*, try-square; *d*, jig-saw; *e*, bevel square; *f*, hammer; *g*, auger bit; *h*, drill bit; *i*, brace and bit.

Ax,
Wedge,
Draw-knife,
Spoke-shave.

Boring Tools: Bits (auger, center, Forstner, expansive),
Drills (single- and double-cut),
Gimlet,
Brad-awl,

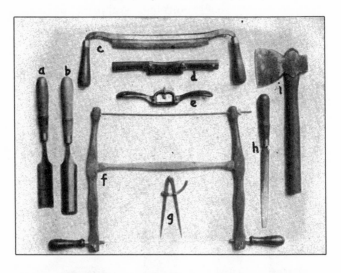

FIG. 5. *a*, gouge (inside ground); *b*, gouge (outside ground); *c*, draw-knife; *d*, spoke-shave; *e*, spoke-shave; *f*, turning-saw; *g*, compass; *h*, wood rasp; *i*, hatchet.

	Reamer,
	Countersink.
Chopping Tools:	Ax,
	Hatchet,
	Adz.
Scraping Tools:	Scraper,
	Rasp,
	Files (single-cut, blunt, flat, bastard, double-cut, taper, half-round).

Pounding Tools: Hammers (claw, upholsterer's,
                                 riveting, veneering),
                               Mallet,
                               Nailset.

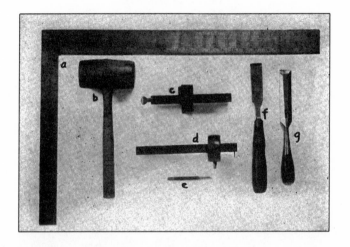

FIG. 6. *a*, carpenter's square; *b*, mallet; *c*, mortise gage; *d*, marking gage; *e*, nailset; *f*, tang chisel; *g*, socket chisel.

Holding Tools: Bench,
                               Vise,
                               Saw-horse,
                               Bench-hook,
                               Handscrew,
                               Carpenter's clamps,
                               Pliers (end-cutting, side-cutting),
                               Pinchers (nippers),
                               Bit-brace.

Measuring and
Marking Tools: Carpenter's square,
                       Rule (two-foot, steel or scale),
                       Try-square,
                       Bevel square;
                       Marking gage,
                       Compass.

FIG. 7. Woodworking bench with the tool rack.

Sharpening Tools: Grindstone,
                    Grinder,
                    Slip stone,
                    Oilstone,
                    Saw-filing machine.
Cleaning Tools: Broom,
                    Brush,
                    Buffer.

CHAPTER III

SAWS AND SAWING

Suggested Projects:

    a) Garden marker (Fig. 8).
    b) Flower trellis (Fig. 9).
    c) Window stick (Fig. 10).
    d) Buggy axle rest (Fig. 11).
    e) Peck crate (Figs. 12, 13, 14).

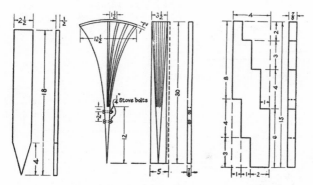

FIG. 8. Garden   FIG. 9. Flower trellis.   FIG. 10. Window stick.
marker.

8. Saws Used. The tools emphasized in this group are the crosscut-saw and rip-saw. Auxiliary tools are the hammer, brace and bit, bevel square, try-square and marking gage.

While there are many saws which constitute a complete equipment, as indicated in the classification of woodworking tools (Sec. 7), there are three only which are used generally—the crosscut-, rip- and back-saws.

27

9. Rip-saws. The formation of the teeth on a rip-saw is shown in Fig. 15. This saw cuts *with* the grain and,

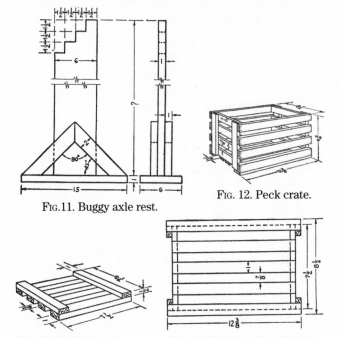

FIG.11. Buggy axle rest.

FIG. 12. Peck crate.

FIG. 13. End of peck crate.   FIG. 14. Bottom of peck crate.

consequently, cuts off the ends of the wood fiber (Fig. 16). The teeth, filed squarely across the saw-blade, form a series of chisels. Alternate teeth are set to one side of the blade, one series being set one way and the alternate series the other way (Fig. 15). The saw-blade is thus made thicker on the tooth edge of the blade than elsewhere, permitting the saw to pass thru the wood without binding while it makes its cut, or "kerf."

The back-saw is a combination of the rip and crosscut in tooth formation and is used for cutting either with or across the grain, particularly where fine sawing is required, as in the making of joints.

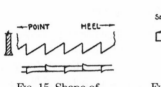

FIG. 15. Shape of ripsaw teeth.

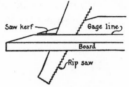

FIG. 16. Position of ripsaw in action.

10. Crosscut-saws. The teeth of a crosscut-saw are filed on both the front and back edges at an angle with the surface of the saw-blade (Fig. 17). This saw cuts *across* the grain, and does its work as it makes its

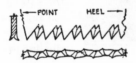

FIG. 17. Shape of crosscutsaw teeth.

forward stroke. The saw is "set" by pushing all teeth outward from the sides which are filed. This results, as in the case of the rip-saw, in forming two series of teeth, those of one series being pushed toward one side of the blade, and those of the other in the opposite direction (Fig. 17).

Working Instructions for Flower Trellis.

Stock: 1 piece, 1″ x 5″ x 32″.
Soft, straight-grained wood. (Drawing, Fig. 9.)

11. Rip-sawing. The chief tool exercise in this project is rip-sawing. It is more difficult to make a series of parallel rip-saw cuts than to make an individual one. In this project, the cuts must be made with great care, that one fan strip may not be weakened more than another. The guide lines must be followed accurately.

There is a possible element of difficulty in sawing each edge of the trellis stock to a taper. The saw must run at an angle with the grain. The piece should be placed in the vise with the end that goes in the ground at the top, and the taper line to be followed by the saw must be in a vertical position (Fig. 18). The saw should run just outside the line in the waste stock.

Fig. 18. Correct position when using rip-saw

12. Squaring and Measuring for Length. Select the best surface *(1)* and the best edge *(2)*, as in Fig. 19. With the try-square blade on one face, called the face side, and its beam on one edge, called the joint edge, square a line across the face side near one end (Fig. 19).

With the beam of the try-square on the face side and the blade on the joint edge, run the try-square with the left hand toward the end of the line squared across the face side until the blade touches the blade of the knife held in the right hand, the point of the knife-blade being on the end of this squared line. With the try-square in this position, square a line across the joint edge (Fig. 20).

Measure the board for length from the squared line on the face side and mark a point with the end of the

FIG. 19. Position of try-square when squaring face side.

FIG. 20. Position of try-square when squaring edge.

knife-blade (Fig. 21). Using the try-square as just described and holding the end of the knife-blade in this point, bring the square up to the knife, square a line

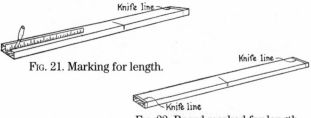

FIG. 21. Marking for length.

FIG. 22. Board marked for length.

across the face side, and then, as on the first end, across the joint edge. The board is now marked for length (Fig. 22).

13. Gaging for Width. Gage two lines on the face side—one 3-1/2″ and the other 4″ from the joint edge.

Set the marking gage so that the width of the board is indicated by the distance from the marker to the stop (Fig. 23). This distance should be measured with a ruler before using the gage (Fig. 24). Inspect the marker before setting the gage to see that it protrudes from the beam of the gage about 1/32″ and that it is filed to a knife edge parallel to the surface of the stop (Fig. 25).

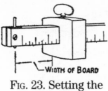

FIG. 23. Setting the marking gage.

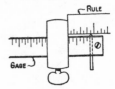

FIG. 24. Testing gage with rule.

Hold the gage on the face side of the wood with the head against the joint edge (Fig. 26), and run the gage from the end of the wood nearest you to the far end,

FIG. 25. Correct shape of point of marking gage.

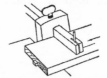

FIG. 26. Position of gage when marking on wide boards.

which, in the case of a long piece, may be rested on the bench (Fig. 27). The relative position of the gage and the wood is shown in Fig. 28.

Do not roll the gage as it is pushed over the surface of the wood, as this will make the marker run too deeply into the wood.

The board is now marked for width (3-1/2″), with another mark to guide the rip-saw in its first cut, and to provide a 1/2″ strip along the edge of the board to be used in fastening the fan strips on the end of the trellis (Fig. 29).

14. Marking Fan Strips. Lay off six points on the fan end of the board, 1/2″ apart. Do this by laying the graduated edge of the ruler across the end of the board on the face side, with the end of the ruler against the joint edge and the graduated edge on the squared knife line, and making a point with a sharp pencil at each 1/2″ graduation mark on the ruler (Fig. 30).

FIG. 28. The correct angle for position of gage.

FIG. 27. Correct method of holding gage and stock.

With a straight edge, connect each one of these points with the center point of the 3-1/2″ strip on the other end of the board. The outside lines only need be drawn the full length.

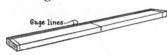

Gage lines

FIG. 29. The board after gage lines have been drawn.

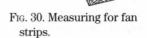

FIG. 30. Measuring for fan strips.

All others should be drawn a distance equal to the depth of the saw cuts for the fan strips (Fig. 31). The bottom of these cuts should be located by a squared pencil line across the face side of the board, as should the position of the center line of each of the bolt holes (Fig. 32).

15. Boring Holes. Place the board edge up in the vise. With a 5/16″ auger-bit in the bit-brace, stand squarely before the board, placed horizontally edgewise in bench vise, with spur of bit on center for one of the holes to be bored for bolts and with bit in a vertical posi-

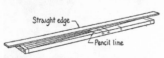

FIG. 31. Laying out rip-saw cuts.

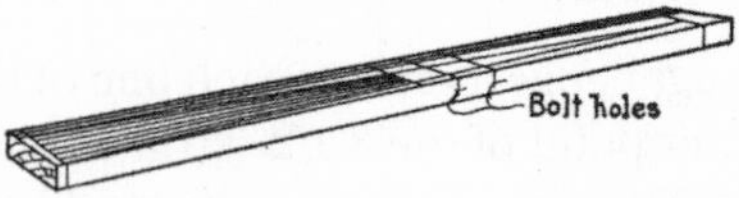

tion (Fig. 33). This position may be tested by the use of the try-square (Fig. 34). With left hand on knob and right hand grasping the handle, turn the handle clockwise until about one-half of the hole is bored. Repeat this operation in boring the second hole. Reverse the board in the vise and bore the second half of each hole. Great care must be taken to make all borings straight to secure holes without shoulders near the center.

FIG. 32. The board marked for bolt holes.

16. Sawing Ends. The saw works at an angle to the surface of the board (Fig. 35). The strokes are taken the length of the saw without exerting more pressure than to guide the saw. The squared line on the face side should be touched by the saw as it goes across the surface (Fig. 35). The squared line on the joint edge should be touched by the saw as it finishes its cut thru the board. In a similar manner saw to the squared lines on the other end of the board.

When sawing, place the board on the top of wooden horse with its end projecting over the end of the horse and with face side up and joint edge toward operator (Fig. 36.) Hold the stock with left knee and left hand, allowing thumb of left hand to guide the saw when beginning the cut. The first stroke should be upward. Very little pressure is used in downward strokes, and none in upward strokes.

17. Ripping Off One-half-Inch Strip. Place the board with long dimension vertical in the vise. Have

FIG. 33. Correct method of using auger bit.

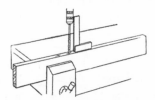

FIG. 34. Testing for squareness when boring.

FIG. 36. Correct position of operator using a crosscut-saw.

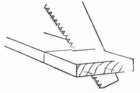

FIG. 35. Position of cross-cut-saw when cutting.

the gage lines 3-1/2″ and 4″ from the joint edge beyond the end of the bench (Fig. 37). Stand squarely in front of the board with right hand grasping the handle of the rip-saw (Fig. 18), allowing the index finger to rest on the side of the handle. Grasp the upper left-hand corner of the board with the thumb and the first two fingers of the left hand, stand in a bracing position, and place the saw on the upper end of the board in a position to

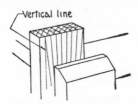

FIG. 37. Stock put in vise for rip-sawing.

draw it toward you. Pull the saw slightly downward without pressure and guide it against the thumb of the left hand. Make the stroke approximately the length of the saw blade. In a similar manner push the saw from you, slightly upward. Continue this backward and forward motion, gradually bringing the saw to a horizontal position, or nearly at right angles with the surface of the board. The saw should always be cutting so that the angle formed between the cutting edge and the board on the operator's side is less than 90 degrees. In this manner

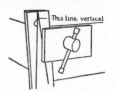

FIG. 38. Rip-sawing at an angle over the grain.

saw on the outside of the vertical gage lines on the left (Fig. 37) in sawing to the 3-1/2″ and 4″ gage lines.

18. Ripping Tapered Edges. Place the board vertically in the vise with fan end downward and marked surface toward the front. One of the lines indicating a tapered edge of the trellis must be vertical (Fig. 38). Saw to this line in waste stock, leaving a sufficient amount of stock to plane finished edge on the board. Reset the board in the vise so that the second line making a tapered edge is vertical. Saw to this line as to the first one.

19. Ripping for Fan Strip. Place the board vertically in the vise and carefully saw *on* each line, marking the dividing line between two fan strips so that one-half of the kerf is taken on each side of the line. The end of each of these cuts must be square with the surface of the board, and must be exactly on the pencil line which limits these cuts.

All sawing on the board is now completed. Plane the two tapered edges and the back of the board.

To secure a definite thickness, the board may be gaged for thickness on finished tapered edges before the back of the board is planed.

FIG. 39. Nailing the trellis.

Insert a stove bolt in each of the holes bored, and fasten in position with a washer under both the head of the bolt and the nut.

Plane the strip which was first sawed from the edge of the board. Saw off 12-1/2" of it, being certain that each end is square. With try-square and sharp pencil, mark a center cross-line on one edge of the strip. This line locates the center position for the end of the middle fan strip. Similarly on this supporting strip locate the center position for each of the other fan strips. With this line at the center of the middle fan strip and with trellis in natural position in the vise, nail the strip to this middle fan strip at the center of its end with two 1" brads, each about 3/16" from the outer surface of the trellis (Fig. 39).

Carefully bend each of the outside fan strips to its proper position, and fasten it with two brads as in the case of the middle fan strip. In like manner, fasten each of the other strips. This work must be done with great care to avoid splitting either the supporting or any one of the fan strips. A wise precaution against such an accident is to bore holes with a brad-awl for each of the nail holes.

Supplementary Instructions.

20. The Buggy Axle Rest and the Measuring Crate require the use of tools not described in instructions for the flower trellis.

Fig. 40. Bevel square used with a protractor.

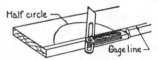

Fig. 41. Bevel square used in geometric construction.

21. The Bevel Square, which is used to lay off the angles of the ends of the braces in the buggy axle rest, is shown in Fig. 4. It has an adjustable blade. It may be set by placing it upon a protractor, as shown in Fig. 40, or

Fig. 42. Setting bevel square to an angle of 45°.

Fig. 43. Laying out with bevel square.

for the more common angles, it may be set on the edge of a board with a geometrical construction made near this edge with compass and straight-edge, as shown in Fig. 41. The angle of 45 degrees is easily secured by placing the edge of the bevel-square blade thru two equal graduations on the sides of a carpenter square, as shown in Fig. 42. A bevel angle should be laid off with a bevel-square, much as a right angle is with a try-square.

Each end of the brace in the buggy axle rest should be completely defined by making bevel-square lines on edges and try-square lines on broad surfaces (Fig. 43).

22. Nailing. The nailing exercise is the principal one in the construction of the measuring crate, aside from the use of the try-square and crosscut-saw, as it is assumed that lath or strips dressed to dimensions will be used as stock.

FIG. 44. Proper use of hammer.

The hammer should be grasped in the right hand near the end of the handle and swung freely from the elbow in a vertical plane with but slight wrist and shoulder movement. The thumb and finger of the left hand should hold the nail (Fig. 44).

FIG. 45. Jig for nailing.

Where a good many operations are repeated, it is often well to use a form, or jig, to secure uniform results and to avoid waste of time in unnecessary preliminaries in making each individual operation.

Fig. 45 shows jig which might be used in locating and driving nails when fastening crate strips on corners. The holes are sufficiently large so that when the jig is placed over the end of a crate strip in position to nail, and the nail is driven thru the jig hole, the jig may be lifted off, the head of the nail being smaller than the hole in the jig.

CHAPTER IV

Planes and Planing

Suggested Projects:

    a) Scouring board for kitchen (Fig. 46),

    b) Bread-cutting board (Fig.47).

    c) Bulletin board to hang on wall (Fig. 48).

    d) Bill board for filing meat and grocery bills (Fig. 49).

    e) Swing board (Fig. 50).

    f) Rope wind (Fig. 51).

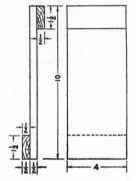

Fig. 46. Scouring board

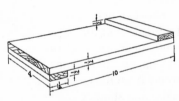

The tool chiefly emphasized in this group of projects is the plane. Other tools needed are the *try-square, ruler, marking gage, crosscut-saw, rip-saw* and, for some of the projects, the *hammer* or *bit* and *bit-brace.*

23. The Plane. There are four principal planes used in a woodworker's kit. They are the jointer, jack, smooth and block. It is not necessary to have all of these in order to do satisfactory work. The jack plane (Fig. 52) shows the plane and its parts.

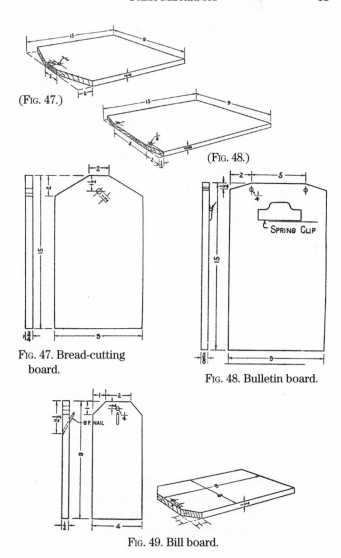

(FIG. 47.)

(FIG. 48.)

FIG. 47. Bread-cutting board.

FIG. 48. Bulletin board.

FIG. 49. Bill board.

24. Care of the Plane. The plane-iron must be kept sharp. Grind it when it is very dull or nicked; otherwise, whet it on an oilstone. Fig. 53 gives the position of the plane-iron on a grindstone as held by the operator. Fig. 54 shows the position of the plane-iron on the oilstone as held by the operator.

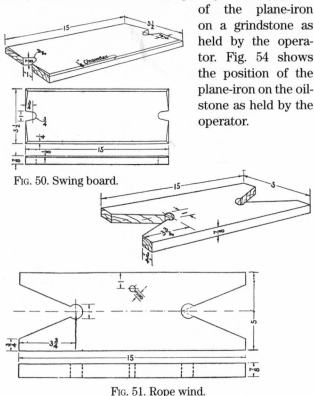

Fig. 50. Swing board.

Fig. 51. Rope wind.

25. Grinding the Plane-Iron. To grind the plane-iron, hold it steady and at such an angle that the proper bevel will be secured. Move it back and forth sideways to account for any unevenness in the stone, but do not raise or lower it.

26. Whetting the Plane-Iron. To whet the plane-iron, hold it so that the bevel formed by the grindstone will be in contact with the oilstone. Use a circular

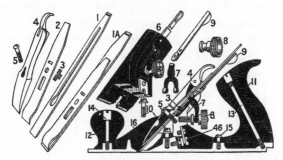

FIG. 52. Parts of jack plane:

1A Double plane-iron.
1 Single plane-iron.
2 Plane-iron cap.
3 Cap screw.
4 Lever cap.
5 Lever cap screw
6 Frog complete.
7 "Y" adjusting lever.
8 Adjusting nut.
9 Lateral adj. lever.
10 Frog screw.
11 Plane handle.
12 Plane knob.
13 Handle bolt & nut.
14 Knob bolt & nut.
15 Plane handle screw.
16 Plane bottom.
46 Frog adj. screw.

FIG. 53. Position of plane-iron on the grindstone.

FIG. 54. Position of plane-iron on the oilstone.

FIG. 55. Difference in angles for grinding and for whetting.

FIG. 56. Whetting the face side of plane-iron.

motion in whetting (Fig. 54). Finally, raise the hands, slightly continuing this motion. This will tend to create a whetted bevel made slightly at an angle with the ground bevel (Fig. 55). The plane-iron should be held in this position for a few moments only, when it may be reversed, laid flat on the top of the stone and given a few circular strokes (Fig. 56).

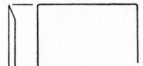

FIG. 57. Shape of cutting edge for general use.

FIG. 58. Shape of cutting edge for jack plane.

The irons for all planes except the jack should be ground at right angles with the edge, with the corners rounded (Fig. 57).

The plane-iron for the jack plane, if used principally as a roughing plane, should be ground rounding on the

FIG.59. Correct way of holding the plane.

edge as in Fig. 58. When used as the only plane in a kit, it should be ground very slightly rounding, if at all. The angle for grinding, except when a plane is used exclusively for very hard wood, should be approximately 20 degrees. The whetted bevel should make approximately 5 degrees with the ground bevel (Fig. 55).

27. Care of Plane. In order to protect the edge of the plane-iron, lay the plane on its side when not in use. Fig. 7 shows the plane and other tools in position on a carpenter's bench, which is a very satisfactory kind to use on the farm.

28. Use of the Plane. To use a plane, the operator stands in front of the bench in a bracing position, the left foot in front of the right and the body turned slightly toward the bench.

Note that the handle of the plane is grasped with the right hand, with the fingers and thumb wrapped about the handle. The palm of the left hand is placed on the knob of the plane (Fig. 59).

The plane is placed upon the board so that its bottom is in contact with the surface to be planed. The left hand presses the plane downward, and the right hand pushes it forward. When

Fig. 60. Planing a wide board.

the plane bottom is fully in contact with the wood, both hands exert an equal pressure. As the plane projects over the end of the board in completing its stroke, the right hand exerts the pressure and the left hand merely serves to hold and guide the plane.

It is customary in planing a surface to begin the planing at the edge nearest the operator, and to finish at the opposite edge. However, if the board is warped or twisted, shavings must first be taken from the high surfaces to establish a flat and true surface. Then the finishing shavings should be taken as suggested.

Working Instructions for Swing Board:

Stock: 1 piece, 1″ x 6″ x 16″.

29. Face Side. Place the board flat upon the top of the bench with one end against the bench stop (Fig. 60).

With the plane set to take a light shaving, proceed to surface the stock, as explained in Sec. 28. The planed surface should be tested with the blade of a try-square or other straight-edge, placed in several positions. When testing, place the blade of the try-square across the surface at different points. The amount of light shown between the board and the try-square blade will indicate the low places in the surface. Continue planing either by taking regular shavings across the board or by planing high places only until the straight-edge test shows approximately the same amount of light for all positions of the try-square. Mark this surface 1.

30. Joint Edge. Place the board in the vise with one edge up. Plane this edge until it tests straight length-

Fig. 61. Testing edge with try-square.

wise by the straight-edge test, and straight and at right angles with the face surface by using the try-square, as shown in Fig. 61. The try-square should be placed on the edge at several points, always having the beam of the square against the face side. When edge tests are satisfactory, mark the planed edge 2.

31. Second Edge. Set the marking gage and gage for width of the board, using the method described in Sec. 13.

Plane the second edge of the board as you did the joint edge. Test frequently with the try-square, and keep the amount of wood to be planed off the same in thickness the entire length of the board. Remember, when the gage line is reached, planing must stop and the edge must be straight and square with the face side.

32. Second Surface. From the face side, gage the thickness of the board on both planed edges. Plane the second surface as you planed the first, testing frequently for straight-ness in width and length so that the surface will be true when the gage lines are reached.

33. First End. Place the board vertically in the vise. First from one edge and then the other, never allowing the plane to take a shaving completely across the end, plane the upper end of the board square with the face side and the joint edge (Fig. 62).

34. Second End. Measure the board for length, and square across the face side and joint edge with knife and try-square. (See Sec. 12 for instructions on sawing and squaring.) Plane the second end according to the directions for planing the first end.

FIG. 62. Planing the end.

Note: The face side, joint edge and first end are surfaces from which all measurements must be taken in securing the dimensions of a board or in making surface measurements.

FIG. 63. Layout for boring.

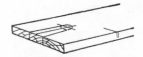

FIG. 64. Lines for rip-sawing.

35. Boring Holes for Rope. With marking gage, make a short, light center line on face side of board from each end thru a point 3″ from each end (Fig. 63). By the use of the try-square and knife, cross each of these center lines with a short knife line at right angles to the joint edge (Fig. 64).

With 3/4″ auger-bit and brace, bore the two holes for the rope, as shown in Fig. 65. A piece of board must be placed on the back of the swing board, opposite the auger-bit, to prevent splintering the fibers of the wood in the swing board, or, the stock must be

FIG. 65. Boring on broad side of stock

FIG. 66. Position of board for rip-sawing.

reversed in the vise as soon as the spur protrudes on the back side of the swing board so that the hole may be finished from the opposite side.

36. Sawing End Notches. On each end, measure 1″ in each direction of the center lines, square across the ends at these points and on the face side join the end of each of these lines with the corresponding side of the hole, to form tangents (Fig. 66).

Place the board in the vise so that one of these lines is in a vertical position (Fig. 66).

FIG. 67. Sandpapering.

As previously instructed, saw to this line in waste stock with a rip-saw.

Sandpaper used over a block and run lengthwise of the grain may be used to smooth surfaces of the swing board and round edges slightly (Fig. 67).

Supplementary Instructions:

The "Working Instructions" for the swing board includes practically all those necessary for any one of the suggested projects in this group. However, in the bulletin board and bill board, the following suggestions should be made:

In cutting off the corners on each of these projects, you

FIG. 68. Layout for corner cuts.

should work from the center line shown in Fig. 68. By measuring on each side of this line, one will be sure to make the end symmetrical. The lines drawn to show where the corners are to be cut should be drawn on the face side and

FIG. 69. Use of hand drill.

from each end of these a line should be squared across the joint edge or end of board (Fig. 68).

To saw each of these lines, put board in vise so that line is in vertical position.

To insert bill-board stake at any particular point on the front of the board, drill or bore a hole slightly smaller than a ten-penny finishing nail thru the board from the front side, as suggested (Fig. 69). If a hand drill is not available, use bit and brace (Fig. 4).

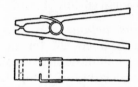

FIG. 70. Clothe spin which may
be used with bulletin board.

From the back side of the board, drive thru the hole a ten-penny finishing nail and set the head under the surface of the board by the use of a nailset or second nail.

The bulletin board may be equipped either with a spring clip, as shown in Fig. 48, or with clothes pin (Fig. 70).

CHAPTER V

ESTIMATING MATERIALS; CONSTRUCTING AN ASSEMBLY PROBLEM

Suggested Projects:

 a) Wash bench (Fig. 71).
 b) Chicken coop (Fig. 72).
 c) Feed bin or wood-box (Fig. 73).
 d) Shipping crate (Fig. 74).
 e) Flower box (Fig. 75).

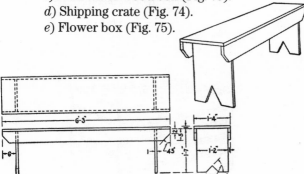

FIG. 71. Wash bench.

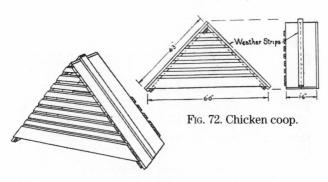

FIG. 72. Chicken coop.

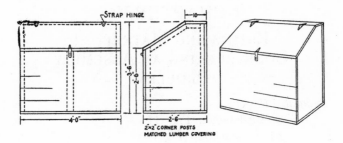

FIG 73. Feed bin or wood box.

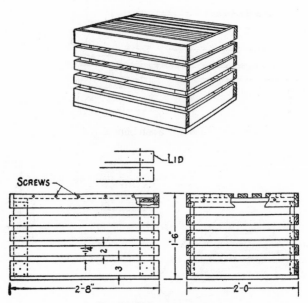

FIG. 74. Shipping crate.

This group of projects does not require the use of tools not already described. It represents, however, a type of project slightly different from any of those included in former groups. The projects in this group are larger and generally include more distinct parts requiring the use of more and larger stock. In a sense, they represent a type of work which is neither carpentry on the one hand nor bench woodwork on the other; they combine the elements of both.

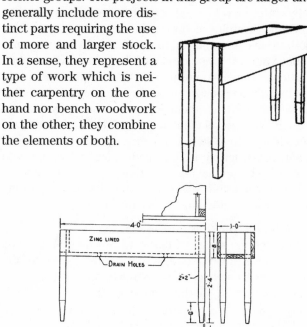

Fig. 75. Flower box.

37. Calculations of Stock. In Sec. 5, rules are given for finding the number of board feet in one or more boards. It is essential that we know how to apply this rule, both to estimate the cost of a project and actually to determine the amount of material that has gone into it. It is equally important to form a judgment of what stock to select before a project is chosen. For example, small pieces of wood may sometimes be used up for the smaller parts of a project, while boards from which pieces for the project may be cut can be selected carefully with a view to wasting as little material as possible.

Think carefully of the means of getting out stock, both to save material and to save time. Be as systematic about your work as possible. When a tool is set for a particular dimension or use, do all that you can with it, not only on one board, but on all which are to have similar work done upon them. Plan ahead so that you know exactly what you should do next, and how you will proceed from step to step. Think thru a problem before you begin construction. If you need to make changes, you can do so better, having once thought out one solution. Whenever possible, make a complete working drawing of the project with dimensions and notes.

Working Instructions for Chicken Coop:

Instructions are given below for making the chicken coop. Use strips of wood, if they can be found, for the slats in the front, and select boards for the roof and back as nearly as possible the desired width. For such a project as this, use old material if available rather than new. Old fence boards are satisfactory for the back.

38. Roof Boards. Secure lumber free from knots which will cut economically to make the roof boards. Example: Two boards, each 9″ wide, or one 12″ and the other 6″ wide, the latter a fence board, possibly. Test the ends of boards for squareness. Use a carpenter's square for this, and in case an end needs to be sawed square, follow the usual method of squaring and sawing, substituting the carpenter's square for the try-square. If two boards are used find the center in length of each one.

Square across thru the center points and saw on lines.
Place one 12″ and one 6″ board, or the two 9″ boards
edge to edge with ends flush. Nail a 2″ or 3″ strip 1/2″

 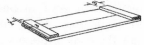

Fig. 76. First cross cleat in place.   Fig. 77. Second cross cleat
                                              in place.

from one end of the pair of boards (Fig. 76), and another
3″ strip flush with the opposite end (Fig. 77).

Place the remaining two boards together in a similar
manner and fasten at one end only with a strip placed
1/2″ from the end.

Nail the unfastened end of one pair to the end of the
other pair which has the strip attached flush with the

 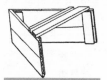

Fig. 78. Roof sections nailed   Fig. 79. Attaching side
       together.   slats.

end (Fig. 78). This joint forms the ridge of the roof of the
coop. The cleat should be on the under side, and nails
should enter it as well as the ends of the boards to which
it is fastened when the two pairs of boards are nailed
together at the ridge.

39. Fastening Front Strips. Place the roof edgewise
on the ground and fasten the lower ends together with a
4″ or 6″ strip of siding to form the lower front board of

the coop (Fig. 79). The lower edge of this board should be high enough to permit the coop to set off the ground at least 1-1/2″. A pan of water can then be placed under it and be held by it when the coop is in use. Before fastening the lower front board in place, set a bevel-square to 45 degrees. Mark and saw the ends of the board to come flush with the outside surfaces of the roof boards. A miter box may be used to saw the ends of this board and other

FIG. 80. Miter box.

boards to be fastened on the front and back (Fig. 80). Stock for remaining cleats may be sawed by using the method of laying out and sawing, as shown in Fig. 81.

In a similar manner, mark, saw and nail the remaining front strips which may be laths or narrow strips of wood. A space of from 1″ to 1-1/2″ should be left

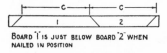

BOARD 1 IS JUST BELOW BOARD 2 WHEN NAILED IN POSITION

FIG. 81. Method of laying out strips.

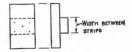

FIG. 82. Gage for spacing slats.

between adjacent strips, all of which should be parallel. The space can easily be determined by the use of a gage made as shown in Fig. 82.

40. Fastening Back Boards. Turn the coop over, cut and fasten the back boards, beginning at the bottom. Alternate boards should be reversed in order to save lumber by taking advantage of the end cut made at 45 degrees (Fig. 81).

41. Trimming. Place the coop on the floor in its natural position. If it does not set squarely on all bottom edges, plane those which are too low until all surfaces rest on the floor. In case the ends of the front or back boards project over the surface of the roof boards, they should be planed flush with these boards.

42. Checking Estimate. Measure carefully all stock used, and determine the number of board feet of lumber in the project. Compare this amount with the original estimate. If this and the planning at the beginning of a project are both done whenever a project is constructed, you will gain in efficiency in making close estimates and in planning to save both material and time.

CHAPTER VI

CHISELING; MAKING COMMON FRAMING JOINTS

Suggested Projects:

  a) Milk stool (Fig. 83).
  b) Combination milk stool and pail rest (Fig. 84).
  c) Harness rack (Fig. 85).
  d) Harness clamp (Fig. 86).
  e) Seed tester (Fig. 87).
  f) Saw horse (Fig. 88).

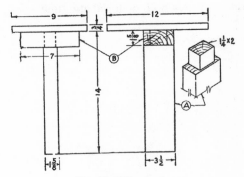

FIG. 83. Milk stool.

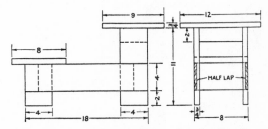

FIG. 84. Combination milk stool and pail rest.

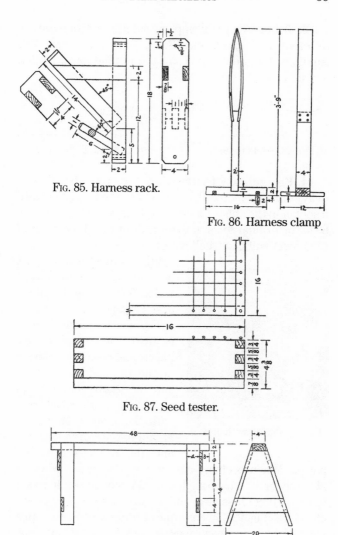

FIG. 85. Harness rack.

FIG. 86. Harness clamp.

FIG. 87. Seed tester.

FIG. 88. Saw horse.

43. Tools. The tools emphasized in this group are the different kinds of chisels. (See Classification, page 29.) Auxiliary tools described are the double gage, mallet, and nailset.

Fig. 89. Socket chisel.   Fig. 90. Tang chisel.

44. Preliminary Instruction. For carpentry work, a heavy chisel is required, one in which the handle fits into a socket in the chisel blade, called the socket chisel (Fig. 89). For ordinary use, however, even tho the handle of the chisel will be struck with a mallet occasionally, but

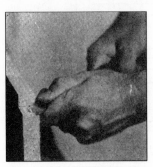

Fig. 91. Paring with chisel.

lightly, a chisel with a spike on the end of it (a "tang") which fits into the handle is used (Fig. 90).

The work a chisel does is divided into two classes, depending upon the relation of the direction of cutting and the grain of the wood cut. When the chisel cuts with the grain (Fig. 91), it is said to pare off a shaving, and the process is called paring. When a chisel cuts across the grain, whether abruptly or at a sharp angle with the wood fiber, it is said to be cutting crosswise, and the process is called cross-chiseling (Fig. 92). In case one cuts across the grain in a vertical position, the process is called vertical chiseling. It is in cross or vertical chiseling that the mallet is much used to force the chisel across the grain. Such work is illustrated by the cutting of joints such as the tenon and mortise joint, in which the mortise is chiseled out.

A chisel is ground and whetted in the manner described for sharpening a plane-iron (Sec. 26).

Working Instructions for Seed Tester:

  Stock: Four 1″ x 3-1/2″ x 6″ hard pine S2S.

    Four 1″ x 6-1/4″ x 6″ matched flooring S2S.

    Six-penny (6d) nails and 1-1/2″ brads.

45. Purpose of Seed Tester. The following instructions are for the seed tester, or germinating box, which is used to test the fer-tility of seeds. As the soil must be moist for this purpose, the box must be made strong to withstand the effect of the moisture, which has a tendency to open up joints and change the shape of boards by warping or winding.

It is for this reason that the corners of the box are made with a lock joint, and that the tongues of the joint are

FIG. 92. Chiseling across the grain.

glued and nailed together (Figs. 87 and 93).

The upper surface of the soil is blocked off into squares that each one may be used for an individual seed or a group of seeds. These squares are determined by a cord strung between the opposite sides of the box (Fig. 87).

46. Rough Cutting. If necessary, rip the pieces for the sides of the box from a board, and square and saw each to the required length. Plane each board to width (3-1/2″) and thickness (3/4″).

FIG. 93. Corner joint for seed tester.

FIG. 94. First step in laying out joint.

47. Laying Out. On the face side of each board, which should become the inside surface of a side of the box, square a knife line 1″ from each end, or a distance equal to the thickness of the stock. Continue this as a fine pencil line around the piece.

FIG. 95. Second step.

FIG. 96. Waste wood marked for removal.

From the joint edge, which should become the upper edge of a side of the box, gage consecutively on the end and on both surfaces of each end of each board a line 3/4″ from the joint edge to form the first dovetail line (Fig. 94). Reset the gage to 1-3/8″ (3/4″ + 5/8″), and in a similar manner gage the second dovetail line on each end of each piece (Fig. 95). Continue this process of gaging, adding 3/4″ and 5/8″ alternately until all cuts are indicated. On the end of each board, mark with a pencil the parts of the joint to be removed (Fig. 96).

48. Sawing. Saw with the rip-saw to each line in the stock to be removed. Saw with a crosscut-saw to the shoulders, those corners to be removed (Fig. 97).

49. Chiseling Joints. Lay each board flat on a wooden surface and chisel out remaining parts of joint to be removed. The chisel should be held at an angle, and the first cut should be made near, but not on the shoulder line (Fig. 98). The last cut should be made by hold-

FIG. 97. Sawing the joints.

ing the edge of the chisel on this line, perpendicular with the surface of the board, the depth of the cut being about one-half the thickness of the board (Fig. 99). Reverse the board, again place the edge of the chisel on the line, and gently tap the

FIG. 98. Chiseling dove tails.

chisel with mallet, or push it with right hand thru to meet the opposite cut already formed. This must be done with great care not to under-cut the joint to any appreciable extent.

The sides of the tongues formed by the rip-saw cuts should not be touched with a chisel unless the saw has not cut to the gage lines. In this case, the chisel should be used to pare off this superfluous stock (Fig. 91).

50. Fastening Corners. Drive a 1-1/2″ brad thru each projecting piece of the joint, as shown in Fig. 100.

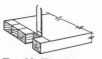

FIG. 99. Finishing the chisel cut.

Before driving the nails in the corners, cover each sliding surface of each joint with cold or hot glue.

51. Nailing Bottom. The bottom boards may be nailed onto the lower edges of the sides of the box with six-penny common nails (Fig. 101). Each bottom board should be squared and sawed to length before it is nailed in place.

A more satisfactory box in appearance and in strength, but one more difficult to make, would have the bottom set inside of the sides of the box and nailed from the outside. Such a bottom would be completely fitted and set in place at one time (Fig. 102).

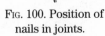

FIG. 100. Position of nails in joints.

52. Marking Edges. With pencil and ruler, divide the length of the inside of each side of the box into equal spaces—1″, 1-1/2″ or 2″—depending upon the distance the strings on the top of the box are to be separated.

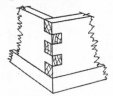

FIG. 101. Method of attaching the bottom.

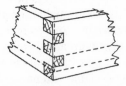

FIG. 102. The bottom fitted inside of side pieces.

Square a light pencil line across the upper edge of each board at the points located, and place at the center of each of these lines a 1-1/2″ brad (Fig. 103). The brad should be driven into the wood to allow the head to project 1/4″ above the surface of the wood. Continue the process until all brads are driven in place.

FIG. 103. Location of nails for stringing.

String the cord continuously back and forth between the opposite sides of the box, the cord running between one pair of sides to be at right angles with that strung between the sides of the other pair.

53. Supplementary Instructions. In the instructions for the seed tester, two operations in which the chisel is a principal tool are not fully described, viz.—(*a*) paring a broad surface and (*b*) cutting a mortise. Examples of

Fig. 104. Making the joints for saw horse.

the former are found in the body of the saw horse, where are-cess for the leg is formed, or in the upright of the harness clamp, where supporting surfaces for

Fig. 105. Method of chiseling joints.

Fig. 106. Layout of joint for milking stool.

the barrel staves are formed; while the latter is used in working out and cutting a mortise and tenon for the joint in the milking stool.

54. Special Operations. After the recess for the saw horse leg is laid out in the body of the saw horse, and the shoulders are cut with a crosscut-saw (Fig. 104), the waste stock must be removed. Fig. 105 shows how the chisel is used in taking paring cuts. The waste stock being removed, the surface is finally tested with a try-square blade used as a straight-edge to determine when the surface is perfectly true.

To lay off the tenon on the top of the upright piece in the milking stool, the try-square and single- or double-marking gage should be used and lines drawn, as indicated in Fig. 106.

Parallel lines can be made at one time with the double gage (Fig. 107). The cross-hatched part of Fig. 106

represents the end of the tenon. Saw to lines a with crosscut-saw in waste stock. Saw to lines b in waste stock with rip-saw.

Fig. 108 shows the joint for the milking stool with the rectangle representing the mortise made up of lines marked b, corresponding to similarly-lettered lines on top of upright. The long lines of this rectangular hole should be made on both top and bottom of board with the marking gage, the short lines with knife and try-square. The

Fig. 107. Gaging joint for milking stool.

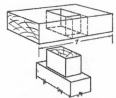

Fig. 108. The joint completed.

extension of the short lines marked b about the edge of the board suggests how the try-square will be used to secure lines on the under side of the board corresponding to those on the top side.

Fig. 109. Boring out the mortise.

The mortise should be bored out by the process illustrated in Fig. 109, the ends of the mortise chiseled as described in Sec. 49, and the sides of the mortise pared out as described in Sec. 44.

CHAPTER VII

USE OF MODELING OR FORMING TOOLS; SHAPING IRREGULAR FORMS

Suggested Projects:

      a) Hammer handle (Fig. 110).

      b) Hatchet handle (Fig. 111).

      c) Neckyoke (Fig. 112).

      d) Singletree (Fig. 113).

      e) Shoulder carrier (Fig. 114).

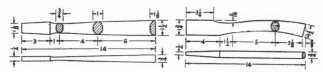

Fɪɢ. 110. Hammer handle.   Fɪɢ. 111. Hatchet handle.

Fɪɢ. 112. Neckyoke.   Fɪɢ. 113. Singletree.

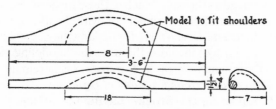

Fɪɢ. 114. Shoulder carrier.

55. Modeling Tools. Under the classification of tools (Sec. 7), are listed those in common use. Among these, but not under a single heading, are those which are used principally for fashioning irregular surfaces. In such a group are found the spoke-shave, the draw-knife and similar tools; the hatchet, ax and adz, and also such miscellaneous tools as the turning-saw, woodrasp and gage (Fig. 5).

Perhaps in no place where woodworking hand tools are in common use are the modeling tools more generally used than on the farm, with the possible exception of the cooper shop. The cutting edge of any one of these tools, except the turning-saw and file, is, in form and use, both a chisel and a knife, yet none of them are used either as the chisel or knife.

Both the draw-knife and the spoke-shave (Fig. 5) are chisels with a handle at either end of the cutting edge. In the case of the spoke-shave, the thickness of the shaving is controlled by a gage in an opening in the shoe or bed-plate of the spoke-shave. There is also similarity in construction between this tool and the plane.

On the other hand, the hatchet, ax and adz are chisels, but controlled differently from either the chisel or the spoke-shave and draw-knife. The descriptive matter under heading, "Working Instructions." in the following pages, suggests the use of each of these tools, and should be studied carefully in connection with the illustrations.

The instructions given are for making the hatchet handle. This project includes the principal modeling exercises for the majority of the forming tools herein listed. *Working Instructions for Hatchet Handle:*

56. Squaring the Stock—Laying Out. Plane stock to over all dimensions, 3/4″ x 1-1/2″ x 14″.

On the face side, sketch the outline of the handle, as shown in Fig. 115. Taper the front end of the handle to 1/2″ thickness on the end, beginning the taper at a point 4″ from this end, as shown in the edge view of the mechanical drawing of the handle.

57. Using the Turning-saw. Place the stock upright in the vise, one-half its length being above the vise. Stand in front of the vise in position to saw (Fig. 115).

FIG. 115. Correct use of turning saw.

Grasp the turning-saw in hands, as shown in Fig. 115, the teeth pointing toward the operator. Move the saw away from you to start the saw cut, or kerf; then toward you without downward pressure, until the saw blade has begun to cut. Continue to move the saw backward and forward the approximate length of the saw blade, holding the frame vertically except when necessary to vary from this position in order not to have the frame strike the stock. Gradually turn the right hand as forward strokes are made to direct the saw blade on the curve.

Where possible, the saw cut should be taken *over* the grain. However, unless the saw can be removed from stock and started in a new place without much difficulty, it is best to complete a saw cut regardless of relation of wood fiber to saw cut. Continue work with the turning-saw until the complete outline of the handle is sawed out.

58. Use of the Spoke-shave. Place the stock in vise, as illustrated in Fig. 116. Stand at end of vise, bending slightly over stock with spoke-shave grasped firmly, but

FIG. 116. Using the spoke-shave.

not rigidly, in both hands. Draw or push it *over* the grain, holding the blade square with the face side, but allowing one hand to lead the other slightly, that the shaving may be cut more readily. It may be advisable to shift the position of the stock in the vise from time to time, that the tool may be used with the least difficulty.

When the spoke-shaved edges are square with the face side, the corners should be taken off to form a cross-section, as shown in Fig. 111 and Fig. 117. Care must be taken to remove no more stock than must be taken off finally to secure a good oval-shaped handle. The oval should be an ellipse.

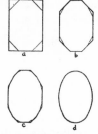

FIG. 117. Steps in modeling handles.

After the first corners are removed, the process of cutting off corners should be continued, as shown in Fig. 117, to secure the closest approximation to an elliptical cross-section. The front end of the handle may now be modeled to fit the hatchet head. This may be done with the spoke-shave or the plane, or partly by the use of each.

59. Scraping and Sandpapering. Finally, all irregular surfaces should be scraped with a piece of glass or a steel scraper, and sandpapered, first using the sandpaper on a block and moving the block slowly around the handle as it is moved back and forth lengthwise with the grain. Finally, with the sandpaper in the hand, continue to move the paper lengthwise to secure the finished surface. Cross strokes with the sandpaper may be taken if followed by strokes with the grain.

Supplementary Instructions:

60. The Wood Rasp. In some cases, it is advisable to use a wood rasp (Fig. 5) separately or in conjunction with the spoke-shave, scraper and sandpaper in modeling a piece of wood to an irregular form and shape. If a spoke-shave had not been available for use on the hatchet handle, the same general procedure could have been followed with the wood rasp in modeling the form for each of the different shapes described, viz.—rectangular cross-section, eight-sided cross-section, etc.

The wood rasp is held like a file. It is pushed forward with pressure for the cutting stroke, and lightly drawn back in contact with the wood, or lifted from the wood entirely on the return stroke. As it is pushed, it is rolled slightly and, also, moved lengthwise on the stock, thus avoiding rutting or gouging the wood.

The hatchet, ax or adz may be used to remove a considerable portion of stock to secure roughly the desired form or shape.

Of the projects listed in this group, little or no difficulty should be experienced in securing the result indicated by the drawings, if the instructions for the hatchet handle are followed as a guide.

61. The Shoulder Carrier. The most difficult project to form is the shoulder carrier. This may be modeled from a straight-grained, well-seasoned piece of cord-wood, or from a heavy plank. It is advisable to cut out with the turning-saw the shape of the carrier shown in the top view, or the one you would get if looking down on the carrier as it is placed on one's shoulders. Next, model the upper surface with a draw-knife, spokeshave and wood rasp. Finally, the under surface should be modeled to fit over the shoulders. This work may be done with an outside ground gouge (Fig. 5). It is pushed into the wood with the grain, and, as the right hand is lowered, the stock is removed and the desired shape is secured. The convex, concave and cylindrical surfaces of the carrier may all be smoothed finally with a wood rasp or sandpaper, or both.

CHAPTER VIII

SUPPLEMENTARY PROJECTS

62. Sheep Rack and Feed Bunk (Fig. 118).

Directions:

1) Frame up each end with 1″ x 10″ boards 3′ 0″ long, cleated together on the outside by the two 1″ x 4″ strips, and on the inside by the 1″ x 3″ strip upon which the trough floor is to rest. Flush with the upper edge of this cleat, and with each of the outside edges of the bunk end, fasten the 2″ x 4″ corner posts or legs.

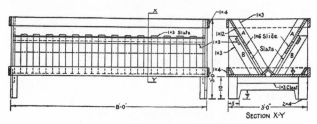

FIG. 118. Sheep rack and feed bunk.

2) Fasten the two bunk ends together by nailing in place the top and bottom bunk rails.

3) Lay the floor, nailing boards to the floor cleats on the bunk ends and to the top of the legs. Fit the middle "V" feed guides and the outside trough edge boards, nailing ends of the same from outside of bunk end.

4) From lower edges of feed guides to upper corners of ends of bunk, draw lines to locate guide boards for the 1″ x 12″ board and rack which hold feed. Construct and nail these guides in place.

5) Cut slats for rack and fasten together at top ends by means of a 1″ x 3″ board, to which all are squarely nailed.

73

6) Nail 1″ x 12″ feed boards in position to the inside guide boards (*A*).

7) Place feed rack in position, supported by outside guide boards *(B)*. Toe-nail the bottom of each rack slat to the "V" feed guide board, and nail the top of the rack

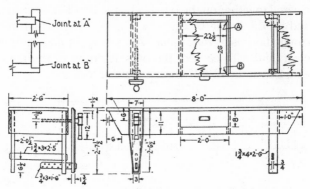

Fig. 119. Woodworking bench.

securely to the 1″ x 12″ feed board, over the lower edge of which they should lap by 3″. This should be done after the 1″ x 6″ slide has been placed in position. This should slide in the openings between the feed rack and the inside end guide boards with difficulty, that it may be held in any particular position by friction, or it should be fastened thru grooves in the end boards of rack by means of wing nut bolts.

63. Directions for Woodworking Bench (Fig. 119).

1) Construct the frame by planing for each end:

         2 oak boards (uprights), each 1-3/4″ x 4″ x 2′ 6″.

         1 oak board (lower crosspiece), 1-3/4″ x 3″ x 2′ 5″.

         1 pine board (upper crosspiece), 1″ x 11″ x 2′ 4″.

2) Lay out and construct the mortise and tenon joints to join the front and back uprights with the lower crosspiece. Tenons may be full width, viz., 3″ wide and 3/4″ thick. Length between tenon shoulders, 2′ 1/2″.

3) Assemble each frame by joining the parts; the lower crosspiece to be fastened to the uprights by gluing joints, and clamped for at least twelve hours, and the upper connecting piece to be nailed in position as soon as clamps are applied to the lower part of the frame.

4) Construct the front of the vise, planing it to dimensions as given in the drawing. Fasten the lower guide piece in with glue and nail it from each edge of the vise board. The staggered holes in guide piece for vise, in which to insert pin to keep the lower portion of the vise board the proper distance from the bench, should each be 1/2″ in diameter, in two rows each about 1″ from the edge of the guide board, holes to be 2″ apart in each row.

5) Nail front and back side boards, or rails, of the bench onto the end sections.

6) Purchase a 1-1/4″ iron vise screw. Bore the holes for this in vise board and bench, and cut the slot for the vise guide board. Assemble vise.

7) Cut the opening for the drawer 2′ x 8″ in upper portion of center of front board, and fasten in place the runner and guide boards for the drawer by nailing or screwing into their ends thru front and back boards.

8) Lay the top boards on. These may be of oak, altho dressed pine will suffice. Joints between boards must be tight. They need not be glued.

9) Construct drawer, as shown by drawer details of joints, and fit in bench to slide freely. The bottom of the drawer may be nailed onto cleats fastened to the lower inside surfaces of the sides of drawer.

64. Working Directions for Dog House (Fig. 120).

1) Cut 2″ x 4″'s to proper lengths for either sills or plates, and one-half the number of studs.

2) Rip all pieces of 2″ x 4″ into 2″ x 2″ strips.

3) Construct sill and plate frames with horizontal half-lap corner joints, and connect sill and plate frames with studs by nailing thru frames into ends of studs.

4) Nail on sheathing (fence boards), lapping side boards over ends of end boards.

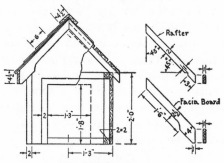

FIG. 120. Dog house.

5) Beginning at the bottom, cut and nail on siding on sides and ends.

6) Cut and nail on roof boards (fence boards), allowing space of 1″ between boards.

7) Shingle roof, beginning at eaves and working toward ridge, breaking joints for every two consecutive layers or rows of shingles.

8) Cut, fit and nail ridge, facing, corner, base, and trim boards.

65. Directions for Corn Drier (Fig. 121).

1) Secure pine lumber 1″ thick, 3-1/2″ or 4″ wide, and 16′-0″ long, dressed.

2) Cut each piece to form lengths for parts of drier with least possible waste. Example: 10′ and 6′ or 6′, 6′ and 4′ (2 braces).

3) Nail, as shown in drawing, nailing two pieces together, surface to surface for ends. Toe-nail in braces. Use six-penny and eight-penny common nails.

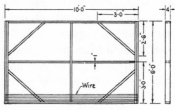

FIG. 121. Corn drier.

4) On end pieces and vertical center piece, with two-foot rule or carpenter's square, lay off points with pencil on front and back for spacing wire. Drive shingle-nail, or 1-1/2″ brad at each point.

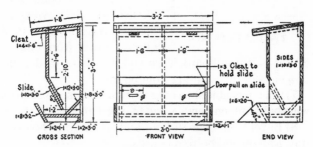

FIG. 122. Self-feeder.

5) Wind wire (1/16″ annealed iron), taking one turn about each nail.

66. Making the Self-Feeder (Fig. 122).

1) Cut ends to dimension—10″ x 3′ 0″—bottom end square and top end tapered toward the front to make it 2′ 10″ long.

2) Cut center partition to overall dimensions of end boards. Bevel front and back edges of lower end to fit to deflector board on the back, and to the front board, to which the adjustable slide is attached on the front.

3) Nail cleat 2″ wide on inside surfaces of end boards at the bottom, upon which floor will rest; also 1″ x 3″ cleat to hold slide, as marked in front view of drawing.

4) Nail to edges of end and center partition boards ship-lap to form vertical portion of front of feeder and all of back of feeder.

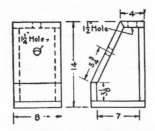

Fig. 123. Egg tester.

5) Lay floor of ship-lap on inside bottom cleats fastened on end boards of feeder. Cut deflector board and toe-nail into position.

6) Cut slanting 12″ board of front of feeder, bore holes for 1/2″ wing-nut bolts and nail board in position to lower front edge of center partition board. Thru end boards of feeder nail into ends of this slanting front board.

7) Cut adjustable slide board to dimensions, cut slots for wing bolts, attach handles and fit board in position.

8) Bevel front edge of floor and attach front board of tray. Nail on end bottom boards.

9) Nail ship-lap to cleats for top. Hinge at rear with two 4″ leaf strap hinges.

10) Paint outside of feeder with brown creosote paint.

67. Making the Egg Tester (Fig. 123).

1) Secure stock from one board 8″ wide, or from more than one board of shorter length, but the same width, to construct the complete box.

2) Cut stock to convenient planing lengths, each to cut finally into a certain number of pieces for the box. A little must be allowed in length for crosscutting and squaring ends.

3) Plane stock to dimensions. Saw to proper lengths and square ends.

4) Taper front edges of side boards.

5) Set bevel-square for angle of front edge of top board and ends of front board. Mark and trim to proper angles.

6) Bore 1-1/2″ hole and 1-1/4″ hole in centers of top and front boards, respectively.

7) Nail box together with six-penny (6d) finish nails in the following order: Back, bottom, front vertical board, top to sides; front slanting board.

68. Constructing a Cow Stanchion (Fig. 124).

1) Select straight-grained hickory, oak or other close-grained, tough wood from 2″ or 1-1/2″ dressed plank.

2) Rip stock to width or thickness to secure strips which will dress to 1-1/2″ x 2″.

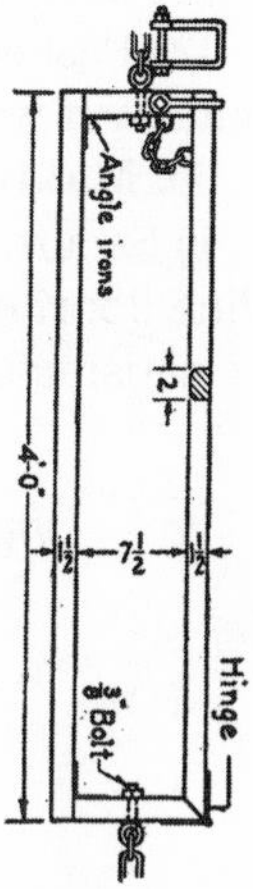

FIG. 124. Cow stanchion.

3) Plane strips to correct width and thickness.

4) Square and saw strips to correct lengths.

5) Bore 3/8″ holes in center of end pieces for chain bolts.

6) Bore 3/8″ hole in upper end piece for corner fastening clamp. (Position of hole depends on length of clamp.)

7) Bevel lower end piece and one side piece, each on one end to 45 degrees for hinged corner.

8) Fasten corner angle iron with 1″ flathead screws to inside of each end piece to draw tightly to side piece when screwed to it. Drill small holes for screws.

9) Fasten remaining side of each angle iron to side piece. First place in position, mark for screw holes, and then drill for them.

10) Fasten strap hinge in manner similar to that used in fastening angle irons.

11) Drill small holes for staples for corner chain.

12) Fasten chain and corner clamp bolts by setting up nuts over washers and burr ends of bolts slightly.

69. **Making Tomato Trellis** (Fig. 125).

1) Secure eleven 12″ strips of pine, 3″ or 3-1/2″ wide. Pine flooring or 6″ pine fence boards ripped in two will be satisfactory.

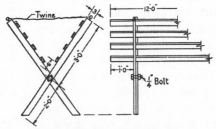

Fig. 125. Tomato trellis.

2) Cut from each of three strips two pieces 5′ 0″ long.

3) Two feet from one end of each of the 5″ strips, bore a 1/4″ hole in the middle of the stock.

4) Fasten two of the strips together with a 1/4″ bolt, using a washer under the head and under the nut, to form one end of the rack. In like manner, make the remaining two supports, one for the opposite end and one for the middle of the rack.

5) Nail four of the 12' strips evenly spaced on the 3' leg of each support, allowing the strips to project 12″ on the end of the frame beyond the end support, and leaving sufficient space below the strip nearest the hinge for the vines. Place the middle support centrally in the frame.

6) Put strong screw-eyes at the top of each bar of each end support, in which to fasten wire or cord to hold the top edges at a fixed position when the frame is in place.

70. Feed Bunk for Cattle (Fig. 126).

1) Construct each trestle or pair of legs by first cutting to length the four legs from 4″ x 4″ stock and connecting each pair with the four 2″ x 4″ cross and brace pieces.

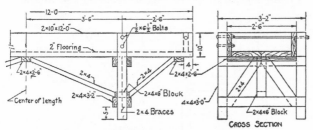

FIG. 126. Feed bunk for cattle.

First nail in position the two crosspieces on one side of a pair of legs, then insert and nail securely to this pair of cross-pieces the two leg frame braces. Now nail on the two remaining crosspieces, both to the legs and the braces. Finally, nail on the 6″, 2″ × 4″ block lengthwise of the bunk in the center of the lower crosspieces, allowing it to rest 1″ on each crosspiece.

2) Construct the frame of the box from 1-1/2″ × 10″ or 2″ × 10″ stock. Bore holes to secure the sides to the end pieces. Note that the end boards of frame rest on floor boards.

3) Turn box bottom side up. Lay and nail floor boards to end boards, and nail on 2″ × 4″ crosspieces.

4) While box is bottom side up, place leg frames in position and bore holes thru legs and side rails of box. Insert and fasten bolts. Fit and nail the four length braces in position, carefully locating center position for them on the crosspieces. The lower ends of these braces butt against the 6″, 2″ × 4″ blocks already nailed to the lower leg frame crosspieces.

5) Place bunk in upright position and insert and tighten end rods.

71. Saw Buck (Fig. 127).

1) Saw legs of frame to lengths from 3″ × 4″ stock.

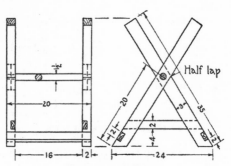

Fig. 127. Saw buck.

2) Locate, lay out and cut half-lap joints for each end of frame.

3) Lay out and cut 2″ notches for thickness and width of connecting braces.

4) Bore holes for center rod in each half-lap joint connecting ends of frame. Do this when each end of frame is halved together.

5) Plane and fit all brace rods.

6) Form center rod, using draw-knife and spoke-shave on center portion of rod, saw to cut shoulder and use chisel and wood rasp to form ends of rod.

7) Nail cross braces on each end frame and trim ends with plane.

8) Place center rod in position and wedge ends with thin wooden wedge.

9) Nail length braces in place and trim edges.

10) Place saw buck upright on level floor. With open dividers, scribe line for bottom of legs; saw to lines.

72. Chicken Feeder (Fig. 128).

1) Cut nine pieces from 6″ fence boards, each 2′ 6″ long.

2) Construct each of the ends and the partition of the feeder by nailing three of these pieces together with a cleat at the bottom and another at the upper edge 2′ 0″ from the bottom.

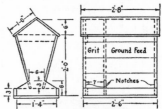

FIG. 128. Chicken feeder.

3) Lay out upon a vertical center line of each end board thus constructed the shape of the feeder end according to dimensions. Saw to shape.

4) Saw 6″ fence boards to lengths of 2′ 6″ for sides.

5) Cut out corners a distance of 3″ on the lower edge of two of the side boards to fit between the end pieces at the bottom of the feed box. Nail each in position for lower board of each side. Nail other side boards on from bottom to top. Dress with plane upper edge of top side boards at roof angle to allow roof boards to fit on same closely.

6) Saw, fit and nail on bottom, roof and side boards for tray. Use 8d. common nails.

73. Garden Marker (Fig. 129).

1) Secure stock as follows:

>One 2″ × 4″ × 4′ 0″.

>One 1″ × 6″ × 8′ 0″.

Short stock for braces and marking pins may be secured from waste from handle.

2) Plane bed board to dimensions; bevel front edge.

3) Locate centers for marker pin holes on top and bottom surfaces of bed board by means of marking gage and try-square. Angle of pin should be about 15 degrees to a vertical line.

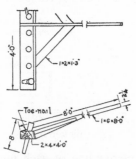

FIG. 129. Garden marker.

4) Bore 1-1/4″ holes for marker pins, working from each side of bed board. With jack-knife, ream out holes on top side to approximately 1-1/2″.

5) Shape handle, nail securely in place, and brace with pieces of stock ripped from handle on under side.

6) From waste stock ripped from handle, or better, from pieces of 2″ x 4″ ripped to 2″ square strips, whittle out marker pins. Drive pins in place and toe-nail from the top, allowing nail heads to project sufficiently so that the nails may be removed with a hammer.

74. Individual Hog Cot (Fig. 130).

1) Frame floor by nailing four 2″ × 4‴'s edgewise across the two 2″ × 4″ runners, one at each end, front and back, and the remaining two evenly dividing the remaining space.

2) Cut rafters, three for each side, each 6′ 6″ long. Toe-nail bottom ends on runners, one at each end and one in the middle. Toe-nail tops of rafters of each pair together.

3) Fasten rafters together on each side by three strips (roof stringers) of 1″ × 3″ stock. These preferably should be set in (housed) to upper edge of rafters. If so, housings should be cut before rafters are placed in position.

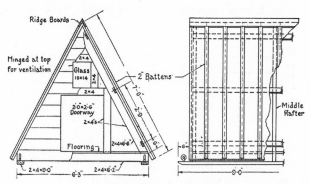

FIG. 130. Hog cot.

4) Lay floor of 1″ × 6″ matched lumber, matching outside floor strips to fit around rafters and come flush with outside edges of them.

5) Erect supports or studs front and back under end rafters to form framing for door and window.

6) Toe-nail window framings between studs.

7) Cover in ends with 1″ × 6″ matched siding, resting bottom edge of bottom boards on top of runners.

8) Cover roof with 1″ × 6″, 1″ × 8″ or 1″ × 10″ boards vertically, nailing each to each roof stringer. Cover each crack with a batten (2″ strip), first placing ridge boards in place.

9) Set window and hinge at top so that it may be opened for purpose of ventilation. Place framing strips around door and window, if desired, to represent casings.

10) Fasten large eyebolt in each end of each runner to serve as connection in dragging cot from one place to another.

75. Feed Bunk for Sheep (Fig. 131).

1) Cut all 2″ × 4″ stocky viz., four corner posts and two horizontal cross-bar supports for the floor, from a 16′ piece.

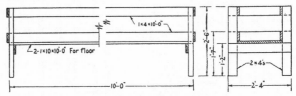

FIG. 131. Feed bunk for sheep.

2) Secure five boards, each 1″ × 4″ × 10′ 0″, and cut from one of them four pieces, each 2′ 4″ long.

3) Frame each end of the bunk.

4) Connect the end frames by nailing the two upper side strips in position.

5) Lay the floor.

6) Nail in position the two lower side strips to form the sides of the feed tray.

76. Plow Doubletree (Fig. 132).

1) Select, from 2″ hickory or straight-grained oak, stock for each of the three parts of the doubletree.

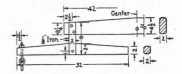

FIG. 132. Plow doubletree.

2) Saw and plane each piece of stock to rectangular shape and to overall dimensions.

3) Plane back edges of each part to the correct taper, first making lines with straight-edge and pencil defining these edges.

4) Bore holes for metal fittings.

5) Secure in stock, or forge out the tug hooks and bolts to fasten same to wooden parts; also the iron straps to fasten the singletree and doubletree together.

6) Attach metal fittings.

77. Wagon Jack (Figs. 133 and 134).

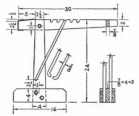

FIG. 133. Wagon jack.

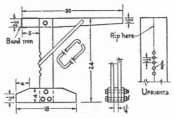

FIG. 134. Another type of wagon jack.

1) Secure hickory, strong, straight-grained oak or other tough wood in following dimensions:

2 pieces 7/8″ × 5″ × 2′ 2″ S2S, uprights.

2 pieces 7/8″ × 4″ × 2′ 6″ S2S, base strips.

1 piece 7/8″ × 3″ × 2′ 6″ S2S, handle.

1 piece 7/8″ × 4″ × 5″ S2S, block at bottom between uprights.

2) Saw and plane each piece of stock to shape and dimensions, as shown in Figs. 133 and 134.

3) Bore series of 5/8″ holes 2″ apart, beginning 12″ from end of handle, each with center 1/2″ from upper edge of handle.

4) Saw notches, as indicated in Fig. 133, saw-cut in each case meeting outer surface of bored hole.

5) With all pieces fastened together in vise or clamp, bore holes for 5/8″ bolts to fasten upright to base strips and handle.

6) Bend 3/4″ band iron around lifting end of handle; drill and countersink holes for 1″ flat-head screws and fasten band iron in place.

7) Assemble all parts of the jack, except the iron rod, to hold the handle in particular positions. The bolts used should be fastened each with a washer under the head and under the nut.

8) Measure with a cord the distance from one hole into which the handle holding iron is to be slipped to

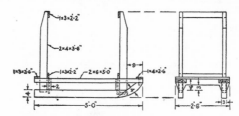

FIG. 135. Farm sled.

the opposite one, thru the first notch from the standard on the handle. In doing this, lower the lifting end of the handle to the lowest desired position. Make allowance for the ends of the handle holding rod, which will slip into the holes in the standard. Add the amount of this allowance to the length of the cord; the total length will be that of the handle holding rod.

9) Cut a 5/8″ round wrought-iron rod to the length of the cord as calculated. Bend the rod to the desired shape by heating portion at bend, and working over end of peen of blacksmith anvil. Cool and spring rod into position.

78. Heavy Farm Sled (Fig. 135).

1) Cut all stock (rough) to overall lengths.

2) Lay out, saw and cut all joints on similar pieces. Example: Two horizontal parts of runners; two front parts of runners; two cross-beams, etc.

3) Frame together the two parts of each runner, driving dowel in with glue, and toe-nailing runner parts together from top and bottom.

4) Put cross-beams in place, driving dowel in place in glue and spiking from under side of runner.

5) Nail long boards of runner frames (bed boards) in position on each cross-beam and on top of runner.

6) Place corner uprights in place, nailing from both sides of bed boards and runners.

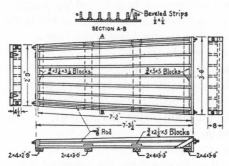

FIG. 136. Fence-post mold.

7) Nail in position all cross-boards—front and rear of sled, inside corners of bed boards and uprights, top of uprights.

Note: Letter each set of boards, and use letter in operation steps.

79. Fence Post Mold (Fig. 136).

1) Lay floor on 2″ × 4″ cleats, as shown in cross-sectional side view. Upper surface of flooring material should be surfaced, and joints between boards made close, but not absolutely tight if lumber is very dry.

2) Prepare ends by planing one board 4-3/4″ wide and 2′ 9″ long, and the other 6″ wide and 3′ 5-3/4″ long, out of 3/4″ stock. On the first board flush with upper edge, fasten a series of 3-1/2″ x 3-1/2″ blocks, leaving 13/16″ between them. Begin to fix these blocks at the center of the board where a 13/16″ space is to be left for the middle partition. On the second board, fasten 5″ × 5″ blocks in a similar way.

3) Prepare seven partition boards dressed on all surfaces, 3/4″ thick, 3-1/2″ wide at one end, 5″ wide at the other end, and 7′ 2″ long. The ends must be square with the center line of the partition board.

4) Place the two end boards and the partition boards in place on the floor and nail into position the 2-1/2″ x 5″ blocks noted on the top view of drawing,

5) Nail in position end blocks on each end board thru which 3/8″ rods pass. These should touch, but not bend, the outside partition boards. Locate position of holes for rods to come in center of space on end boards outside of the last partition board. Locate position of hinges on small end board.

6) Remove end boards, bore holes for rod, fasten on hinges and replace end boards, inserting side rods and fastening hinges to floor (see end view).

7) Remove partition boards, prepare beveled strips (section *A*), fasten them to the floor, and replace the boards.

8) Cover inside surfaces for each individual mould with linseed oil.

CHAPTER IX

Wood-Finishing and Painting

80. Purpose of Wood-Finishing. With few exceptions, all woodwork, whether exposed to the weather or used under cover, is given some sort of surface finish. The object of wood-finishing is twofold, viz.:

First, to preserve the wood. All wood is porous and, consequently, absorbs moisture. With the change of temperature and amount of humidity in the atmosphere, the quantity of moisture taken up by wood will vary. The change in the moisture content of wood causes a change in its shape, known as *warp* (the word used for buckling) and *wind* (the word used for twisting).

The absorption of moisture by wood is accompanied by swelling. As wood dries, it shrinks, thus causing checks and cracks.

Second, to decorate the wood. Decoration may be natural or artificial. Any substance such as oil or wax which, when applied to the surface of wood, brings out its natural markings and colorings, is regarded as a natural decorative agent. Any substance such as colored stain or paint, which covers the grain of wood when applied to it, may be made a decorative agent, but is considered artificial, as it changes the natural appearance.

81. Method of Preservation. Both the natural and artificial wood-finishing agents serve to seal the pores of the wood. All stains have a tendency to enter the wood fibre and to close the pores, but not to fill the cells or larger holes and openings. Paint, on the other hand, covers the entire surface of the wood and, consequently, fills all openings—both pores and cells, as well as such artificial openings as cracks and checks. It must be evident, therefore, that for exposed woodwork, paint is the most satisfactory physical preservative covering. However, besides the fact that it obliterates the material appearance of the wood, it has the possible disadvantage of checking and peeling. On the other hand, when a stain has some inherent preserving quality, chemical or otherwise, it results in both protecting the wood and preserving its natural beauty.

82. Classification of Wood Finishes.

A. Non-covering agents may be divided as follows:

1. Oil.
2. Wax.
3. Stain
   a. Water.
   b. Oil.
   c. Chemical.
   d. Creosote.

B. Covering agents may be divided as follows:

1. Shellac
   a. White.
   b. Orange.
2. Varnish
3. Filler.
4. Paint.

83. **Oil Stain** is used on work which does not require a high finish, but which, to present the full effect of the natural grain, needs a light coat of finishing material. Raw linseed oil is generally used for this purpose. That it may penetrate to the greatest extent, the oil should be applied when hot. A soft cloth, cotton waste or a brush may be used. When the oil has evaporated, or has set in the wood, a brisk rubbing will secure a dull polish, which, however, will not long continue except by repeated rubbing, which may be done on inside work in the process of dusting.

Oak, when used outside, as for garden furniture, is protected somewhat from the weather when given coats of hot linseed oil two or three times annually.

84. Wax. This may be secured in cans as "prepared wax." It is frequently used to give a natural finish of low gloss. This material is a substitute for oil and serves not so much as a preservative by means of penetration as by virtue of its filling up openings. When rubbed with a soft cloth, it gives a velvet-like polish. Wax hardens with time and, therefore, makes a very satisfactory wood finish, especially if new coats are added from time to time and if the waxed surfaces are rubbed occasionally.

85. Water Stains are the simplest of all liquid finishes to apply. They are sold both in powder and liquid form. A water stain is applied with a brush and, before dry, is rubbed with a cloth or with waste. If care is taken in mixing and applying, there is little difficulty in securing a uniform color. Wax or one of the class B finishes may be used after the stain has dried.

Before applying a water stain, the wood should be thoroly scraped and sandpapered, and then "wet down" with water. Water raises the grain as would the water stain if applied first. When the wood surface has dried after the application of the water, it should be thoroly sanded. The application of the water stain will raise the grain slightly, but not sufficiently to require sanding, which, of course, would injure the appearance of the stained wood.

86. Oil Stains; Chemical Stains. These are applied in the manner described for water stains, except that the previous washing is omitted. An oil stain will strike into the wood more freely than will a water stain, and, consequently, because of the variation in the porosity of the average piece of wood, and especially of different pieces of wood assembled in one unit, difficulty is sometimes experienced in getting a uniform color. It may be necessary on particularly porous woods to dilute an oil stain, or to apply a thinner coat than would be used on a less porous part or piece of wood. Wax or one of the class B finishes may be used after an oil stain has dried and the surface oil has evaporated thoroly.

Chemical stains, which now constitute the largest part of those to be secured in the open market, are prepared to overcome the disadvantages of poor penetrating qualities of water stains and the uneven penetration of oil stains. They prove quite satisfactory in giving a uniform and well-set color on wood of fairly uniform quality. They may be covered either with wax or the finishes under class B.

87. Coal-tar Creosote Oil. The preservation of wood on the farm cannot always be most satisfactorily accomplished by the use of wood finishes already described. Wooden fence posts, bridge and trestle supports, piles or posts used to support roofs for grain and hay stacks, timbers used in silos, wooden shingles for roofs, etc., are neither stained nor painted as a rule; they are frequently left unprotected. Moisture, air and temperature are natural weather elements which permit the development of fungus growths which cause rot and decay. All wooden structures exposed to the weather should, therefore, be protected.

Toxic mineral salts or coal-tar creosote oil is used to protect outside woodwork which it is not desirable to decorate as the common stains and paints do. Coal-tar creosote oil eradicates fungus organisms or suspends their destructive growth. It is insoluble and, therefore, is impervious to moisture. Present practical results of treating wood with it have justified its use.

The two general methods of treatment are known as the pressure processes and the non-pressure processes. The former are used extensively by large corporations, and the latter by small consumers, in which class the farmer would be placed. Of the non-pressure processes, there are two, viz., the open-tank system and the brush method.

88. The Brush Method is the one which the conditions of the average farm make entirely possible. It consists of painting refined coal-tar creosote oil, heated to approximately 150 degrees F., on the wood in the same manner as is done with paint, or pouring the heated creosote over the lumber, catching the drippings in pans or basins, or applying the heated creosote with a mop instead of a brush. It is current opinion that in order to make effective the use of coal-tar creosote oil, it must be applied under pressure; nevertheless, the fact remains that the brush method of surface treatment results in a most surprising increase in the life of the material treated, and in a most satisfactory reduction in the annual cost of maintenance of structure.

Two or three coats of coal-tar creosote oil are necessary, and all surfaces exposed or in contact with moisture-collecting materials, such as concrete, should be covered. Particular attention is directed to the covering of surfaces of joints, such as the sides of mortises and tenons, etc.

89. The Open-Tank Process, while not feasible under ordinary farm conditions, is here briefly described, that it may be used where conditions permit. It consists of alternate hot-and-cold treatments of wood with refined coal-tar creosote oil by immersion and continuous soaking in open tanks without artificial pressure, requiring no mechanical apparatus other than tanks, hoist (in some cases), and means of heating the oil.

The procedure is as follows: Season the lumber sufficiently to expel any excess of moisture. When cut for sizes, construction, etc.—that is, when completely framed—immerse lumber in a bath of coal-tar creosote oil maintained at a temperature of from 150 to 210 degrees F. for a period determined as follows: For close-grained wood (naturally resistant to impregnation), one hour in the hot and one hour in the cold, or cooling, bath for each inch of the largest cross-section. For species more susceptible to treatment, one-quarter of an hour for each inch of the largest cross-section, and milled lumber from ten to thirty minutes in each bath; or, if the stock is in the form of boards, an immersion of a few minutes is sufficient. Frequently, heavy-milled stock is not subjected to the cold-bath treatment, but allowed to remain in the hot bath after the source of heat is removed and while the oil cools. On the other hand, boards are not subjected even to a "cooling" bath as suggested by the use of the word immersion above.

A project in creosoting may be selected from the buildings or structures already erected or to be erected.

In some cases, the possibility of creosoting is suggested in the instruction given for woodworking projects.

90. Shellac is a gum preparation prepared from the secretion of the lac bug. It is procurable in the market in dry flakes, and is dissolved in alcohol. The consistency for satisfactory use should be that of thin syrup. It is applied with a brush, which should be of good quality. Shellac evaporates rapidly; hence, unusual precaution is necessary in applying it to avoid streaking the surface. Long, single strokes with a well-filled brush will produce the best results. The brush should not make a second stroke over the same surface until the first coat of material is dry.

A dry shellacked surface may be sandpapered and again shellacked. By repeated coats and careful sandings, a very smooth and highly-polished surface may be secured which can be improved by a final light rubbing with a piece of felt or burlap wrapped over a piece of cork or wood, and first dipped in a shallow dish of rubbing oil, and then into pumice stone.

91. Varnish acts very similarly to shellac. It is the customary finishing material for highly-polished woodwork. It is applied and treated the same as shellac, but dries much slower.

92. Wood-Filler is used to fill the pores of the grain of wood. When shellac or varnish is used, both as a filler and as a finish, many coats are required before the grain is filled and a finishing surface is built up. Wood-filler is, therefore, used to fill holes and level up the surface for the finishing material, which, ordinarily, is varnish.

Wood-filler is silex mixed with linseed oil, japan and turpentine. It should be thinned with turpentine or benzine to the consistency of paste and applied by means of a brush. When it begins to "gray," a sign of its drying, it should be rubbed across the grain with a handful of excelsior, shavings or waste. Before applying shellac, varnish or other finishing material, the filler should dry at least forty-eight hours. Colored fillers are common to produce particular color effects. The white filler may be mixed with dry pigment colors to secure the color desired. In case wood is both stained and filled, the stain should be used first.

93. Paint is made from white lead and linseed oil. It may be secured in the market prepared ready for use after being thoroly stirred. It may be made by mixing white lead and linseed oil with a coloring material. The surface of wood to be covered with paint should be clean and smooth. Paint is applied with a brush with the grain of the wood. The brush should be run back and forth over the same surface several times to work the paint into the grain of the wood. Two or three coats are usually necessary to cover the surface properly. Each coat may be sanded carefully when dry before the succeeding coat is applied. Unless a paint has considerable drier in it, or is a cheap substitute for white lead and oil, it needs at least three or four days to dry before it can be smoothed with sandpaper, or a second coat of paint can be applied.

The projects in wood-finishing and painting should be worked in approximately the order given in the "Classification of Wood Finishes" in Sec. 82. The projects may be those given in the several groups under "Woodworking." Upon the completion of a woodworking project, the proper finish may be applied, or all woodworking projects may first be completed and then finished. In this case, there will be an advantage in concentrating attention upon the work, both of using woodworking tools and of applying wood-finishing materials.

Paint is regarded as easier to apply than shellac or varnish; hence, the project in painting may well precede that in shellacking or varnishing.

Always keep a "full" brush of finishing material; that is, have the lower half of the bristles full of the finishing material, but do not allow the upper part of the brush to be covered. As one removes the brush from the material, it should be drawn upward against the edge of the receptacle on each side, that not too much material may be left in the brush, and also that the upper part of the bristles shall be free from material and the brush kept clean.

Brushes when not in use should be kept hanging in the material in which they are used so that the ends of the bristles will be clear of the bottom of the receptacle. Receptacles should be covered to prevent accumulation of dust and dirt. Any wide-necked bottle or fruit jar may be used as a receptacle for brushes, the stopper being made of wood.

The projects given in the woodworking section of this book suggest the finish which each may be given. It is suggested that the finishing of these projects in the order presented be regarded as the desirable wood-finishing projects to secure the necessary knowledge and practice in this subject.

CHAPTER X

GLAZING AND SCREENING

94. Definition. Glazing consists of cutting and setting glass in frames. The chief use of this art is in cutting, tacking and puttying panes of glass in window sash, hotbed frames, etc.

95. Precautions. Window glass may be secured in single- or double-strength thicknesses. Double-strength glass is thicker and stronger than single-strength. Glass also is manufactured in a variety of qualities. That known as common is used for ordinary purposes. Whatever the strength or quality, sheet glass should be handled with care, both to prevent breaking it and to provide against being cut by it. It should be grasped by thumb and fingers of both hands, each taking hold of one of opposite edges. When working upon a pane of glass, it should be laid flat on a plain wood surface, such as the top of a bench or table.

96. Cutting Glass. Clean off a flat wooden surface and lay the glass on it, preferably by sliding the glass upon the surface rather than placing it upon the surface from above. If an irregular piece of glass is to be used, place a straight-edge, preferably of wood, but the edge of a carpenter's square may be used, near one edge and run a glass cutter across the glass and against the edge of the straight-edge with one firm stroke, using moderate pressure. If the glass cutter is sharp and the single operation is done carefully, a cut will appear at all points on the glass where the cutter has run. Slide the glass into a position so that the waste stock projects over the edge of the wooden surface, table or bench top, on which it is placed, and so that the line cut in the glass is directly above this edge. With the left hand placed flat on the surface of the glass which is on the table, and with the thumb and fingers of the right hand grasping the edge of the glass projecting over the edge of the table, gently press downward with the right hand.

The glass should crack or make a clean break on the line made with the glass cutter, thus giving one edge of the piece of glass desired.

Place one leg of carpenter's square against this edge and the other in a position to secure an adjacent edge of the piece of glass being prepared. Repeat the operation of cutting and breaking off the waste.

In a similar manner, secure the opposite edges. First, measure carefully for the desired width or length at two points near each end of an edge already formed, and mark in each measurement by a short line—1/4″ is sufficient—made with the glass cutter. Connect these points by the edge of the blade of a carpenter's square or wooden straight-edge against which the glass cutter is run as before.

97. Setting a Pane of Glass in a New Frame. Place the pane of glass in the frame and very gently fasten it in position with three-cornered pieces of tin (glazier's points) used by glaziers, which may be secured when purchasing putty. Lay a triangular piece of tin flat on the glass as it rests in the frame on a bench or table top. With a finger or thumb, press one corner of this tin into the frame near a corner of the pane of glass. With the end of the putty-knife blade resting on the pane of glass as the knife is held in the right hand, or with a square-edged chisel, very carefully drive the point about 3/16″ into the wood by letting the edge of the putty-knife or chisel blade gently strike the point three or four times.

Likewise, insert other points, locating them so as to have one come near the corner of the frame on each edge of the pane, and others placed to make the distance between consecutive tins about 8″ or 10″. In case of a small pane, at least one point should be placed near the middle of each edge of the pane.

If a pane is being set in a vertical frame, as in a window sash in a window frame, care must be taken to hold it firmly in position with the left hand while the right hand is used to drive the points into the frame. Care must always be taken to have the pane well seated; that is, firmly resting against the frame on which the flat surface of the pane rests.

98. Applying the Putty. In order to seal the pane in the frame, making the joint waterproof, putty is pressed into the corner between the pane and the frame. Putty as it comes from the stock receptacle, may need to be mixed with a little boiled linseed oil to soften it. The oil should be mixed thoroly with the putty. Unless the putty is quite dry, the oil need not be added to it, as kneading it in the hands will make it soft.

In applying putty, one should practice the following method: (1) After having beaten and kneaded the putty to an even consistency, cut off a small amount and form it roughly into the shape of a ball. (2) Put this putty into the palm of the left hand and hold the putty knife in the right. Set the frame to be puttied on an easel or on some similar device so that the glass slants away from the operator. (3) Now, with the left hand preceding the right hand, and with the putty knife in position against the glass, feed the putty with the thumb and the first two fingers of the left hand from its position in the palm of the hand and under the corner of the putty knife. Move both hands slowly from right to left, feeding enough putty under the knife to fill the triangular opening formed between the knife and the wood and the glass. (4) When one complete stroke is made, go back and fill in any imperfect spaces, and also clean off any surplus putty which may be left. A little practice is necessary before a perfect job is made with the first stroke. Care should be taken not to allow the putty to get smeared on the glass more than is necessary. The putty should not be high enough to show above the wood on opposite side of the glass.

If a broken pane of glass is being replaced or the opening in an old frame is being filled, care must be taken to clean thoroly the corner into which the pane fits of all dirt, especially old putty. Use broken panes of glass as far as possible in re-glazing windows.

The projects in glazing should consist both of replacing an old pane or panes of glass, and setting the glass in a new frame. After the putty is thoroly dry and hard, it should be painted with the frame in which it is set.

99. Screening. Every farm home should be screened as a protection against the house fly, rightly called the typhoid fly. Screens for doors and windows of standard sizes can be bought in stock from most lumber dealers. One who is handy with tools can easily construct screens.

During the winter months, the screens should be removed from the windows and doors and stored away in a dry place. During spare time, they should be cleaned and painted. Paint especially prepared for this purpose is obtainable at most paint stores. Painting the screens keeps them from rusting and will increase their life many years.

PART II

CEMENT AND CONCRETE

CHAPTER XI

HISTORY OF CEMENT

100. Preliminary. The fact that concrete is now being used so universally, both on the farm and in the city, makes it desirable, if not necessary, that every one should study its possibilities and learn at least the first principles of correct concrete construction. There are too many poor jobs of concrete work, the failure of which is due to lack of knowledge on the part of the man doing the work. Concrete, when properly made, has too many good qualities to be condemned merely because of lack of information and judgment on the part of the man who uses it.

The main reasons concrete is being used to such a great extent are:

a) It is permanent.

b) It is more nearly fireproof than any other building material.

c) It is rat-proof.

d) It is attractive.

e) It is sanitary.

105

> *f)* With the aid of steel, it can be used for most any purpose in building.
>
> *g)* It can be used with success by the average farmer with less special training than is required with other available materials.
>
> *h)* It is economical.

101. Pre-historic Uses of Concrete. Altho we now find concrete being used in nearly all types of construction work, it is only of recent years that the cement industry has been developed. Some form of cement was used thousands of years ago. The ruins of Babylon and Nineveh show traces of it, as does the Pantheon of Rome. It is said that the prehistoric people of America—the Aztecs and Toltecs—used a cement mortar that has been so durable that the mortar joints are projecting where the adjacent stones have been worn away by the weathering action during the ages.

There is little evidence of the use of cement during the intervening period from three or four thousand years ago up to the beginning of the nineteenth century. During this period, the art of making cement seems to have been lost and the builders of the Middle Ages had to resort to the use of lime and silt mortars, which were not very durable, as evidenced by the ruins of this age.

102. Re-discovery of Cement. The re-discovery of the method of manufacture of hydraulic cement, a cement that will set or harden under water, was made by John Smeaton, an English engineer, in 1756. He discovered that limestone containing clay, when burned and then ground until very fine, produced a material which would not only set under water, but also resist the action of water. This we call natural cement. The manufacture of this natural cement on a commercial basis is credited to Joseph Parker, who established a factory in 1796 and called his product Roman Cement. Other factories were established in Europe about the same time.

103. Natural Cement in America. In 1818, Canvass White established a factory at Fayetteville, New York, for manufacturing natural cement on a commercial basis. Other plants sprang up along the canals in New York state; also in Ohio, and a plant was established near Louisville, Kentucky. The output for a number of years was very small—about 25,000 barrels per year. After the Civil War, during the reconstruction period, an impetus was given to the cement industry, and the production of natural cement reached its maximum in 1899, when 10,000,000 barrels were produced. Since then, the production of cement from natural stone as found in the quarries has been on the decline. At the present time practically all cement used in America is artificial cement, or Portland cement.

104. Portland Cement. The process of making artificial cement, or Portland cement, was discovered by Joseph Aspdin, an Englishman, in 1829. The cement was given its name because it resembles the Portland rocks near Leeds, England. In the United States it was first manufactured in 1870 at Copley, Pennsylvania. Its use has increased so rapidly that now the output amounts to about 100,000,000 barrels per year. Portland cement manufacturing plants can now be found thruout the country. Wherever there is an abundance of suitable limestone and shale, or clay, and a supply of fuel and labor, a cement plant can be successfully operated. Portland cement is different from natural cement, in that the materials of which it is made are carefully proportioned and artificially mixed. The essential components of Portland cement are silica, aluminum and lime, with small quantities of other materials. The silica and aluminum are in the clay. The material is first ground, then mixed in proportion of three parts of limestone to one of clay; it is then burned to a clinker and re-ground to proper fineness. While there are a great many brands of Portland cement on the market, the composition is practically constant and the buyer can feel safe in buying any recognized brand.

CHAPTER XII
PROPERTIES AND USES OF CEMENT

105. Properties. The properties of cement with which every builder is most concerned are those of strength and permanence. The requirements ordinarily mentioned are proper fineness, proper setting qualities, purity, strength in tension, and soundness. A cement that is fresh, free from lumps, properly packed and stored, is nearly always first-class.

106. Mortar. Mortar is a mixture of (1) cement or hydrated lime, or both, (2) sand, and (3) water. It is a plastic mass, the water content being varied with its use. Lime mortars are little used at present because they set slowly, will not set under water, are not very strong, and will deteriorate, due to weathering action. A small amount of lime, 10 to 20 per cent, is usually added to cement mortar to make it work well with a trowel and to make it more adhesive.

107. Definition of Concrete. Concrete is often defined as an artificial stone. It is made by mixing cement with sand and gravel, or broken stone, and water; or, in other words, it is a mixture of (1) cement, (2) a fine aggregate, (3) a coarse aggregate, and (4) water. The addition of water causes the cement to undergo chemical changes forming new compounds that develop the property of crystallizing into a solid mass. The strength and durability of plain concrete (that is, concrete without reinforcing) varies with:

      a) The quality and amount of cement used.

      b) The kind, size and strength of aggregate.

      c) Correctness of proportioning.

d) Method and thoroness of mixing.

e) The amount of water.

f) Method and care of placing.

g) Method of curing.

h) Age.

108. Aggregates. As ordinarily employed, the term aggregates includes not only gravel or stone—the coarse

Fig. 137. Gravel bank.

material used—but also the sand, or fine material, which is used with the cement to form either mortar or concrete. Fine aggregate is defined as any suitable material that will pass a No. 4 sieve or screen (having four meshes to the linear inch), and includes sand, stone screenings, crushed slag, etc. By coarse aggregate is meant any suitable material, such as crushed stone or gravel, that is retained on a No. 4 sieve. The maximum size of coarse aggregate depends on the class of structure for which the concrete is to be used.

The fact that the aggregates may seem to be of good quality and yet prove totally unsuitable (Fig. 137), shows that study and careful tests are necessary if the best results are to be obtained. The idea that the strength of concrete depends entirely upon the cement, and that only a superficial examination of aggregates is necessary, is altogether too prevalent. The man who recognizes the quality of his aggregates, who grades them properly, sees that they are washed if necessary, then mixes them in proportions determined by thoro testing, study or actual experience, is the one who will make the best concrete.

In the selection and use of sand, more precaution is necessary than for the coarser aggregate, due to the physical condition of sand and a wider variation in properties. A knowledge of these properties and of the method of analysis to determine the suitability of sand for use in mortar and concrete, may easily be applied to an analysis of the coarse aggregate.

109. Presence of Rotten or Soft Pebbles in the Gravel. In many cases, gravel from the old glaciers has been used, which have been so badly weathered that the pebbles can be crushed between the fingers. In other cases, small lumps of shale or sandstone are mistaken for gravel. These lumps are not strong at best, and, under the action of water, especially alternate wetting and drying, they go to pieces. No pebbles which can be scratched with a thumb nail, or crushed in the fingers, are suitable for concrete. If there are only a few of them in gravel which is otherwise good, they will not seriously weaken the concrete, but it is a good deal better not to use them at all, since a hard concrete cannot be made from soft materials.

110. Presence of Dirt in the Aggregate. Most gravels and sands contain some clay, but clay in amounts up to three per cent by weight is not especially harmful. More than three per cent is harmful. Where gravels contain organic matter of any kind, the concrete made from them is very likely to go to pieces, and they should not be used unless the dirt can be washed out. Clay may also be removed by washing. To test for amount of dirt, shake up four inches of sand or gravel in a quart fruit jar, three-fourths full of water, for four or five minutes. Then let it stand three hours. If there is more than 1/2″ of dirt on top of the material, it is too dirty to use without washing.

111. Vegetable Matter in Sand. A coating of vegetable matter on sand grains appears not only to prevent the cement from adhering, but to affect it chemically. Frequently, a quantity of vegetable matter so small that it cannot be detected by the eye, and only slightly disclosed in chemical tests, may prevent the mortar from reaching any appreciable strength, Concrete made with such sand usually hardens so slowly that the results are questionable and its use is prohibited. Other impurities, such as acids, alkalis or oils in the sand or mixing water, usually make trouble.

Where limestone is used in an aggregate, it is well to see that the pile of limestone is thoroly wet down before using. This is for two purposes—(1) to remove the coating of dust which would otherwise prevent the formation of a bond between the cement and the stone, and (2) to allow the stone to absorb water before the mixing process. Limestone will absorb a great deal of moisture, and, if mixed dry, it is liable to take up part of the water needed in the process of setting or crystallizing.

CHAPTER XIII

PROPORTIONS AND MIXTURES; HANDLING OF CONCRETE

112. Proportions. The theory of proper proportions is to use just enough sand to fill the air spaces or voids in the coarse aggregate, and enough cement to fill the air spaces in the sand, and also to coat each particle and thus serve as a binder. The small contractor in actual practice rarely attempts to carry this out; in fact, he seldom accurately measures the materials that go into the job. He uses a little cement, some sand and gravel, and, under average conditions, may get fair results. It is no wonder, however, that we find sidewalks going to pieces, foundations of buildings cracking and disintegrating when the work is done in such a haphazard fashion.

To make a concrete that is strong as well as economical, it is essential that the materials be well graded from the larger to the smaller-sized particles so that the voids around the particles are reduced to a minimum. The absolute elimination of voids is an ideal condition which we should strive to obtain. However, the densest concrete is not always the strongest. In some cases, a rather porous mixture with a small amount of fine aggregate is stronger than another piece of concrete with a great deal of fine aggregate and a small amount of coarse material, although the latter mixture would be the denser of the two.

113. Requirements of Good Concrete. The proper proportions to use, under practical conditions, will depend on the use to which the concrete is to be put.

The three properties which are most often required are: (1) Strength, as in bridges, buildings, etc.; (2) resistance to wear, as in concrete sidewalks and roads; (3) water-tightness, as in water tanks, silos, etc. The practical mixtures that are ordinarily used for different kinds of concrete work are as follows:

114. Standard Mixtures. Rich mixture of 1 part cement, 1-1/2 parts sand, and 3 parts broken stone, or gravel, commonly called a 1:1-1/2:3 mixture, is used for columns of reinforced concrete buildings, for thin water-tight walls where very dense, strong concrete is required, and under all similar conditions.

A good, standard mixture of 1 part cement, 2 parts sand, and 4 parts broken stone, commonly called 1:2:4 mixture, is used for reinforced concrete work of all kinds, for water tanks, thin walls, etc.

Medium mixture of 1 part cement, 2-1/2 parts sand, and 5 parts broken stone, commonly called 1:2-1/2:5 mixture, is used for all plain concrete, that is, concrete without reinforcing—for foundations, walls, floors, etc. When the walls are to be water-tight, a 1:2:4 mixture should be used instead.

Lean mixture of 1 part cement, 3 parts of sand, and 6 parts broken stone, commonly called 1:3:6 mixture, is used for very heavy mass concrete where the loads are wholly compressive. Still leaner mixtures are sometimes used for very heavy foundations and abutments, but are not recommended for general use.

115. Common Errors in Proportioning Concrete. A rather common error that is made by the inexperienced concrete worker is to assume that when mixing one cubic foot of cement, two cubic feet of sand, and four cubic feet of gravel, he will secure seven cubic feet of concrete. This is an entirely erroneous idea, as the sand would simply fill the voids in the coarse material, and the cement would fill the voids in the sand and coat the particles of sand and gravel or stone. Since the amount of cement and sand used is more than enough to fill the voids in the gravel, the resulting concrete will be slightly more than four cubic feet, about 4.25 under average conditions. The same error is often made when unscreened, bank-run materials are used. In attempting to secure the equivalent of a 1:2:4 mixture, the contractor will use one part of cement to six parts of bank-run material, when, in reality, he should use only about 4-1/4 cubic feet of bank-run material to one cubic foot of cement to get the equivalent of a 1:2:4 mixture. This is assuming that the bank-run material is of the correct proportion of one part of fine aggregate to two parts of coarse aggregate, which should be accurately determined by testing. The only safe method of using bank-run materials is to screen them before using. Then when the materials are used, the proportions can be definitely secured.

116. Determining Quantities for a Job. In determining the quantities of material for a job, one must remember that the volume of concrete is only a little greater than the volume of coarse aggregate; in fact, this is often taken as a basis for estimate of materials needed. For example, suppose it is required to make 54 cubic feet of concrete of a 1:2:4 mixture. It is assumed that 54 cubic feet of coarse aggregate, 27 cubic feet of sand, and 13-1/2 cubic feet, or 13-1/2 sacks of cement are required. Another rule which may be used for all standard proportions is to take the sum of the proportions and divide into the number 11; the quotient will be the number of barrels of cement required to make one cubic yard of concrete of the particular proportion. For example, $\dfrac{11}{1+2+4} = 1-4/7$ barrels of cement, or 6-2/7 sacks (4 sacks to a barrel) for one cubic yard of concrete. Since 54 cubic feet, or 2 cubic yards, of concrete is required in the above job, it will take 2 x 6-2/7, or 12-4/7 sacks of cement, 25-1/7 cubic feet of sand, and 50-2/7 cubic feet of coarse aggregate. For a small job, the first method may be used, but with the larger job, the latter method, which is more accurate, should be adopted.

117. Requirements of Good Mixing. The requirements of good mixing are: (1) That every particle of sand and stone is coated with cement paste, (2) that the sand and stone are evenly distributed through the mass, and (3) that the whole mixture is of a uniform consistency. A poorly-mixed concrete may be known by its lack of uniformity in color and the separation of fine and coarse material. It is just as important to have materials thoroly and carefully mixed as to have them properly proportioned. It is considered so important by well-informed concrete contractors, that they require the materials to be mixed for a definite period of time, if mixed by machine method, or turned a definite number of times if mixed by hand. Up to a certain limit, it has been found that the strength of the concrete is directly proportional to the length of time it has been kept in the mixer. (In the specifications for the construction of some concrete work, the time of mixing is definitely stated.)

118. Hand-Mixing. A water-tight platform is the first requirement for successful hand-mixing. In mixing by hand, there is always a tendency to mix in small units, which is sometimes a mistaken idea. It is usually best to mix at least enough so that one sack of cement or one cubic foot can be taken as a unit because, if the sack is emptied and only a part of a sack is taken, the cement will fluff up and form more than one cubic foot.

119. Procedure in Hand-Mixing. In the actual process of mixing, it is usually best to spread the sand on

Fig. 138. Spreading cement on sand.

the mixing board, and on top of this spread the sack of cement (Fig. 138); then two men using square-pointed shovels turn this sand and cement over several times until the streaks of color are merged into a uniform shade throughout the entire mass. The coarse aggregate is then added (Fig. 138-a), and during the first turning, water is added by means of a hose or from a bucket (Fig. 139). Care must be observed to prevent washing the cement out of the mass. It is best to turn the materials several times (Fig. 139-a), adding a small amount of water each time until it reaches the proper consistency. The only objection to the hand method of mixing is that a great deal of labor is involved, and this, in some

FIG. 138-*a*. Measuring coarse aggregate.

FIG. 139. Adding water to mixture.

cases, reduces the quality of the concrete because of the fact that the materials are not mixed as thoroly as when mixed in a mixing machine.

Fig. 139-*a*. Turning the mixture.

Fig. 140. Batch mixer.

120. Machine-Mixing. There are two types of machine mixers in use—the batch mixer (Figs. 140 and 140-*a*) and the continuous mixer. The latter type is not as satisfactory as the batch mixer and is seldom used except on small jobs. Better results can be obtained with the batch mixer, because a definite quantity of materials is added and thoroly mixed before any concrete is discharged from the mixer. By allowing the materials to remain in the mixer for a definite period of time, they are more completely mixed, and all parts are of uniform proportion. In the continuous mixer, the dry materials are fed automatically from a hopper into a mixing trough where water is added and where the entire mass is mixed and carried along by blades to the discharge end, where the concrete is discharged continuously

Fig. 140-a. Another batch mixer.

121. Consistency of Mixtures. The amount of water used in making concrete will depend on the use for which the concrete is intended. There are three consistencies ordinarily referred to in discussing concrete. They are generally called the "dry," "quaky" and "wet" mixtures. The dry mixture is of about the consistency of damp earth and is used where the concrete is tamped into place. The quaky mixture is so named because it is wet enough to quake when it is tamped. It is used in molded products requiring reinforcing, such as fence posts, beams, columns, etc. It is also used in sidewalks, floors and foundations. The wet mixture contains enough water to permit its flowing from the shovel or conveyors from elevators to various points in the construction of large buildings. There is a tendency on the part of some contractors to make the mixture very wet so as to make it flow more easily. This will cause the separation of the coarse materials from the finer and reduce the quality of the concrete. One main point to remember in connection with the proper consistency is that the materials must not be too dry nor too wet; either condition will cause the separation of the coarse material from the mortar.

122. Placing of Concrete. No time should elapse between the "mixing" and the "placing." One's judgment must be used in placing; the method adopted will depend on the particular job. The essential feature in placing is to prevent the separation of the stone from the mortar.

123. Three Methods of Placing Concrete.

1) A dry mixture of concrete is placed by thoro tamping or by pressure. The density and the final strength of a dry mixture will depend on the extent of tamping. This method of placing concrete is used in making concrete products that are not reinforced, such as blocks, bricks and jardinieres. The material must be carefully tamped as the mold is being filled, either by hand or by power machines.

2) A quaky mixture can be placed by agitation or slight tamping. This method is used in making reinforced products, such as posts, large tile and tanks; also for slab work, such as floors and sidewalks. Some forms are designed so they can be vibrated to settle the concrete into place.

3) A wet mixture is simply deposited into place, and requires no tamping. A spade or board should be used for working large stones back from the forms and leveling the surface so that no large stones are left uncovered (Fig. 141). This mixture and method of placing is used in nearly all reinforced structures where the reinforcing is put in place before the concrete is poured. For large structures, special apparatus is used for elevating the material.

Fig. 141. Working stones away from surface.

124. Handling Concrete. There are three common ways of conveying the mixture:

    a) It may be shoveled off the board directly into the work.

b) It may be shoveled into wheelbarrows and wheeled to position and dumped.

c) It may be elevated by buckets and hoisting apparatus.

Where the concrete is mixed by hand, it is usually transported by wheelbarrow (Fig. 142). For machine-mixed concrete where the work is of some magnitude,

FIG. 142. Moving concrete with wheelbarrow.

some flexible method of handling it is best, usually a tower with elevating equipment. Derricks and bucket elevators are also used. The one objection to the use of tower and chutes is the tendency, in order to secure easy flow, to use too much water, causing a separation of the fine and coarse aggregate.

CHAPTER XIV

FORMS FOR CONCRETE; CURING CONCRETE

125. Necessity of Forms. The plasticity of concrete, and the readiness with which the material can be adapted to all shapes and sizes of construction, which are two of the chief merits of the material, make necessary the use of forms in connection with it.

126. Importance of Form Construction. The design and construction of forms is one of the most serious problems of concrete work. As a rule, on small work, the expense of the forms is from one-fourth to one-half of the total cost of the work in place. Many people do not appreciate this fact and neglect the forms with the result that the finished work is of poor quality, or else the forms have cost too much. The shape, dimensions and finish of the work all depend on the forms, and it is not possible to do good concrete work without good forms.

127. Earth Forms. In foundation walls, where care has been observed in excavation and the earth stands up properly, it can be used. Earth can be used also in making well tops, etc., where the work can be fashioned out in the clay. The earth must be wet down thoroly to keep it from absorbing too much moisture from the concrete. A combination of wood and clay can be used. Molds of wet sand are used in ornamental work. Frequently, colored sands are used for this purpose, providing both the finished surface and color to the concrete.

128. Cast, Wrought or Galvanized Iron Forms. These are used where a smooth surface is desired without further treatment after removal of forms. In construction work, where the same type of form is used a great number of times, it is economy to have a material which will not go to pieces, warp, swell and crack, even tho the first cost may be higher. Steel forms, if strongly built, will meet these conditions. Forms made of iron are more easily cleaned, and can be used a great number of times. Rusty iron is not good for forms; the concrete will stick badly. There are steel forms on the market for concrete posts (Fig. 143), water tanks, silos, etc.

FIG. 143. Commercial post mold.

129. Wood Forms. Wood forms are most common, and are used most for concrete work on the farm. The chief reason for this is that lumber can be obtained easily in small quantities, and there is always a certain amount of old lumber around every farm.

130. Requirements of a Good Form.

a) One that can be used a number of times.

b) One that is strong so it will not bulge or crack.

c) One that is tight and free from leaks.

d) One that is true and properly aligned.

e) One that is made of good material suited to its use.

Soft woods are better than hard because they (a) are cheaper, (b) do not crack so badly, (c) are an easier material to work. Spruce and yellow pine make good forms; the boards used should be sound and free from knot holes. Partly green lumber is better than either green or kiln-dried, because it will swell just enough to make tight joints without buckling. Dressed lumber has several advantages over undressed: (a) It makes truer work, (b) tighter joints, (c) smoother surfaces, (d) forms are easier removed, and (e) forms are easier cleaned.

131. Use of Old Lumber for Forms. Where old lumber is to be used, it should be sorted and listed so that new lumber can be ordered of proper sizes that will work in best. Care must be observed in the use of old lumber to see that it is strong enough to support the load put on it by the concrete. A great deal of expense can be avoided by taking advantage of old lumber.

132. Sharp Corners in Forms. Sharp corners should be avoided as much as possible in concrete work. It is best to bevel the corners by setting strips in the forms, especially on inside angles. This gives both greater strength and better finish to the work.

133. Removing Forms; Care of Forms. Forms should not be removed until the concrete is thoroly set. The time of setting varies with the wetness of the mixture, and with the weather. Concrete sets much faster in warm, dry weather than in cold or damp weather. On foundation walls or similar work, where the concrete is used in direct compression, the forms may be removed in a few days. Under floors or beams, which are subjected to bending, the forms should be left two weeks or longer.

Care of Forms:

Forms for concrete posts, etc., should be oiled with a heavy oil before they are used. As soon as they are removed, they should be thoroly cleaned with a stiff wire brush. Oiling metal forms or molds after using is better practice than to wait, as a coat of oil prevents rust. In removing wooden forms, care must be observed to avoid splitting boards. All boards should be cleaned, the nails pulled, and boards stacked to prevent warping.

Curing Concrete:

Proper curing of concrete is very essential to success. It must not be allowed to dry out too rapidly. If freshly made and exposed to the intense heat of the summer's sun, it must be protected. The drying out not only produces check cracks, but hinders the setting action of the concrete, making it weak. Floors and walks that are protected and kept moist for some days will harden into a very dense and almost dustless material, while those not adequately protected will wear rapidly and be dusty.

CHAPTER XV

REINFORCING CONCRETE; CEMENT-WORKING TOOLS

134. The Principle of Reinforcing. Plain concrete is strong in compression, but will not resist a very great load when in tension. Steel is a material that has a great tensile strength, as well as compressive strength, so, by combining the two, we have a resultant material which is strong in both tension and compression, and can be adapted to most any use.

The design of reinforced concrete structures is quite technical and has no place in a text of this character. For simple types of construction, such as reinforcing for a silo, water tank, retaining wall, fence posts and well tops, the student can refer to tables in hand-books, or use his best judgment, bearing in mind that the amount of reinforcing will vary from 3/4 to 1-1/2 per cent of the cross-section of the member being reinforced.

135. Compression and Tension in Beams. A consideration of the basic principles underlying simple reinforced concrete construction may be of interest. Consider a simple beam of uniform cross-section like a 2″ x 4″, supported at each end, with a load applied at the center (Fig. 144). It will be found that the upper part of the beam will be in compression, or tending to crush together, and the lower part will be tearing apart, or in tension. It will be noted that there is a plane perpendicular to the force applied and cutting the beam in half where there is neither tension nor compression. This is called the neutral plane or neutral axis.

129

Now, since the lower part of the beam is in tension, and since concrete is weak in tension, it is apparent that to make the lower part of the beam as strong as the

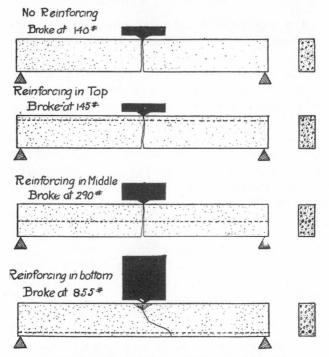

FIG. 144. Results of different placing of reinforcing.

upper part, we must imbed some material in the beam that is high in tensile strength. Steel is not only high in tensile strength, but its co-efficient of expansion is the same as that of concrete, so a strong bond between the two can be maintained. It must be kept in mind that the steel must be placed as far as possible from the neutral axis to be most effective. It must not be placed too near the surface of the concrete. It must be kept in mind, further, that in any reinforcing job, the steel must be placed where it will be under a tensile strain. Fig. 144 shows the relative strength of a concrete beam with reinforcing placed in various positions.

136. Kinds of Reinforcing. As to the kinds of reinforcing, probably square twisted steel rods, or the deformed bars, are best. Round rods are sometimes used, but they should be carefully anchored to give the best results. Some engineers specify either the twisted or the deformed rods, since a better bond is secured between the concrete and the steel with this type of reinforcing. Some contractors claim that a small amount of rust on the reinforcing is advantageous. A very small amount of rust may be of some value in forming a bond between the concrete and the steel. However, if the steel is left outside until it has become pitted with rust, the resultant piece of work would be weakened, as the bond between the steel and concrete would be a poor one.

137. Use of Scrap Iron for Reinforcing Concrete. It is thought by some that scrap iron will make good reinforcing. It is seldom true that as good a job can be secured by using scrap iron, old gas pipe, etc., as by using regular reinforcing steel. Gas pipe that is of value as pipe is expensive reinforcing material.

138. Tools for Concrete Work. Very inexpensive tools are required for concrete work; in fact, few tools that are not found on the average farm. For special work, special tools will be required, which may be secured from any good hardware supply house. A panel containing many of such special tools is shown in Fig. 145.

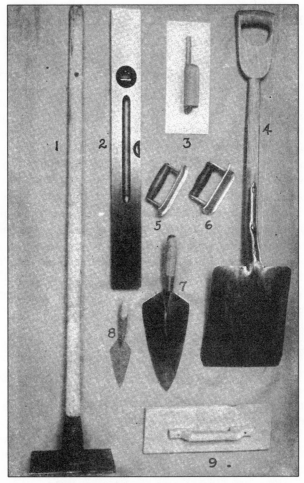

FIG. 145. Tools used in concrete work.

1, Tamper; 2, level; 3, finishing trowel; 4, shovel; 5, groover; 6, edger;
7 and 8, trowels; 9, hand float.

The tools commonly used in farm concrete work and such as will be needed in the following projects are as follows:

a) For screening aggregate—a moulder's riddle for small work, or a screen, as shown in Fig. 138.

b) For washing aggregate—a trough in which dirty aggregate can be freed from clay.

FIG. 145-a. Mixing concrete.   Measuring boxes and other equipment.

c) For mixing and placing—a platform, as shown in Fig. 142; shovel, spade, hoe, tamper, striking board and wheelbarrow (Figs. 142, 145 and 147).

d) For measuring ingredients—a measuring box, as shown in Fig. 145-a.

e) For finishing—trowel, edger, groover, hand float, etc. (Fig. 145).

f) Water container—a barrel or, for large construction, a water tank, to which is attached a hose.

Tools for wood construction—carpenter's square, hammer, saws (rip and crosscut).

Note: To these tools there may be added a mixer, either hand or power, depending upon the extent of the work to be undertaken.

CHAPTER XVI

Projects in Concrete Construction

Project No. 1

139. Study of Concrete Construction and Concrete Materials (Figs. 146 and 146-*a*).

Requirements: To investigate as many types and classes of concrete work as are available and as time

Fig. 146. Defective concrete walk.

will permit. The following are suggested: Concrete tanks—one circular and one rectangular—sidewalks, feeding floor, foundation wall, retaining wall, fence posts, roads, tile, and block. These should be studied with the idea of noting the results obtained by use of poor materials and poor workmanship, and the use of good materials and careful workmanship, and also to determine quantity of material

135

needed for certain jobs. Make a written report on results obtained.

Tools Needed: Rule for taking dimensions.

Preliminary Instruction: Carefully read the preceding paragraphs. Keep in mind the general principles of concrete construction. Remember the

FIG. 146-*a*. An attractive walk.

requirements for well-made concrete, good aggregate, proper proportions, careful mixing and placing, and correct reinforcing.

Working Instructions:

a) Examine at least one of each of the different types of concrete work listed under requirements and report on the following:

1) General condition of the job.

2) If cracks are forming, to what are they due?

3) Where cracks have formed, note if there is a clear fracture, or, if the aggregate is pulled out of the mortar.

4) Was the coarse aggregate worked back from the form when placed?

5) Was a dry, quaky or a wet mixture used?

6) Does the job indicate that the forms were well made?

7) If the forms were not well made, what was wrong with them?

8) Why do poor foundations often cause cracks in concrete walls?

9) Examine the foundation and note if care was observed in its preparation.

10) Is the foundation well drained?

11) What is the effect of poor drainage under a foundation wall? Under a sidewalk? Under a road?

12) What precaution should be taken in constructing an earth-retaining wall?

13) If cracks have formed, were they due to lack of or insufficient reinforcement?

14) Where should reinforcing steel be placed in such a wall? Why?

15) Why should a wet or quaky mix be used where the concrete is reinforced?

16) Write a brief statement about each piece of work, giving your opinion as to what should be done to make a first-class job.

b) Examine concrete material, note the quality, etc.

1) Examine a sack of cement. See if it is free from lumps and is fresh.

2) Note the brand of cement examined.

3) Note the condition of the bag.

4) Why is it important to take care of the bags and not allow them to get wet?

5) Examine available sand. See if it is clean, free from clay, coal or other organic matter.

6) Test a small quantity of sand for clay by putting about 4″ or 5″ in a fruit jar, adding water and shaking until clay is in solution. Set aside and let clay settle on top of the sand. Determine the per cent of clay present.

7) What per cent of clay is allowable in average concrete work?

8) Examine available gravel or broken stone. See if it is free from clay, organic matter or soft particles.

9) Can you scratch the stone with your thumb nail?

10) What would the effect be to use soft stone in making concrete?

11) Is the coating of fine dust ordinarily found on lime-stone detrimental in making concrete?

12) Examine some bank-run sand and gravel as in Nos. 5 and 8.

13) Why is it poor practice to use ordinary bank-run material for making concrete?

14) Suppose it is required that a piece of concrete work be made of bank-run material that has 50 per cent as much sand as gravel, and that it is to be equivalent in strength to a 1:2:4 mixture where the sand and gravel are graded. How much would be required for each sack of cement?

c) *Problems:*

Assume a 1:2:4 mixture and determine the amount of materials needed; also cost:

1) To make a circular tank 6′ 0″ inside diameter at the top and 5′ 4″ diameter at the bottom, and 2′ 0″ deep. The wall of the tank to be 4″

thick at the top and 8″ thick at the bottom, the bottom of tank to be 5″ thick.

2) To make a rectangular tank with the same capacity as No. 1, to have same thickness, walls and bottom, and to be 4′ 0″ across inside at the top.

3) To make a sidewalk 40′ long, 3′ wide, and 4″ thick.

FIG. 147. Making block.

4) To make six concrete fence posts. Assume 20 cents a post for steel.

140. Molded Concrete (Figs. 147 and 147-*a*).

Project No. 2

Requirements: To make tile of different sizes, block, flower boxes, and other pieces of concrete work requiring a dry mixture.

Tools Needed: Shovels, bucket, measuring box, screen, mixing platform, trowels, and suitable molds.

Material Needed: Cement, sand and water.

Preliminary Instructions: The principles outlined in the discussion on selection of sand must be kept in mind. Only the best sand should be used. In the kind of work outlined in this project, a relatively dry mixture must be used, one about as wet as damp earth when plowed; with such a mixture the molds may be removed immediately. Good results cannot be obtained if the materials are either too wet or too dry. The product will stand up due to the adhesiveness of the concrete, and it must be

FIG. 147-*a*. Making flower box.

allowed to set thoroly before handling. Careful measurement of materials is an essential requirement of all concrete work.

Working Instructions:

1) Use a 1:3 mixture; that is, one part of cement and three parts of sand, for the various jobs outlined. Where coarse aggregate is available, the block may be made of a 1:2:4 mixture with a 1:2 face.

2) After measuring the sand, spread it out in a thin layer on a water-tight platform; then spread the cement on top of the sand and mix together dry, continuing the turning until the color is uniform and without streaks. Water is then added slowly

from a sprinkling can or by a hose, the mixing being continued until all parts of the mass are of the same color and wetness.

3) Carefully clean the molds and apply a thin film of oil after using, so they will be ready for the next job. See that they are absolutely clean before placing any concrete.

4) Tile, block, etc., are made by thoroly tamping or pressing the concrete in the molds to be used. Any dry mixture must be thoroly tamped to make dense concrete.

5) Extreme care must be observed in removing the molds to avoid cracking the product or causing it to get out of true shape. Tapping the mold slightly will often prevent failures.

6) After the product has set for twenty-four hours, sprinkle it carefully with water, repeating this frequently for ten days. It should not be used for one month or more. Where such products are made on a commercial scale, they are often cured in a steam kiln.

7) Write a report on each product made. Give the general method of procedure and why. Carefully determine cost of materials in each.

Project No. 3

141. Sidewalk and Floors (Figs. 148 and 148-*a*).

Requirements: To prepare foundation, construct forms to proper grade and position, and construct sidewalk and floors of various kinds requiring a quaky mixture.

Tools Needed: Shovels, buckets, measuring box, screen, mixing platform, trowels, edger, groover and float. Woodworking tools suitable for constructing forms.

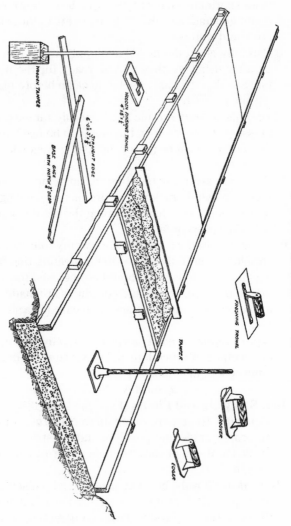

FIG. 148. Sidewalk construction.

Materials Needed: Enough cement, sand and gravel or broken stone, and water to complete the job. For a 1:2:4 mixture, 1 sack of cement, 2 cubic feet of sand and 4 cubic feet of gravel should make 4-1/4 cubic feet of concrete, or 13 square feet of walk or

FIG. 148-*a*. Boys constructing sidewalk.

floor 4 inches thick. Material for forms must also be provided— 2 x 4's with suitable stakes are very satisfactory.

Preliminary Instructions: The general principles of proper proportioning, mixing and placing should be carried out in constructing sidewalks the same as in any other type of concrete construction. In work of this class, a quaky mixture should be adopted. A walk should not be made by putting down coarse material and pouring over it a cement-sand mortar. Because of the close resemblance between other types of floor constructions, such as feeding floors, barnyard pavements, basement floors, garage floors, etc., and concrete walks, only a detailed description of the construction of one type will be given. The location and drainage of any walk or floor must be considered.

Working Instructions:

1) In laying out a walk, the first consideration is its location with reference to buildings and the road. If it is to be located with reference to a certain building, either parallel or at a right angle, it should be definitely located by careful measurement. Stake out the position of the walk and draw a tight string so that the surface may be properly leveled to a uniform grade. This surface should be thoroly tamped to prevent any settling after the walk has been placed.

Under certain conditions, where there is a tendency for water to collect under a walk, cinders or gravel may be used as a sub-base. Ordinarily, the concrete will be placed directly on the well-tamped soil.

2) Make the forms of 2″ lumber, either 4″ or 5″ wide, depending on thickness to which walk will be made; 4″ is satisfactory for most conditions. Place the forms carefully to grade, and fill in with earth and tamp any low places before placing any concrete. Proper and careful alignment of the forms is the most important feature to insure a good-looking job. Definite measurements must be taken to locate carefully the position of the forms. A level should be used in order to see that the forms are properly leveled.

To support the forms, drive stakes every 3' or 4'. It is considered good practice to put in alternate sections of the walk, and, after this has set, remove the end form and fill in the section not built. For short pieces of walk, however, this is unnecessary. If it is desired to give the walk a slight slope to one side, this can be done by use of a level and straight-edge, placing one of the 2 x 4's lower than the other—1/4" to 1' is a good side slope for a walk, and will cause it to shed the water very quickly. To make such a slope on a walk to be 4' wide, the form in the direction of the slope will be set 1" lower than the upper one.

3) For a one-course walk, nothing leaner than a 1:2:4 mixture should be used; that is, one part of cement to two parts of sand and four parts of broken stone. Both sand and gravel, or broken stone, should be clean and free from clay or other foreign material. If bank-run materials are used, careful screening to get the proper proportions is necessary.

4) After measuring the sand required for one batch, spread it out in a thin layer on a water-tight platform; then spread the cement on top of the sand and mix together dry, continually turning until the color is uniform and mixed together without streaks. The cement and sand is then spread out and the coarse material placed on top. It is then again mixed and water is added until it is of a quaky or jelly-like consistency. Such a mixture can be quickly spread about in the forms and easily leveled with a strike-board resting upon the top of the forms. Avoid using too dry a mixture for floor construction.

5) The concrete may be shoveled directly from the mixing board into the form, or handled by means of a wheelbarrow.

6) Level the material off and tamp it enough to force the coarse material in from the surface, and bring enough cement-sand mortar to the surface to make a smooth finish. Slight tamping is also done to remove any air or water bubbles from the material. A spade or board should be pushed in along the side of the form so that all coarse material will be worked back from the edge of the walk.

7) If the walk is to be 50′ or more in length, an expansion joint should be placed approximately every 50″. This expansion joint can be provided by putting in aboard 1/2″ thick at intervals of 50′, which should be removed after the concrete has properly set, and the groove filled with heavy asphalt or Tarvia. To leave the board in place is worse than no expansion joint. This practice is sometimes followed.

8) If the material has been mixed to the right consistency, the surface can usually be given its final finish within one-half hour after placing. The first part of the finishing should be done with a wood float, merely to level off the surface and make a smooth job. If it is desired to make a very smooth surface, continue the finishing by using a steel trowel. The troweling process tends to bring an additional amount of cement and fine sand to the surface, making it very slick. Ordinarily, this practice is not desirable. The edges of the walk must be finished with the edger to give a rounded corner. To line the walk off into sections, use a straight-edge and

groover. This must be done before the concrete has begun to set because it is sometimes necessary to force coarse material farther below the surface to make a good groove. Lay off the walk so that the length of the sections will be about one and one-half times the width; that is, a walk 2′ wide should be divided into sections 3′ long, or a walk 3′ wide into sections 4-1/2′ long.

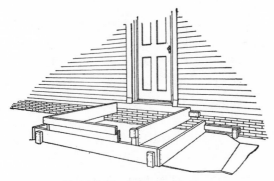

Fig. 149. Form for double step

9) If the walk is built where it is exposed to extreme drying from the sun, it is well to protect it until it has set. The protection may be in the form of moist sand or a tarpaulin of some sort. The hot sun and dry winds will tend to remove the moisture from the concrete and prevent it from hardening. Sprinkle the surface for a week or ten days, after which the walk may be put into use.

142. Constructing a Doorstep (Figs. 149 and 149-*a*).

Requirements: To prepare foundation, construct the form and place the concrete for a step and platform at some door, or a step at the curb, walk or drive-way entrance to the house. A 1:2:4 mixture should be used for such a job.

Tools Needed: Same as in Secs. 140 and 141.

Material Needed: Enough cement, sand or gravel or crushed stone, and water to complete the job, using a 1:2:4 mixture. A sufficient quantity of fence

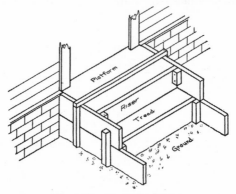

FIG. 149-*a*. Doorstep form.

boards and 2 x 4's to construct the form. Boards 1-1/2″ thick are preferred to light ones.

Preliminary Instructions: In constructing each piece of concrete work, the requirements of good concrete must be ever kept before you. In a job of this kind, the construction of the forms deserves a great deal of attention. If a carriage step or small step at curb is to be built, it will require little foundation; the ground should be leveled and well tamped. For a doorstep, a sub-base should be provided, and if it is a large one, the central portion may be tamped full of clay to serve as a filler; in this case, not less than 6″ of concrete should surround the filler.

Working Instructions:

1) Follow general instructions given for concrete construction. Carefully prepare the form for the step to secure correct dimensions as planned—the proper area of platform, the correct width of tread and the correct height of riser. It is suggested that the riser be 8″ high and the tread 10″ wide; then stock 8″ boards can be used as the part of form for riser. Have each part of form properly braced so there is no danger of its bulging.

2) For a solid step, a 1:2-1/2:5 mixture is adequate. If the step is to be made from one level to another without backing, and is to be reinforced, a 1:2:4 mixture should be used; in fact, for small jobs, such a mixture is best.

3) Carefully mix the concrete to a quaky consistency as outlined in Sec. 141. Place the material in the form and tamp it lightly, working the coarse aggregate back from the surface to secure a smooth finish.

4) The finishing coat of one part cement to two parts sand for the platform and the treads should be placed immediately after the surfaces have been leveled off. Where it is not desired to give an extremely smooth finish, enough fine material can be worked to the surface by troweling, and this can be leveled off. The risers and sides of the steps can be finished only after the form has been removed. Forty-eight hours should elapse for the ordinary job to allow for setting. To finish the risers and sides of steps, remove all marks made by forms by the use

of a stiff brush. If care has been observed in working the coarse material back from the form and no air pockets have been formed, this method of finishing is sufficient. If the wall is left quite rough on removal of the forms, they should be wet down and a cement mortar of the same proportions as used

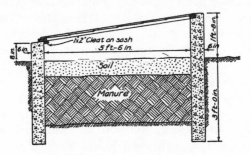

FIG. 150. Section thru hot-bed.

on the treads should be applied with a brush. Keep the step moistened for a week or ten days until ready for use.

143. Hot-bed, Foundation Wall, or a Similar Type of Construction (Figs. 150, 151, 152).

Requirements: To build a form such as needed for the walls of a hot-bed or the foundation for a small building. Determine the quantity of material required. Prepare and place the concrete, remove the form in due time, and finish the job. A mixture of wet consistency should be used.

Tools Needed: Same as in Secs. 140 and 141.

Materials Needed: Enough cement, sand and gravel or crushed stone to complete the job, using a 1:2:4 mixture, a sufficient quantity of boards, and 2 x 4's to make and brace the form, and pieces of wire with which to fasten it together.

Preliminary Instructions:
Concrete is the best material available for foundation wall construction. The super- structure may be built of some other mate- rial, but usually con- crete will be used for the foundation. The particular location of

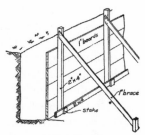

FIG. 151. Foundation wall form.

hot-bed or foundation wall should be definitely decided so the work will not be held back at the beginning of work period. To lay out a rectangular foundation, one should be careful to have all intersections of walls exactly 90 degrees. This can be easily checked by the "3, 4, 5" method. This rule is applied by measuring along one wall a distance from the cor- ner equal to 3 feet; then

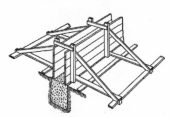

FIG. 152. Wall form above ground.

measure from the same point along the other wall a distance of 4 feet; then, if the two lines form an exact right angle, the distance between the ends of the 3- and 4-foot lines will be exactly 5 feet. For convenience and accuracy, any multiple of 3, 4 and 5 may be used.

Working Instructions:

1) Carefully excavate all soil to proper depth. If the soil is firm, it may be used as the outside form up to the surface, above which a double form will be necessary. For all kinds of foundation walls, it is always essential that the footing be wider than the wall proper, and that it be carried deep enough to be below the frost line. If double forms are necessary, due to the soil caving it will have to be excavated to a greater width.

2) Construct the form with care, duplicating the inside and outside wall dimensions as desired. See that corners are square, walls are well braced, vertical, and carefully aligned. If walls are to be more than six feet high, tie wires should be used in addition to the supporting braces (Fig. 152). If this precaution is not followed, a bulged wall is likely to be the result. The inner form on a hot-bed or other small piece of concrete work, may be supported by braces on the inside, running from one wall to the opposite one.

3) For thin walls up to six inches, a 1:2:4 mixture should be used. Walls more than eight inches thick may be made of a 1:2-1/2:5 mixture.

4) For a job of this kind, the concrete may be mixed to a slightly wet consistency. Care must be exercised to avoid the separation of the coarse material from the fine, which is possible in a wet mixture. Shovel the concrete into form and force the coarse aggregate back from the surface of the wall by means of a spade or a thin board. When the job is a fairly large one, do not mix less than the amount produced when a sack of cement is taken as a

unit. It is desirable to complete the job without interruption after it is started. In case it is necessary that the work be discontinued for a period, see that the surface of the dry concrete is cleaned and thoroly wet down before fresh concrete is poured.

5) To insure against cracks in a concrete wall, a few reinforcing rods bent at right angles and placed at succeeding heights of 12 to 18 inches in the corners, will be invaluable. Reinforcing placed around openings is also recommended

6) The forms on a wall of more than six feet in height should stay on several days. The forms on walls only two or three feet high may be removed in forty-eight hours. As to finishing the surface of the wall, follow instructions given under this heading in Sec. 142.

Note: When a wooden superstructure is to be built on a concrete foundation, it is advisable to set some bolts in the concrete at intervals of every five or six feet, to which the sills may be fastened.

144. Constructing Fence Post (Fig. 153).

Requirements: To construct line fence posts and corner and end posts requiring quaky or wet mixtures and reinforcing.

Tools Needed: Shovels, buckets, measuring box, screen, mixing platform, straight-edge, flat trowels and suitable forms or molds, or woodworking tools suitable for constructing same.

Materials Needed: Cement, sand, gravel or broken stone and reinforcing steel. For line posts, provide 1/4″ to 3/4″ stone; and corner and end posts, 1/4″ x 1-1/4″ stone.

Preliminary Instructions: There is nothing that adds
more to the appearance and usefulness of a fence

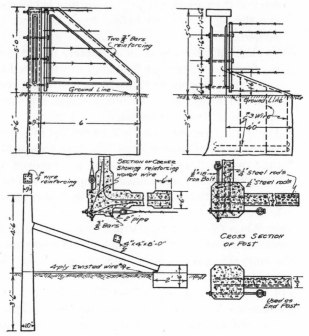

Fig. 153. Corner post.

than a good line of uniform posts, and there is probably nothing that adds more to the appearance of
a farm lay-out than a good, serviceable fence. A
good fence is a real necessity on every farm.

Many of the early concrete posts were failures
because they were not properly made. People tried
to make posts without knowing the first principles

of correct construction. Posts were made of poor material, lean mixtures, and incorrectly reinforced. To make good, uniform posts, provide well-made forms. There are a lot of good patented forms on the market, but homemade forms are just about as good. A very satisfactory form for posts is outlined under woodworking projects, Sec. 79 and Fig. 136.

The chief difference between the construction of line posts, and corner and end posts is that the corner and end posts are usually made right in place, as shown in Fig. 153. The hole is excavated, the form built over it, and the steel tied in place, and the concrete then poured. The method of constructing line posts will be definitely outlined.

Working Instructions:

1) Place forms so they will be level. Clean them with a brush, and apply a thin film of oil before placing concrete.

2) For corner and end posts, a 1:2:4 mixture may be used. For line posts, use a 1:2:2-1/2 mixture, the stone not to be larger than 3/4″. Where materials are not screened, use one part of cement to three parts of sand and pebbles for line posts.

3) Mix the materials to a quaky consistency and fill the form half-full of material. Tamp until the material is free of water and air bubbles.

4) Press two 1/4″ twisted steel rods into the concrete so that they will be within 1/2″ of the corners of the post. Then fill the form full of the mixture, tamp it lightly, and smooth off the surface. Press two more rods in place at each corner about 1/2″ under the surface; then smooth the surface to proper finish.

5) Leave the posts in the molds at least forty-eight hours under most conditions. If the weather is extremely dry and hot, they may be removed earlier. To take out the post, turn down the hinged end of the form, lift the dividing boards between the posts, then grasp the post and slide it on the bottom of form by a pulling motion; after it is loosened, it may be lifted out. Handle the posts with care when green as they are liable to be broken.

6) Set the post on end in sand to cure. Sprinkle daily in dry weather for a week. Do not use until the posts are one month old.

145. Constructing a Circular Stock Tank (Figs. 154, 154-*a* and 155).

Requirements: To construct form according to plan, prepare foundation and construct a circular stock tank to be provided with inlet pipe with float control and outlet. Plumbing work is outlined in Sec. 351 under head of "Plumbing."

Tools Needed: Same as in preceding projects.

Materials Needed: Enough cement, sand and gravel, or crushed rock, to construct tank of a 1:2:4 mixture according to plan. Enough heavy hog wire 30″ high to extend around tank and lap 30″, and enough to extend twice across the bottom and up the sides. For a tank 6′ inside diameter. it will require about four rods of fence.

Preliminary Instructions: A stock tank is a needed piece of equipment on every farm. It should be carefully located with reference to lots for convenience; in fact, it may be placed between two lots or where four lots corner. A drainage outlet must be provided which must be given consideration when the tank is located. Extreme care must be observed in mixing, placing and reinforcing to insure a strong water-tight construction.

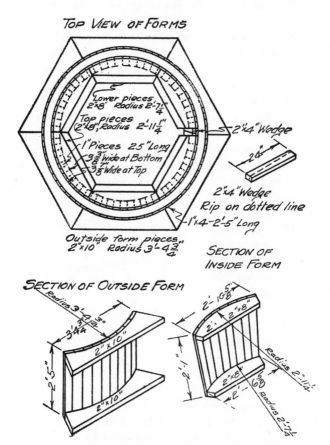

FIG. 154. Circular tank forms.

Working Instructions:

1) Like every other piece of concrete work, the water tank must be constructed on a solid foundation. The soil should be firmly tamped before the form is set in place. In the preparation of the foundation, the proper placing of the outlet and inlet pipes must be given consideration, since both should be brought thru the bottom of the tank. For a large tank, it is well to excavate and form a sub-base of cinders or gravel.

2) The form for a circular tank is the most difficult part of the tank to build. However, by carefully studying the plan, one should experience little trouble. It will be noted on plan that both the inside and outside forms are made in six sections.

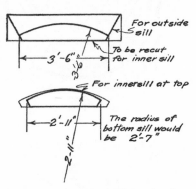

This makes the length of each section equal to one-

FIG. 154-*a*. Detail of sill for circular form.

half the diameter of the tank. If the inside diameter of the tank is to be 6′, the outer length of section for inside form will be 3′, less the thickness of boards used. The inner length of outside section will be equal to one-half the diameter of tank plus the thickness of wall and thickness of boards used. If the wall is to be 5″ thick at the top and boards 1″

thick are to be used, then the sections would be 3′ 6″ long at the inner length. Since the outside surface of the wall of the tank is vertical, both sills for the outside form will be cut the same. The inner surface should be given a slope of 2″ to the foot, or, for a tank 2′ deep, the bottom sill for the inner form will be 4′ shorter than the top one. The sills for this form are best cut from a 2″ x 10″ or 2″ x 12″ timber when a jig or band saw is available, making it possible to get both the inner and outer sills from the same piece, as shown in Fig. 154-*a*.

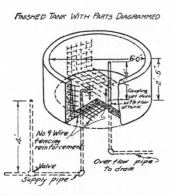

FINISHED TANK WITH PARTS DIAGRAMMED

FIG. 155. Circular tank complete.

After sills are cut, the boards must be carefully fitted to make a tight wall. The boards for the inner section are made 5″ shorter than those for the outer section to allow for thickness of floor. After these sections are completed, they are assembled in place and can be fastened together by strips across ends.

3) For concrete construction such as this, never use a mixture leaner than 1:2:4. For small tanks, a 1:2:3 mixture is better.

4) Follow method of mixing and placing as outlined in Secs. 141 and 143, with the exception of the following: Set the outer form in place and put in floor first. Spread about 3″ of concrete on floor; then put

two or three strips of the hog wire fence across the floor and extend it up the sides. Place balance of concrete for floor and tamp in place. Put a strip of hog wire in place for wall reinforcing, lap the ends, and wire to it the strips that were placed across the floor. Then place inner form in position, carefully center it, and fasten in place with boards nailed across the top. Pour the rest of concrete, keeping the reinforcing near the center of wall. It is desirable to provide a concrete box in the center of the tank to provide protection for the inlet pipe and automatic float. A form for this box should be constructed, set in place and the concrete poured as the tank is being completed. For the outlet pipe, a drain with a 1-1/2″ coupling should be set in a low place in the floor. A short piece of pipe screwed into the coupling and extending to a height that it is desired the water should stand, will act as an overflow.

5) Remove the forms from the tank in about forty-eight hours, and, after wetting it thoroly, apply a cement paint to the entire surface. Allow this coating to set, then wet down again, after which the tank may be filled with water. It should not be put into use for a week or ten days, as the green concrete can easily be broken by stock.

CHAPTER XVII

Supplementary Concrete Projects

146. Constructing Garden or Lawn Roller (Figs. 156, 157).

Requirements: To make a garden or lawn roller, as illustrated, complete with handle for pulling or pushing it.

Fig. 156. Garden roller.

Instructions:

1) Secure a length of drain tile of size desired. If drain tile is not available, an old carbide can or other cylindrical can may be used.
2) Secure lengths of 1/2″ pipe and fittings for axle and handle.
3) Construct a platform on which to make the roller.
4) Lay out a circle on platform slightly larger than tile.
5) Bore a hole in platform, the diameter being equal to outside diameter of pipe for axle.
6) Make cross-frame of two 1″ x 4″ pieces.

7) Center and bore hole in cross-frame as has been done with platform. Nail blocks on ends of cross pieces to hold in place when assembled.
8) Place tile on platform and center axle in place with cross-frame. Nail blocks on platform to hold tile in place.

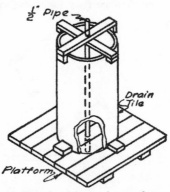

FIG. 157. Details of forms for garden roller.

9) Axle should extend out of the tile at least two inches.
10) Mix concrete 1:2-1/2:4 proportion to a quaky consistency
11) Place concrete in tile around axle. Leave in place for a week or more before using.
12) For handle, assemble pipe and fittings, as illustrated in plumbing project. This makes a very good elementary pipe-fitting exercise. See Sec. 334.

147. A Hog Trough (Fig. 158).

Requirements: To construct form, mix and place concrete, and properly reinforce a trough that will be suitable for feeding slops to hogs.

Instructions:

1) Construct form as illustrated. The inner part of form may be made of heavy clay if it is desired to make the bottom of trough with a curved surface.

2) Provide reinforcing. If the trough is to be more than 4′ in length, 1/4″ rods should be used in addition to the wire netting.

3) Use a 1:2:3 proportion, and mix to a wet consistency.

4) Place concrete and reinforcing.

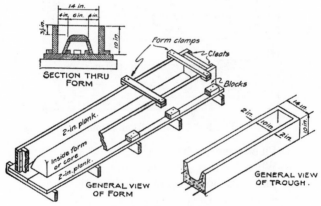

Fɪɢ. 158, Hog trough.

148. Engine or Machine Foundation (Fig. 159).

Requirements: The requirements of this job will depend on the particular machine. A machine subject to a great deal of vibration should have a heavy foundation. The proper-sized foundation can best be determined by the maker of the machine. The structural details would be about the same for all machines.

Instructions:

1) Excavate and prepare footing for foundation.
2) Construct form according to plan.
3) Provide bolts to fasten machine to base.
4) Provide a template to locate bolts in base.
5) Provide pieces of 1″ gas pipe for bolts.
6) Mix concrete of 1:2-1/2:5 proportions to a quaky consistency.

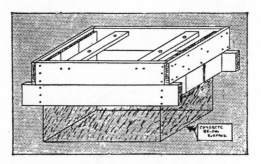

FIG. 159. Machine foundation.

7) Place concrete in form.
8) When form is practically full, set pieces of gas pipe with bolts approximately in place.

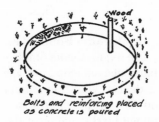

FIG. 160. Earth form for cistern or well top.

9) Fill forms, finish surface, and adjust bolts into correct position by aid of the template.

10) When initial set has been taken, remove template and trowel to a smooth level finish.

11) Remove form after several days and finish surface.

12) Bolt machine in place.

149. Cistern or Shallow Well Top (Fig. 160).

Requirements: To make a circular top for a well or cistern.

Instructions:

1) Describe a circle the exact size of top desired on a smooth level place on the ground.

2) Carefully excavate inside of the circle to a depth of 4″.

3) Cut out a cylindrical wood block and locate where pump pipe will pass through.

4) Provide four bolts to fasten pump to top.

5) Cut two pieces of hog wire for reinforcing across top, and two pieces of large, smooth wire for the edge.

6) Mix concrete of 1:2:3 proportion to a quaky consistency.

7) Sprinkle form so it will not absorb water from concrete.

8) Place concrete in bottom half of form.

9) Place reinforcing and set bolts in place.

10) Fill form with concrete.

11) Build up concrete where pump is to stand.

12) Finish surface with slight slope toward one side so water will drain off.

13) Sprinkle top from day to day. Remove at the end of a week or ten days.

150. Manure Pit and Cistern (Fig. 161).

Requirements: To excavate for manure pit and cistern, construct form, and place the concrete and reinforcing where needed.

Note: Refer to project in Sec. 143.

Instructions:

1) Excavate for both pit and cistern, and prepare foundation.
2) Construct outside form if needed.
3) Construct inside form for pit in place.
4) Provide tile from pit to cistern.
5) Arrange reinforcing for cistern. Heavy hog wire may be used instead of rods.
6) Construct inner form for cistern.

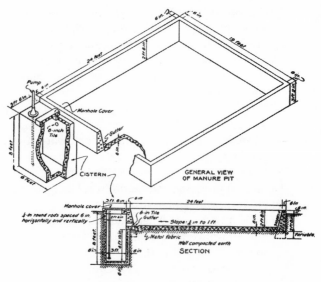

FIG. 161. Manure pit and cistern.

7) Use 1:2:4 proportion and mix concrete to a wet consistency.

8) Place concrete in walls.

9) Remove wall forms.

10) Place concrete in floor.

11) Construct form for cistern top, providing place for pump, also for manhole cover.

12) Place concrete for top with reinforcing, and also for manhole cover.

13) Remove form from cistern thru manhole.

14) Remove form from manhole top.

151. Feeding Floor (Fig. 162).

Requirements: To prepare foundations, construct forms and place concrete for feeding floor for ten hogs. It takes 12 to 15 square feet of space per hog.

Note: Refer to Sec. 141.

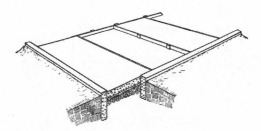

FIG. 162. Feeding floor.

Instructions:

1) Prepare foundation by leveling the spot where floor is to be built. Remove all vegetable matter and have soil thoroly tamped.

2) Construct form to grade so that floor will be at least 4″ thick. Have a slope of 1/4″ to 1′ in one direction.

3) Mix concrete of 1:2:4 proportion to a wet consistency.

4) Place concrete in floor; complete one section at a time.

5) Remove forms.

6) Excavate for curb around floor.

7) Construct form for curb.

8) Place concrete in curb.

9) Remove curb form.

152. Constructing a Scale Pit (Fig. 163).

Requirements: To excavate for scale pit, construct form, and place the concrete.

Note: Refer to project under Sec. 143.

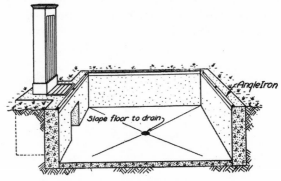

FIG. 163. Scale pit.

Instructions:

1) Excavate pit to dimensions to be determined from size of scale.

2) Provide drain for center of pit.

3) Construct outside form if needed.

4) Construct inner form so wall will be at least 6″ thick.

5) Use 1:2:4 proportion and mix to a wet consistency.

6) Place concrete in wall.

7) Provide bolts at top of wall to fasten angle iron.

8) Remove forms from wall.

9) Place concrete in floor with slope toward center drain.

153. Vault for Privy (Fig. 164).

Requirements: To construct a sanitary vault for privy with partition as illustrated in plan. As many sections can be made as desired. This is a dry type of vault, dry earth or ashes being used to absorb liquids.

Instructions:

1) Prepare footing for vault so its lower level will be no lower than the surface of the ground.

2) Construct form in place.

3) Provide pieces of wire for reinforcing to insure against shrinkage cracks.

4) Mix concrete of 1:2:3 proportion to a wet consistency.

5) Place floor and wall of vault as one unit.

6) Remove inner form after twenty-four to forty-eight hours, and paint up any holes with 1:2 cement-sand mortar.

7) Paint inner surface with a cement wash.

8) Remove outer form after several days.

9) Finish outer surface.

154. Milk-Cooling Tank (Fig. 165).

Requirements: To excavate, construct form and place concrete and reinforcing for a milk-cooling tank to be 2′ 6″ wide, 20″ deep, length as needed, bottom 8″ lower than floor of milk room. The bottom to be corrugated to allow free circulation of water with drainage outlet.

Note: Refer to Sec. 143.

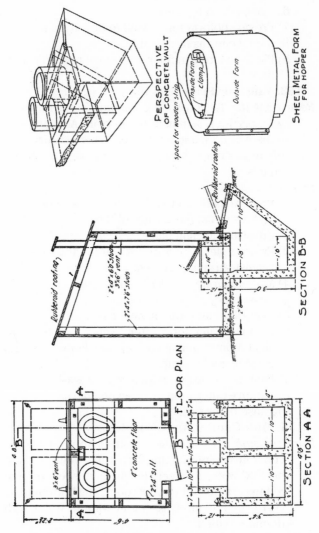

FIG. 164. Plan of sanitary privy.

Instructions:

1) Excavate to a depth of 14″ below floor level.
2) Construct outside form in place.
3) Construct inner form, to be put in place after floor is made, so wall will be 4″ thick.
4) Provide reinforcing material, either rods or heavy hog wire.
5) Put drain in place so coupling will be at surface of low place in floor.
6) As a protection to top of inner wall, provide a 4″ channel iron with 3/8″ by 6″ anchor bolts threaded into it, as illustrated.
7) Mix concrete of 1:2:4 proportion to a quaky consistency.
8) Place concrete in floor to a depth of 6″ with reinforcing in place.
9) Form corrugations in bottom of tank sloping toward outlet.
10) Adjust inner form in place, with reinforcing extending up from floor and entirely around wall of tank.
11) Place concrete in walls of tank.
12) Firmly seat the channel iron with anchor bolts on inner wall, hammering it into place with a wood maul.
13) Remove inner form at the end of twenty-four to forty-eight hours, and finish surface with a cement wash. If there are any holes, use a 1:2 cement-sand mortar to fill them.
14) Remove outside form after several days, and finish surface with a stiff brush.

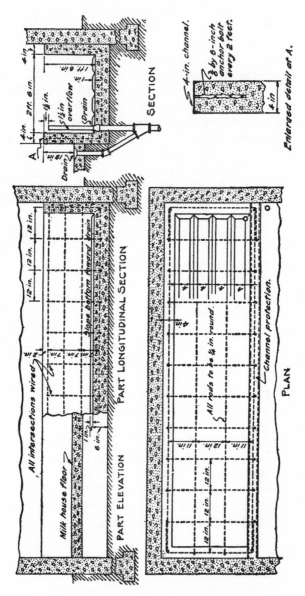

Fig. 165. Milk-cooling tank.

155. A Rectangular Water Tank (Figs. 166, 167, 167-*a*).
Requirements: To construct form according to plan, prepare foundation and construct a rectangular water tank, to be provided with inlet pipe with float control and outlet pipe.

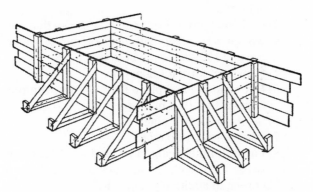

Fig. 166. Outside form for rectangular water tank.

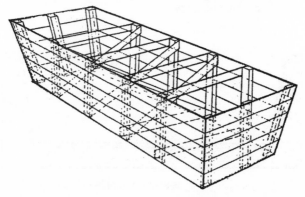

Fig. 167. Inside form for rectangular water tank.

Note: This project is quite similar to that described in Sec. 145, with exception of shape.

Instructions:

1) Prepare foundation. Set inlet and outlet pipes in place.
2) Construct outside form in place 38″ high (Fig. 166).
3) Construct inner form ready for use so that tank will be 2-1/2′ deep with 6″ floor, and 5″ wall at top and 10″ wall at the bottom.
4) Construct form for float box.
5) Mix concrete and place floor of tank with reinforcing. Use 1:2:4 mixture.
6) Set inner form in place.
7) Place form for float box.
8) Put wall reinforcing in place. Use at least four twisted 1/4″ rods 3′ long, bent at right angle at corner, in addition to hog wire.
9) Place balance of concrete.
10) Remove forms after concrete is thoroly set.
11) Finish surfaces.

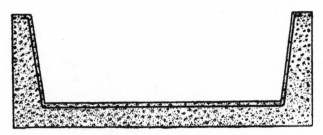

FIG. 167-*a*. Cross-section of finished tank.

156. Potato and Fruit Storage Cellar (Figs. 168,169).

Requirements: To excavate, construct forms, and place concrete and reinforcing for storage cellar as illustrated.

Note: Refer to project under Sec. 145.

Instructions:

1) Lay out, excavate and prepare foundation for storage cellar according to plan.

2) Construct end forms in place.

3) Construct inner side form in place. The sills supporting top section of inner form to be divided into three parts, the length of each part to be equal to one-half the width of cellar.

4) Provide wedges at bottom of form support to make the form easily removed.

5) Construct outer side form so wall will be 10″ thick at bottom of arch and 6″ thick at top. When excavation is carefully done, the outer form will be required only above the ground surface. Be very careful in bracing both the inner and outer forms to get best results.

6) Mix concrete 1:2:4 proportion to a wet consistency, except for the top of arch, which should be to a quaky consistency.

7) Place concrete in lower side wall.

8) Place heavy hog wire from lower walls over arch to insure against shrinkage cracks.

9) Place concrete over arch.

10) Keep concrete from being exposed to extreme heat of sun and moisten each day.

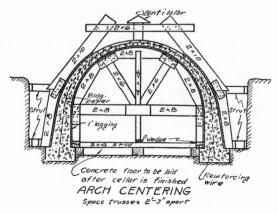

FIG. 168. Fruit storage cellar.

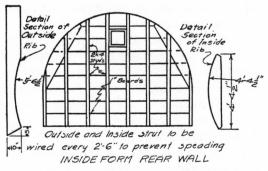

FIG. 169. Inside form of fruit storage cellar.

11) Remove forms after a period of about one week.

12) Finish job by smoothing off rough places with a brush and by plastering where necessary.

157. Hog Wallow (Fig. 170).

Requirements: To excavate for hog wallow suitable for 20 or 30 hogs weighing 200 pounds each. To construct form and provide overflow drain and inlet pipe similar to drain and inlet for tank described in Sec. 145. To reinforce floor and side walls of wallow with wire mesh.

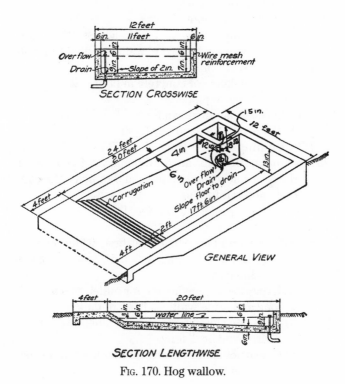

FIG. 170. Hog wallow.

Note: Refer to project under Sec. 145.

Instructions:

1) Excavate and prepare foundation for wallow.
2) Place inlet and overflow pipes.
3) Construct inner form ready for use.
4) Construct form for box to protect inlet and overflow pipes.
5) Use a 1:2:4 proportion and mix concrete to a quaky consistency.
6) Place floor about 4″ thick.
7) Put reinforcing in place across the floor and extending up the wall, as illustrated in cross-section.
8) Place balance of concrete in floor.
9) Put forms in place.
10) Place concrete in walls.
11) Remove forms and finish job.

158. Dipping Vat for Hogs (Fig. 171).

Requirements: To excavate for dipping vat, to construct the forms according to plan, and to place the concrete and reinforcing.

Note: Refer to project under Sec. 143.

Instructions:

1) Excavate main part of vat first, which is 8′ 6″ long and 2′ 10″ wide. With a 5″ wall, this gives 2′ in the clear.
2) Excavate sloping incline for outlet, this to be 8′ long and same width as body of tank.
3) Excavate sloping "step-off" incline; this to have 2′ drop and be 2′ long.
4) Construct an inner form so that floor and walls will be at least 5″ thick.

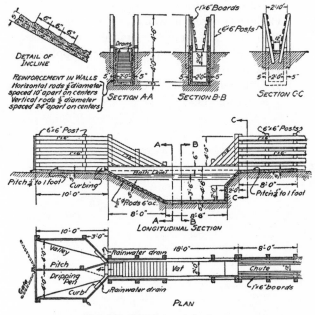

FIG. 171. Dipping vat for hogs.

5) Outside form should be unnecessary if care is observed in excavation.

6) Construct form for curb and floor of dripping pen.

7) Provide reinforcing to insure against cracks. Heavy hog wire is sufficient.

8) Use a 1:2:4 mixture and mix to a quaky consistency.

9) Place concrete in floor first with reinforcing in place.

10) Put in inner forms.
11) Place concrete in walls.
12) Form treads on incline before concrete sets.
13) Place concrete floor and curb in dripping pens.
14) Remove forms.
15) Coat surface with cement wash to insure water-tightness.

PART III

BLACKSMITHING

CHAPTER XVIII

MANUFACTURE OF IRON AND STEEL

159. Preliminary. There is quite as great need on the farm and in the house for a knowledge of metalwork and facilities to carry it on as for woodwork and cement work. The house-owner and home-maker is more independent if he can do the ordinary things about the home premises which demand the use of metalworking tools for the simpler constructions and repairs.

Under the general head metalwork, we shall consider, under separate parts, the following special branches of metalwork: Forging, Sheet-metalwork, and Farm Machinery Repair and Adjustment.

Under the sub-heads given above, the first will deal chiefly with steel, wrought iron or cast iron, while under the second, tin, zinc or lead, or sheet iron, will be the material handled.

160. Iron Ore. The commercial varieties of iron are pig iron, wrought iron and steel. Iron is found in the ground in natural deposits as "ore," which consists of metal imbedded in mineral and extraneous matter of no value. If the ore contains 50 per cent or more of metal, it is called "rich." It cannot be worked commercially with profit if it contains less than 30 per cent of metal. The valuable ores are oxides, hydrates or carbonates of iron. Ores appearing as sulphides are poor, as it is difficult to remove the sulphur. However, weathering ore—allowing it to stand in the open—will change sulphides to sulphates, which are largely dissolved out by rain.

One of the richest ores is magnetite, or black ore, which, when pure, contains 72.4 per cent iron and 27.6 per cent oxygen, Hematite, or red ore, when pure, contains 70 per cent iron and 30 per cent oxygen.

161. Pig Iron is made by crushing ore to uniform size and heating it in a blast furnace until it can be drawn off at the bottom in a molten condition. The blast furnace is a long, vertical cylindrical shaft which is fed from the top with ore, fuel or flux. Air under pressure is introduced at the bottom for purposes of combustion. The metal when molten is drawn off at the bottom, usually twice during twenty-four hours, and run into sand molds or iron chilled molds to form "pigs" of cast iron. Cast iron has 4 or 5 per cent of impurities such as carbon, sulphur, phosphorus, manganese and silicon. The amount of carbon present determines whether the iron is gray or white. If the greater part of the carbon is free as graphite, the iron is known as gray. If the greater amount of carbon present is combined, the iron is known as white. White iron, or iron with low combined carbon, is soft.

162. Wrought Iron is pure carbonless iron produced in a pasty condition. It is the converse of cast iron, as it is fairly tenacious and extremely ductile. When heated, it can be welded better than any other iron or steel. When heated to full red and quenched in water, it will not harden.

Wrought iron may be produced from iron ore in one operation, but this is costly, as the yield is low. Commercially, it is produced by indirect methods, by purification of pig iron, removing impurities by oxidation. This can be done in an open hearth or reverberating furnace, the methods being known as the open-hearth and Bessemer processes, respectively.

163. The Open-Hearth Process oxidizes the impurities of the pig iron by means of adding iron ore to a bath of molten pig iron. The fuel is, therefore, in contact with the metal, and the oxygen of the blast combined with the impurities are eliminated as oxides. This is a comparatively slow process of refinement, taking from seven to twelve hours to complete. Its advantages are a fine quality of iron produced and a large amount of material which can be handled at one time.

164. The Bessemer Process also is an oxidizing one, but the metal and fuel are not in contact. The oxygen is furnished by means of a large volume of compressed air blown thru a bath of molten pig iron. The oxygen combines with the carbon to evolve as gases while it combines with other impurities to form slag. The process requires but a few minutes—from twelve to twenty.

165. Steel. It is practically impossible to define steel accurately. It is an alloy of iron and carbon, but as alloys of iron and carbon include cast iron, this definition is not a technical one. Ordinary steel may be said to be iron containing from 0.1 to 2.0 per cent of carbon in combined form which has been subjected to complete fusion and poured into an ingot or mold for the production of forgeable metal. Such a metal—steel—has the composition of wrought iron, but it has been produced in a steel-melting furnace.

166. Tempering Steel. The greater the amount of carbon in steel, the harder it is, but the more ductile. The amount of carbon in steel practically determines the purpose for which steel may be used. Steel is hardened when heated to redness and quenched in water or oil.

When steel is heated and allowed to cool, naturally, it softens. It is upon this fact that tempering, which is the process of getting the proper combination of hardness and ductibility, is based. As the hot steel cools, surface oxides are formed which range from faint yellow thru straw, full yellow, brown, purple and full blue to dark blue. The lightest of these colors indicates the highest degree of hardness.

Machine and Tool Repairs

Under this heading is considered such work as one may be called upon to do in constructing tools and machines made of iron or steel, and which does not require the heating of the metal. For the most part, such work will be done with hand tools, as hammer, chisels, files, drills, taps, dies, rivets, etc. Work which requires the careful shaping or fitting of cold metal will need to be done in a machine shop and is not considered here.

Took and Equipment

For general use about the premises, a small out-building or room should be equipped with the following:

One wooden bench made of well-braced 2″ x 4″ uprights and stringers covered with plank and fitted with a spring screw vise, or machinist's vise.

One hand forge and anvil, with the common forge tools.

One grindstone, hand- or foot-power type, about 24″ in diameter and 3-1/2″ thick.

One bench hand emery grinder and oilstone.

In the room should be stored:

Wooden horses, wooden and metal blocks, skids, a block and tackle, and, possibly, a chain block, crowbars, pinch bars, rollers and pieces of iron pipe, rope and rope lashings with ends tied or wound, and jacks of the adjustable top and simple and heavy erecting types. The bench tools should consist of a simple equipment, some of which can be made and others purchased, such as:

> Machinist's hammers with ball, straight and cross peens, each weighing from 1 to 1-3/4 pounds.
>
> Hand hack-saw and blades.
>
> Center and prick punch.
>
> Machinist's chisels, principally the flat or cape chisels.
>
> Files, handled, of the rough and middle-cut grades principally, and both single- and double-cut in flat, round, halt-round, square and triangular shapes. The total number need not exceed 12.
>
> Drills in sizes ranging from 1/16″ to 3/4″ graded to sixteenths.
>
> One drilling ratchet and one breast drill.
>
> A variety of wrenches, including a good pipe wrench, monkey, alligator and a variety of single-end and solid or closed wrenches such as those included in a first-class automobile tool kit. A variety of socket wrenches will also be found very handy.
>
> Two or three sizes of outside and inside calipers and dividers.
>
> One scriber.
>
> One surface gage.
>
> One surface plate, about 2′ square.
>
> One 2′ rule, carpenter's folding.

One 6″and one 12″ steel scale.

One 6″ screwdriver, one 12″ screwdriver.

A variety of machine screws, bolts and nuts, washers, rivets and cotter pins.

One small set of taps for cutting machine thread.

One small set of dies for cutting machine thread.

One pair 6″ end pliers.

One pair 6″ side pliers.

One 4″ spirit level.

One carpenter's level.

One plumb bob.

One gasoline blow-forge.

Such supplies as the following should be accessible:

Waste, cotton wick, emery, emery cloth, lard and machine oil, and cup grease.

It will be well to keep in stock a small supply of bar strap and sheet iron.

FIG. 172. Arrangement of forge and tools, showing position of blacksmith at the forge.

CHAPTER XIX

EQUIPMENT FOR BLACKSMITHING; FUNDAMENTAL PROCESSES

167. Use of the Forge on the Farm. The village black-smith shop has always been a place of both first and last resort in helping to solve the many construction problems of a community. Likewise, the blacksmith's forge on the farm may be made the means of developing and repairing many tools and machines. The farmer who would save both time and expense may very well, therefore, be familiar with the work of the blacksmith.

It is suggested that the forge be a part of the equipment of the farm shop and occupy one end of a room, along one side of which may be placed the metalworking bench, thus bringing the vise near the anvil. It is frequently desirable to grasp a hot piece of metal in the vise when it is taken from the forge.

168. The Forge and Anvil. The forge which will be as serviceable as any on the farm, is one of the hand-operated fan, or bellows, type (Fig. 172). In front of it should be placed the anvil at easy-turning distance from the forge (Fig. 172). It may be mounted on the end of a heavy hardwood block or piece of the trunk of a tree, or it may be mounted upon a concrete pillar, to which it should be lagged. The height of the face of the anvil from the floor should be approximately 30″. It should weigh from 150 to 200 pounds.

169. Blacksmith's Tools. In addition to the forge and anvil, the following general equipment of tools should be at hand:

One each 1 to 3-pound cross-peen, straight-peen and ball-peen hammer.

One sledge, 5 to 10 pounds.

One pair of flat-jawed tongs for general work.

One pair of hollow-bit tongs for holding rod stock.

One pair of anvil or pick-up tongs for holding short pieces of heavy work while upsetting.

One bell tong for flat or scroll work.

One short-piece tong.

One handled top and bottom swage.

One handled top and bottom fuller.

One handled punch.

One handled flatter.

One handled hardie or hot chisel.

One heading tool.

One hardie for anvil.

Figs. 173 and 174 show photographs of a number of these tools. Tools other than those listed above and ordinarily included in a blacksmithing kit, are listed under the head of "Farm Machinery."

Of these, the most essential are:

Carpenter's square.

Calipers.

Dividers.

Scriber.

Folding steel rule.

Tire measurer.

Vise (solid box blacksmith).

Cold chisels, one on handle in shape of hammer.

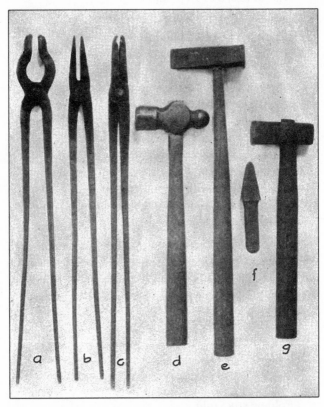

FIG. 173. Forge tools: *a*, hollow-bit tongs; *b*, flat-jawed tongs; *c*, pick-up tongs; *d*, ball-peen hammer; *e*, handled chisel; *f*, hardie for anvil; *g*, chisel.

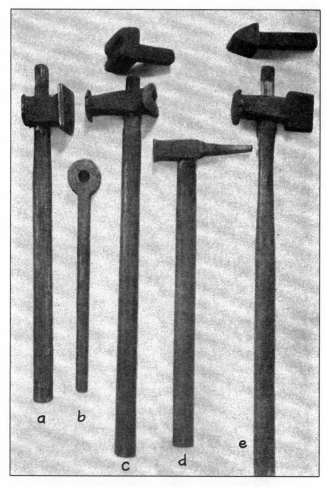

FIG. 174. Forge tools: *a*, flatter; *b*, heading tool;
c, swages; *d*, handled punch; *e*, fullers.

170. Supplies for Forge Work. It is well to carry in stock a small supply of wrought iron and steel in the following sizes:

3/8″ rods.

5/8″ rods.

1/4″ x 5/8″ bars.

1/4″ x 1″ bars.

(Note: Also material for buggy tires, bolts and rivets.)

171. Use of Wrought Iron. Wrought iron will be used chiefly. It can be worked either hot or cold. When worked cold, it becomes denser, harder, more elastic and brittle, but can be brought to its original condition by heating to red and cooling slowly.

The ordinary processes of tool construction are described in the instructions for projects. For ordinary work, a "red" heat is given the stock. When pieces are to be joined to form one solid piece by welding, however, the stock is brought to a "white" heat.

172. The Fire. The blacksmith's forge is a pan with a grate at the bottom which admits the air pumped for the purpose of creating a draft. The pan, or fire pot, contains the coal. This must be bituminous, or soft, coal of the very best quality. It is very important that it be free from sulphur and phosphorus.

To build the fire, remove all clinkers, slate, stone and other foreign material. Push the coal and coke to one side to expose the grate, tuyere, or wind box. Upon this, place a few shavings, some straw or paper, and cover with a little kindling as the match is applied.

Use a very light blast at first. As the fire burns, add green coal. When the fire gets strong, surround it with a ring of green, dampened coal, except toward the front, which should be kept open for the insertion of the iron to be heated and used to hold the iron while being heated, and for the tools. These should be kept in a horizontal position. As the work proceeds and the fire extends into the ring of green coal, it may be dampened to hold the fire to a limited area. Green coal may be added at the rear and the sides, but the fire should not be disturbed by pok-

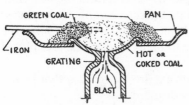

FIG. 175. Cross-section of forge.

ing it. As it burns from underneath, cinders should be raked out to keep the fire clear, and the coal should be gently patted down with a small shovel. Continuing this process will keep a clean, well-confined and fresh fire. Fig. 175 shows a cross-section thru the fire-pot.

As iron heats in the fire, the following shades of color will appear, indicating the proper condition of the iron for certain classes of work:

a) Dark blood red (block heat).

b) Dark red, low red (finishing heat).

c) Full red.

d) Bright or light red (scaling heat).

e) Yellow heat.

f) Light yellow heat (good forging heat).

g) White heat or welding heat (beyond this, iron will burn).

173. Welding. Upon continued heating of wrought iron or mild steel, the temperature increases, the metal becomes increasingly soft, and, if another piece equally soft is touched to the first, the two will stick; light tapping will complete the weld. The greater the range of temperature thru which the metal remains pasty, the more easily may it be welded. The greatest trouble in welding is in heating the metal properly. The fire must be clean and bright; otherwise, small pieces of cinder, etc., will stick to the metal. The heating must be slow enough to get the metal heated thru. Have all tools in place before taking a piece of metal from the fire. Hold the tongs on metal so that pieces can be easily placed in position without difficulty. When "stuck," first tap the thin parts of the pieces to be welded, as these cool first and most rapidly.

Do not have an oxidizing fire in welding; that is, not too much oxygen going thru fire.

In the welding process, the oxide formed is really a flux. In welding, steel will burn before the oxide becomes white hot; hence, a flux is used made of sand and borax; this is put in at yellow heat and protects the surfaces to be welded, preventing the forming of oxide. The oxide melts at a much lower heat when combined with the flux. This is the principal object of using a flux. Sal ammoniac seems to clean the surface, so a flux is sometimes made of one part sal ammoniac and four parts borax.

The following typical welds should be familiar:

a) Fagot or pile.	*f*) Chain-making.
b) Scarfed.	*g*) Butt.
c) Lap (flat).	*h*) Jump.
d) Lap (round).	*i*) Split.
e) Ring (round stock).	*j*) Angle.
k) "T" (round stock).	

CHAPTER XX

PROJECTS IN BLACKSMITHING

Problem No. 1: Drawing and Bending of Iron.
Projects Suggested for this Group:

 a) Staple (Fig. 176).

 b) Gate hooks (Fig. 177).

 c) Hay book (Fig. 178).

 d) Eyebolt (Fig. 179).

 e) Stove poker (Fig. 180).

174. Tools to Be Used. The tools needed to make projects in this group, aside from the forge and anvil, are a blacksmith's hammer (light) and a pair of flat-jawed or hollow bit tongs.

175. Maintaining the Fire. Every operation at the forge requires the maintenance of a good fire, the heating of iron to the proper temperature, and the proper handling of the blacksmith's tools to accomplish satisfactory results. Before beginning work on this project, read carefully the instructions on preparing the fire (Sec. 172). While work is progressing, green coal should be added from time to time, but always on the rim or edge of the fire, not on the live fire. The fire should be prevented from running into the green coal farther than desired by occasionally dripping water on the inside edge of the rim of green coal. This coal should be kept well packed down, thus forming a wall around the fire to be kept confined to the grate only. As the coal in the fire is consumed, remove clinkers and draw in fresh coal from the rim.

The operator must at all times keep his tools in good order and near at hand. The hammer to be used may well

FIG. 176. Staple.

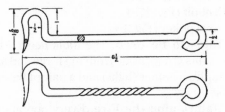

FIG. 177. Gate hook.

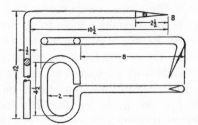

FIG. 178. Hay hook.

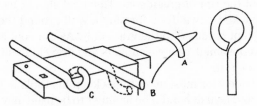

FIG. 179. Eyebolt, showing steps in construction.

be laid in position on the anvil (Fig. 172) to be grasped by the right hand immediately when the iron taken from the forge reaches the anvil. The tongs may be laid on the

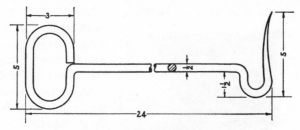

Fig. 180. Stove poker.

top of the forge at the left side of the fire, so that they may be handled by the left hand in removing the iron from the fire.

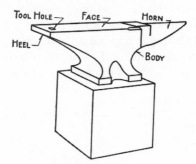

Fig. 181. Anvil and block.

Working Instructions for the Gate Hook:

Stock: One piece of 1/4″ round wrought iron 10″ long.

176. Bending Iron. Place one end of the rod in the tongs held in the left hand, with which place the opposite end of the rod in the front and at the base of the fire in a horizontal position. Heat this end of the rod for 3″ to a light yellow or lemon color.

Withdraw the iron with tongs in the left hand and place on anvil with the heated end projecting over the

FIG. 182. Method of bending iron.

horn 2-1/2″. Fig. 181 shows anvil with parts named. Grasp the hammer well toward the end of the handle with the right hand. Raise the hammer above the iron and strike it a light blow just beyond the point where. it is in contact with the edge of the anvil (Fig. 182). Continue this process until the iron assumes the form shown in solid lines at the right in Fig. 183. This form should be made without reheating the iron.

Reheat the same end of the iron, again to lemon color. Grasp as before with the tongs, but with the iron turned

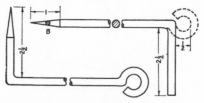

FIG. 183. Steps in making gate hook.

over in the tongs so that the part made at a right angle with the rod in the first operation is upward. Place on the horn of the anvil, as shown in Fig. 184, and, by striking the end at an angle with the hammer, shape this end to a complete circle centrally located on the end of the rod. The dotted lines at the right end of the rod (Fig. 183) show this finished shape. The hole in the ring should be 1/2″ in diameter. It can be made the right size and circular by forming it over the end of the horn (Fig. 184).

177. Drawing Iron. Heat the opposite end of the rod to lemon color, and form 1″ of it to a cone (*B*, Fig. 183).

FIG. 184. Making the eye on gate hook.

The cone is formed by resting the heated end of the rod at the angle of the cone on the face of the anvil and gradually rolling it from side to side while the hammer strikes the iron lightly a repeated number of blows. This end of the rod is to form the hook.

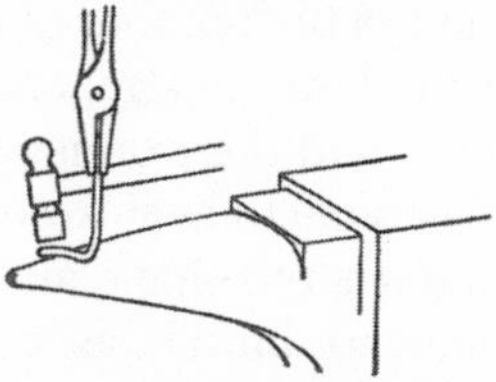

Reheat the hook end of the rod for 3″ to lemon color

FIG. 185. Forming the point on gate hook.

and bend it over the horn of the anvil to form a 2-1/2″ right-angle shoulder. This operation is the same as the first one described in forming the ring end of the hook.

Now, grasp the tongs, as shown in Fig. 185, and proceed as in forming the ring of the hook to bend the hook end in the middle of the 2-1/2″ portion of the L-shaped end to a half-circle (Fig. 177). Bend the point of the cone outward slightly over the end of the horn of the anvil. Lay the hook flat on the face of the anvil and straighten with a few light blows of the hammer.

If it is desired to have the hook twisted in the center (Fig. 177), heat the central portion of the hook to a light yellow color, grasp the hook end with the tongs, place the ring end in the vise, and twist or turn in one direction until the desired number of twists are formed and until the hook and the ring are in the same plane.

Each of the projects in this group is made so nearly the same as the gate hook, that they require no special instructions. The handle both for the hay hook and the stove poker is formed of two half-circles joined by straight portions of the handle. A little care on the part of the operator after making the gate hook will enable him to make either of these handles. The iron may need to be heated a few more times, but this will not be serious unless the number of heatings is sufficient to weaken it or unless the temperature approaches that for welding heat and the iron is burnt in consequence. It is always desirable to heat iron as few times as possible to secure the desired shape and form in order not to weaken the metal or burn it, as well as to save as much time as possible in the work.

Problem No. 2: Upsetting and Punching.
Projects Suggested for this Group:

    a) Open-end wrench (Fig. 186).

    b) Punched screw clevis (Fig. 187).

    c) Machine bolt (Fig. 188).

    d) Log hook (Fig. 189).

178. Tools Needed for Upsetting and Punching.
The same tools as those named for the group of projects

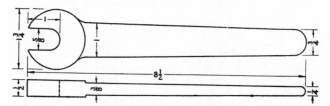

FIG. 186. Open-end wrench.

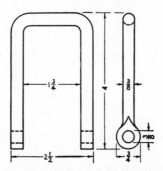

FIG. 187. Punched screw clevis.

in Problem 1 will be required in this group, and in addition, the upsetting tool and punch.

179. Upsetting and Punching. It is frequently necessary to enlarge some portion of a piece of iron. This

is done by upsetting. To upset stock, heat it at the point
to be enlarged, place it on end on the anvil, and pound
it on the other end with a hammer. Repeat this process
for each reheating until the stock is of the desired size
where it is to be upset.

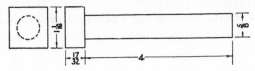

FIG. 188. Machine bolt.

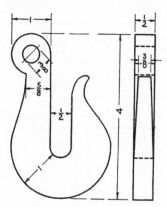

FIG. 189. Log hook.

At times, to make a hole thru a piece of iron with
forge tools, it is only necessary to drive a punch thru it
when hot. At other times, the stock will need to be bent
around to lap back on itself, when it must be welded as
described in the next group in this section, or the hole
will have to be drilled.

**180. Working Instructions for the Punched Screw
Clevis.** One piece of 1/2″ wrought iron 12″ long will be
used for this project.

1) Heat one end of the iron to light red and bend 1-1/4″ of it to right angles with the rod over the back edge of the anvil (Fig. 190).

2) Reheat the same end to lemon color, place on the face of the anvil with bent end upwards and upset by pounding on this upturned end with quick, sharp blows of the hammer. Roughly shape to approximate circular form by working the cylindrical surface on

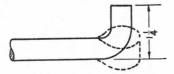

FIG. 190. Upsetting for clevis.

the surface of the anvil and over its corner. Reheat and continue to upset and shape until thickness of flattened end is approximately 3/8″.

3) Reheat to welding, or white, heat, using extreme care not to burn the iron. Remove the iron from the forge the moment it becomes white. Place it quickly on the face of the anvil in former position for upsetting, and strike quickly with the hammer two

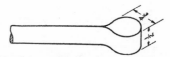

FIG. 192. Appearance of finished job of upsetting.

FIG. 191. A flatter.

or three times. Finish flat surfaces with the flatter (Fig. 191). Turn the iron on edge over the corner of the anvil, and strike quick, sharp blows to form circle. If the iron is at welding heat and the work with the hammer is done quickly, the iron will weld or become a solid mass. Any seams which may have formed in the upsetting process will be obliterated. Fig. 192 shows the finished end. In a similar manner, as described up to this point, forge the other end of the rod.

4) Reheat each end separately to yellow color, mark center with prick-punch and punch 1/4″ hole one-half way thru iron on this prick-punched mark with punch, shown in Fig. 193. Reverse the stock, place the end over the hardie hole, and drive the punch thru from the other side. Reheat the stock, if necessary, and drive the punch thru from each side to enlarge the hole to 3/8″ (Fig. 194).

5) Punch a hole in the other end in a similar manner (Fig. 195).

6) Heat the stock in the center for a space of 3″, and bend it over the horn of the anvil to the shape shown in Fig. 196. The central portion of the curved end of the clevis should be straight.

FIG. 193. Handled punch.

7) By laying the clevis on the face of the anvil with the punched ends hanging over the edge of the anvil,

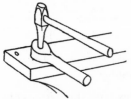

FIG. 194. Using the punch,

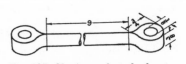

FIG. 195. Clevis ready to be bent into shape.

and striking the two legs of the clevis with light hammer blows, it may be straightened. The two punched holes must be in line. Fig. 187 shows the finished clevis.

Supplementary Instructions:

181. Open-End Wrench. Heat 4″ of one end of 1-1/2″ x 7/16″ soft steel to lemon color, and draw it out to shape and dimensions shown at A, Fig. 197. Mark the stock 1-1/2″ from the point where the forging of the handle was begun, as shown by the dotted line (A, Fig. 197). Cut the stock off on the anvil hardie (Fig. 198), or with the handled hardie

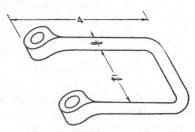

FIG. 196. Completed clevis.

(Fig. 173), cutting, first, from one side and, then, from the other, and, finally, breaking off over the edge of the anvil by striking the stock not to be used a sharp blow with the hammer just beyond the anvil edge.

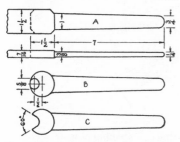

FIG. 197. Steps in making wrench.

Heat the stock to lemon color and forge to shape, as shown in *B*, Fig. 197. The wrench end should be rounded up, keeping stock to original thickness, by first forming an octagon, then a sixteen-sided figure, and, finally, a circle. This work should be done over the corner of the anvil and by moving the edge being formed into different positions as the hammer strikes the iron. Reheat the metal and punch a hole 1/2″ out of center toward the wrench end and expand it until it is 5/8″ in diameter (B, Fig. 197). Cut the end out with a hot chisel or handled hardie to 60 degrees, keeping same centrally located, as shown at

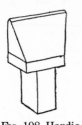

C, Fig. 197. Reheat and forge to shape and dimensions, shown in Fig. 186. This should be done by holding the wrench edgewise on the face of the anvil with the handle held downward at an angle and striking the wrench end an angle blow on the end of each prong of this end, finally flattening the inside of jaws and their surface on the heel end of the anvil. Smooth up with flatter.

FIG. 198. Hardie for anvil.

The wrench should be hardened to make it serviceable. Heat it to lemon color and plunge it in water for a few moments, This cools the outer surface. When

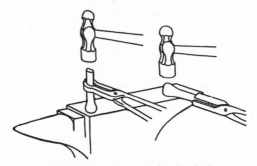

FIG. 199. Upsetting and shaping bolt.

the metal is withdrawn from the water, the heat of the center will draw out toward the surface. While still quite warm, put in water to completely cool.

182. Bolt Head. The construction of the square bolt head involves upsetting (Fig. 199). Care must be taken

not to upset too far, however. When the approximate
dimensions given in Fig. 200 have been secured, heat
the upset end to lemon color and place the bolt thru
the hole in the heading
tool (Fig. 200-*a*) and
into the hardie hole
in the anvil, as shown
in Fig. 200-b. Proceed
to upset the head and

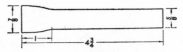

FIG. 200. The upsetting completed.

to keep it circular in form by occasionally removing it
from the heading tool, and, by rolling it in the tongs on
the face of the anvil (Fig. 199), hammer the head into a
true cylindrical form. When the diameter of this cylinder
is slightly less than the distance between corners of the

FIG. 200-*a*. Heading tool.

FIG. 200-b. Using the
heading tool.

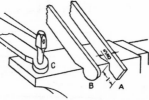

FIG. 201. Steps in making log hook.

finished head, reheat the stock on the head end to lemon
color and forge the square head (Fig. 188).

183. Log Hook. Heat 2″ of one end of 1/2″ x 1″
wrought iron stock 5-1/2″ long to a yellow glow; place
over outside edge of anvil with 1″ overhanging, and
forge to shape, shown in A, Fig. 201. Reheat and forge to

shape, shown in B, Fig. 201. Reheat and punch hole, as shown in *C*, Fig. 201. Round corners of hole over horn of anvil to shape, shown at C, Fig. 202.

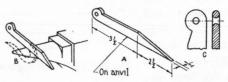

FIG. 202. Further operations in construction of log hook.

Heat the other end and taper to shape and dimensions, shown in *A*, Fig. 202. Bend the point slightly over horn of anvil. Reheat center of stock and form over horn

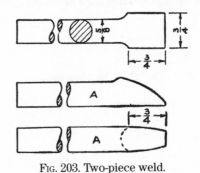

FIG. 203. Two-piece weld.

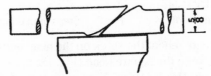

FIG. 204. Position of pieces when welding.

of anvil, as shown by dotted lines at *B*, Fig. 202. Finish to dimensions given in Fig. 189.

Problem No. 3: The Process of Welding.
Projects Suggested:

    a) Two-piece weld (Figs. 203 and 204).
    b) "T"-weld (Fig. 205).
    c) Welded clevis (Fig. 206).
    d) Wagon wrench (Fig. 207).

184. Preparation for Welding. The same tools as those named for the previous groups, in addition to which

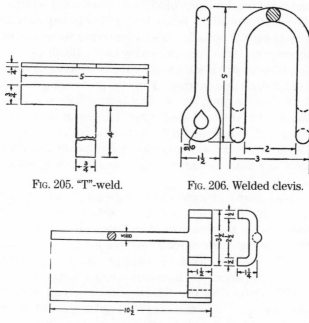

FIG. 205. "T"-weld.

FIG. 206. Welded clevis.

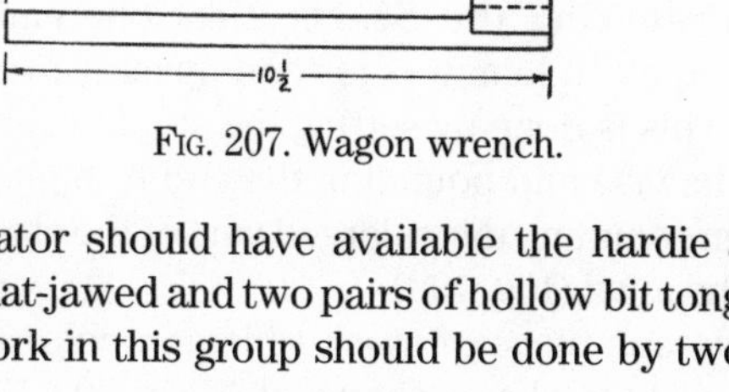

FIG. 207. Wagon wrench.

the operator should have available the hardie and two pairs of flat-jawed and two pairs of hollow bit tongs. Some of the work in this group should be done by two people working together at the forge; hence, the desirability of two pairs of tongs. The top and bottom swage, the flatter and the top fuller will be needed for some projects.

While it is true that the punched screw clevis required a welding heat, the exercise of welding on it was comparatively simple. Welding is probably the most difficult forge work. It requires a perfectly clean fire, exactly the proper temperature of heated metal, and both accurate and rapid manipulation of tools. The end weld is one of the simplest of all the welds. It should be practiced until it can be made upon first trial, when other welds will be accomplished with comparatively little difficulty.

It is necessary always to have the two pieces of metal to be welded first hammered into the proper shape. Both must then be given the welding heat at the same time, taken out of the fire together, quickly placed one on the other, and then immediately hammered with light, quick blows, while the stock is changed in position on the anvil to permit the hammer to strike all portions which must be joined.

Just before taking the iron from the fire, it is well to put some kind of flux on each of the surfaces to be placed together. Sal ammoniac or rosin is generally used.

Working Instructions for Two-Piece Weld:

    Stock: Two pieces of wrought iron or soft steel, each about 5/8″ in diameter and 4″ long.

185. Preparing the Scarfs. Heat one end of each piece of stock to lemon color and upset it to 3/4″ from the end. This is done by setting the stock on end on the face of the vise and pounding the end to be upset (Fig. 208), then rounding the enlarged part of the stock on the face of the anvil (Fig. 199).

Reheat each piece of stock to lemon color and scarf the up-set end to shape, shown at A, Fig. 203. Each scarf should be one and one-half times the diameter of the stock.

186. Making the Weld. Place scarfed surfaces of each piece of stock down in the fire and heat to white or welding heat. Grasp one piece with the hollow-bit tongs in the left hand, and the other with the flat-jawed tongs in the right hand. Take both pieces from the fire, quickly turn the one held by the right hand as it is moving toward the anvil, so as to place it quickly on the anvil under the scarf of the piece held with the left-hand tongs, as shown in Fig. 204. Instantly drop the right-hand tongs and pick up the hammer which should be lying near at hand. Strike quick, sharp blows on the ends to be welded, at the same time turning the pieces with the left-hand tongs. Continue until the two pieces are thoroly joined, then until the diameter is reduced to that of the original stock and the surfaces of the stock at the weld are smooth.

FIG. 208. Upsetting for two-piece weld.

FIG. 209. One member of the "T"-weld.

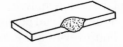

FIG. 210. The second member shaped for welding.

Supplementary Instructions: To form and weld the parts of the other projects listed in this group, a few special instructions are needed beyond those given for the two-piece weld.

187. "T"-Weld. The center of one piece and the end of the other must be upset, as shown in Fig. 209. Fig. 210 and Fig. 211 show how these pieces must be swaged to form the welded joint. A difference of 1/8″ between the thickness of stock and the upset portions of stock will

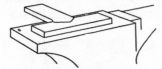

FIG. 211. Position of weld on anvil.

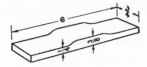

FIG. 212. Effects upon piece of iron from upsetting.

be sufficient to form the welded joint to the thickness of stock, as shown in Fig. 211.

188. Wagon Wrench. The preliminary steps of heating and upsetting the two pieces of stock for this project are similar to those already described. A little more difficulty may be experienced because of the dimensions of the stock and the lengths of the upset portions of same. When the rec-

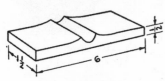

FIG. 213. Preparing wagon wrench for welding.

tangular stock is fully upset, it must be laid flat on the face of the anvil and pounded on the upper surface near each end to flatten the lower surface (Fig. 212). This will make the additional thickness of the upset portion of the stock offset on the top surface. Heat this part of the stock and make a groove 1/4″ deep with a 5/8″ fuller (Fig. 213).

The remaining exercises involved in making this project should be clear by a study of Figs. 214, 215 and 207.

Mark the points where the bends are to be made on the rectangular stock of the wagon wrench with prick-punch or hardie before heating to make either bend over the edge of the anvil.

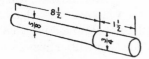

FIG. 214. The handle of the wrench prepared.

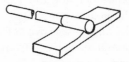

FIG. 215. The pieces ready for welding.

189. Welded Clevis. The drawings for this project show in detail the succeeding steps in forming one end of the clevis. The offset at *C*, Fig. 216, should

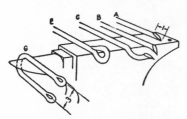

FIG. 216. Operations in making welded clevis.

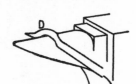

FIG. 217. Bending iron for welded clevis.

be made by striking light blows just over the edge of the anvil with the peen of the hammer 1-1/2″ from the end of the stock. The form shown at *E*, Fig. 216, is made over the end of the horn of the anvil, as shown at D, Fig. 217. When the ring for the end is nearly completed, the stock should be reversed on the horn, placed over

FIG. 218. Dipping iron in water preparatory to welding.

the end and rounded up carefully with the hammer, leaving the joint to be welded in perfect condition. One end of the clevis should be welded before the other is formed. Before taking the welding heat, dip the end to be welded in water, as shown in Fig. 218, and then, when the heat is completed, make the weld over the edge of the anvil, as shown at *E*, Fig. 216. Reheat and drive a 5/8″ punch in each eye from each side (Fig. 219). Finish over end of horn (Fig. 220).

FIG. 219. Punching the clevis.

FIG. 220. Finishing eye of clevis.

Problem No. 4 : Welding and Tempering Steel.
Suggested Projects:

    a) Butcher-knife (Fig. 221).

    b) Punches (Fig. 222).

    c) Cold chisel (Fig. 223).

    d) Sharpening cultivator shovel (Fig. 224).

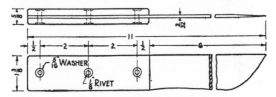

FIG. 221. Butcher knife.

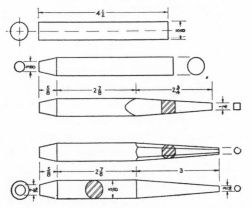

FIG. 222. Punches.

190. Forging Steel. Tools needed are those required for ordinary work at the forge, including flatter and swage.

The forging of tools which are not of unusual shape demands only the use of simple exercises in forging. The new exercise is that of tempering.

191. Working Instructions for Cold Chisel.

Stock: One piece 3/4″, six- or eight-sided tool steel, 7-1/2″ long.

1) Heat 1″ of one end of stock to lemon glow and round to cone shape, leaving 3/8″ flat on end in form of

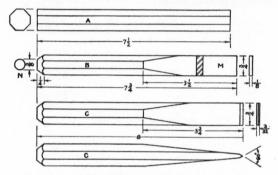

Fig. 223. Cold chisels.

Fig. 224. Cultivator shovel.

circle. Keep circular flat end centered on axis of stock (*N*, Fig. 223).

2) Heat 3″ of opposite end of stock to lemon glow. Forge to shape and dimensions, as shown at *M*, Fig. 223. Care must be taken to keep taper uniform on both sides and to keep width of stock unchanged.

3) Reheat chisel end of stock to bright red and smooth with hand hammer, and, if necessary, finally with flatter.

4) Heat entire stock to dull red, plunge each end for entire length of forged part in water for a few moments. Remove stock from water and allow color to run to light

blue at extreme end, then plunge in water to harden completely. It may be well to temper each end separately.

5) Grind chisel end of tool to a cutting edge, with ground surfaces making angle of about 60 degrees. If the flattened surfaces forming the chisel end and the conical end are rough, grind them smooth. All grinding should be done on an emery wheel if available; otherwise, on a grindstone. Keep

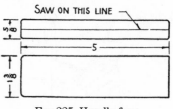

FIG. 225. Handle for butcher knife.

the tool from overheating and, possibly, burning if it is ground on an emery wheel running dry, by frequently plunging tool in water.

192. The Butcher-Knife. The butcher-knife is made from 1/16″ or 3/32″ tool steel, forged thin on one edge to form cutting edge of knife. The handle should be made in two halves, or, better, in one piece (Fig. 225), to be cut in halves. The two halves of the handle should be held in place on knifeblade when holes are drilled thru both knife-blade and handle. Soft-steel rivets placed in each hole can be riveted down on each side of the handle over a rivet washer, to fasten the knife-blade and handle securely together. The knife-blade is tempered by heating to dull red, plunging in water, or, better, oil, and almost instantly withdrawing and allowing a light blue color to draw to edge. The knife-blade can then be ground for use.

193. A Cultivator Shovel. This is sharpened by heating, forging and tempering in the general manner described for the cold chisel or the butcher-knife.

More difficulty may be experienced, however, in forging to shape. Fig. 226 suggests the position of the cultivator shovel on the face of the anvil. Position, as shown at A, is the one taken after first heating when point of

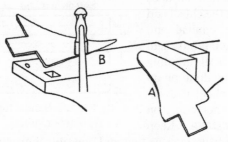

FIG. 226. Steps in sharpening cultivator shovel.

shovel is drawn to a sharp point by quick blows of the hammer. Position, as shown at B, is the one taken after a second heating when the side of the shovel is drawn to an edge. Care must be taken to keep the surface of the shovel free from hammer marks.

CHAPTER XXI

SUPPLEMENTARY PROJECTS IN BLACKSMITHING

194. Directions for Making Wagon-Box Stake-Irons (Figs. 227 and 228).

1) Secure 1/4″ strap band iron of proper width, or use as substitute old wagon-wheel tire.

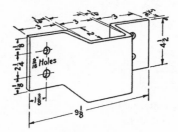

FIG. 227. Wagon-box stake-iron.

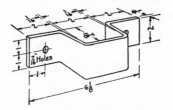

FIG. 228. A lighter stake-iron.

2) Cut to length as per dimensions with cold chisel or over anvil hardie.

3) Heat in center portion and make inside bends over corner of anvil

4) Heat between center and end, and make each outside bend over corner of anvil.

219

5) Prick-punch for center of holes, and drill or punch, heating metal in latter case.

6) Straighten on surface of anvil with hammer and flatter.

195. Making a Ring (Fig. 229).

1) Cut calculated length from band or rod iron.

2) Heat one end to light red and draw out, as shown in *A*,

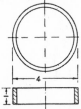

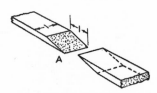

Fig. 229. A ring constructed from rectangular stock.

Fig. 230. Ends of metal prepared for welding.

3) Repeat operation on second end, making drawn-out taper on reverse side.

4) Reheat entire rod to light red and round over horn; bring ends together on face of anvil (*B*, Fig. 231), ready for welding heat.

Fig. 231. The ring shaped for welding.

5) Heat ends of ring to welding temperature, and weld over horn of anvil.

6) Reheat welded part to light red and smooth up over horn and on face of anvil.

196. Constructing a Chain (Fig. 232).

1) Cut to link lengths 1/4″ round, soft steel or wrought iron.

2) Heat and swage ends of link, forming same roughly, as shown in perspective In *A*, Fig, 233, and *B*, Fig. 234.

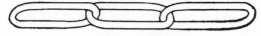

FIG. 232. Chain links.

3) Put link into last one welded, heat and form carefully on face of anvil (C, Fig. 235), ready to weld.

FIG. 233. Preparing the weld.

FIG. 234. Link ready for welding.

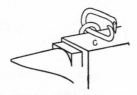

FIG. 235. Link inserted in chain.

4) Heat to welding heat, weld on face of anvil, and smooth over end of horn.

197. Making Ice Tongs (Fig. 236).

1) Cut to estimated length two pieces 3/8″ x 3/4″ rectangular rod.

2) Heat one end and form handle.

3) Heat center and flatten, and form portion for joint.

4) Heat remaining portion of hook end, form over horn of anvil to semi-circular shape, and forge end over corner of anvil to shape of blunt-pointed spur.

5) Heat flattened portion to light red and punch for 3/8" bolt.

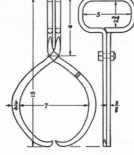

FIG. 236. Ice tongs.

6) Straighten and smooth on face and horn of anvil.

7) Insert bolt and burr-end over nut.

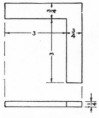

FIG. 237. Right-angle weld.

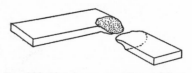

FIG. 238. Metal prepared for welding.

198. A Right-Angle Weld (Fig. 237).

1) Heat both pieces 1-1/2" on one end to lemon color. Upset 1/8" thicker than rest of stock 3/4" in length.

2) Scarf both pieces, using peen of hammer (Fig. 238).

3) Heat both pieces, scarfs down, to welding temperature (white heat). Lay together and weld with quick, hard blows.

4) Finish to perfect right angle. Round inside corner and keep outside corner square (Fig. 237).

199. Forge Tongs (Fig. 239).

1) Heat one end of stock, 18″ x 3/4″ x 3/8″, to lemon color.

FIG. 239. Forge tongs.

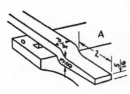

FIG. 240. Bending iron for forge tongs.

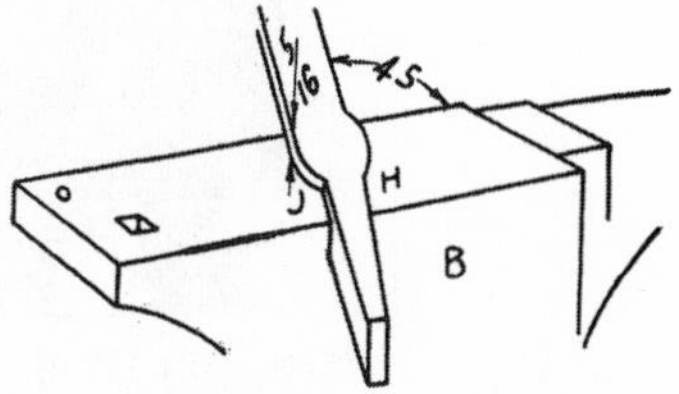

FIG. 241. Shaping the joint of forge tongs.

2) Lay flatwise over round corner right angle to anvil, forge jaw 2″ long, 3/4″ wide, and taper from 3/8″ to 5/16″ to dimensions, as in A, Fig. 240.

3) Reheat to lemon color. Place on anvil at an angle of 45 degrees, as in B, Fig. 241; finish to 7/8″ wide by 5/16″ thick. Place stock edgewise and use fullers (Fig. 242) as shown in Fig. 243, to secure shape, as at H and J, Fig. 241.

4) Reheat to lemon color. Place over anvil 7/8″ from shoulder, jaw down, as in C, Fig. 242; strike at D, forging shank to E, Fig. 244.

FIG. 242. Top and bottom fullers.

5) Heat other end of forging to lemon color. Forge to 5/16″ round to form the handle; cut to 18″ over all.

6) Reheat the jaw to lemon color. Put 1/4″ fuller lengthwise on inside of jaw and fuller 1/8″ deep (*F*, Fig. 245).

Fig. 243. Using the fullers.

7) Reheat eye at G, Fig. 245, to lemon color. Punch 5/16″ hole for rivet in center of eye.

8) Repeat operations for other half.

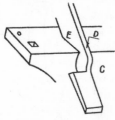

Fig. 244. Another step in construction of tongs.

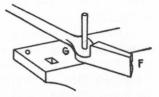

Fig. 245. Punching for rivet.

9) Heat one end of piece cut from handle to lemon color. Cut off 1″ for rivet. Reheat and insert rivet and rivet with hammer (Fig. 239).

Tongs for special uses are shown in Fig. 246. Bottom and top swages (Fig. 247) may be used to finish handles, as at *A*, *B*, *C*, and *D*. Fig. 246.

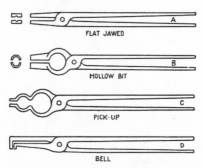

FLAT JAWED

HOLLOW BIT

PICK-UP

BELL

FIG. 246. Several types of tongs used in forge shop.

200. Repointing Cultivator Shovel (Fig. 248).

1) Mark new stock for lines A and *B* under shears (Fig. 249).

2) Heat to bright red. Cut on lines *C* with hot chisel (Fig. 249).

3) Reheat to bright red; scarf inside edges *(C)* to dimensions in drawing.

4) Heat old shovel to bright red. Straighten shovel.

5) Reheat shovel, place borax on back side of section to be welded; leave it there until dissolved.

6) Place new point on shovel (Fig. 250), allowing it to project 1/2″ beyond old point.

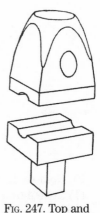

FIG. 247. Top and bottom swages.

7) Rake coke (good supply) in fire hole, place shovel on it, add more coke on top of shovel,

then spread a shovelful of wet coal on top of this. Heat slowly to welding temperature.

8) Remove to anvil and strike series of blows all over new point.

9) Reheat other side to welding temperature.

FIG. 248. Cultivator shovel.

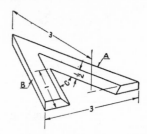

FIG. 249. New piece of stock for cultivator shovel.

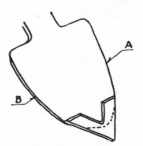

FIG. 250. Correct placing of new points.

10) Remove to anvil and weld this side onto point.

11) Reheat to lemon color, hammer on edges at B, Fig. 250, until sharp. Grind off irregular edges.

12) Reheat to bright red. Bend shovel over horn to shape as at the beginning (Fig. 248).

13) Draw color to straw and plunge in water to harden.

201. Sharpening Plowshare (Fig. 251).

1) Place share on floor and mark around outside lines with chalk.

2) Heat 4″ of share, starting at A, Fig. 251, to a bright red.

3) Place on anvil, as shown in Fig. 252, and forge to sharp edge.

4) Reheat 3″ or 4″ at a time, and forge to sharp edge until share is finished from A to B, Fig. 251

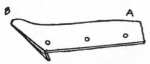

FIG. 251. Old plowshare.

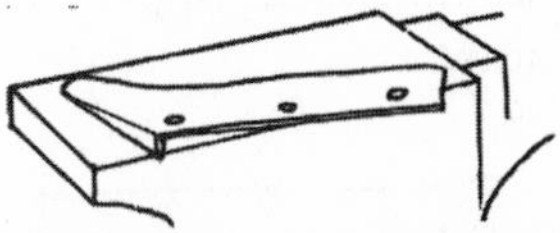

FIG. 252. Position of plowshare on anvil when sharpening.

5) Heat point to bright red, place on anvil and forge to sharp point.

6) Grind off irregularities.

7) Reheat point and set share so it will have correct suction and landside, which are 1/8″ and 1/4″, respectively.

8) Reheat to bright red and case-carbonize with potash.

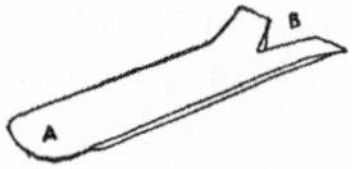

FIG. 253. Piece of steel for new point.

FIG. 254. Steel for point shaped for welding.

9) Share should fit as nearly as possible to outline on floor.

202. Pointing Plowshare (Figs. 253 to 256).

1) Heat 3″ of new stock on one end to lemon color.

2) Scarf end, as shown at *A*, Fig. 253.

3) Heat other end to lemon color. Scarf and split, as shown at B, Fig. 253.

4) Heat center of stock to lemon color. Bend into shape of V, as in Fig. 254, having bottom, or split, side 1″ longer than top side.

5) Heat old share (Fig. 255) to re*d* heat.

6) Place on anvil and apply borax on both sides of share.

7) Heat new point to red heat.

8) Place new point on share, as in Fig. 256. Reheat to welding heat. Apply a little borax to share while it is heating.

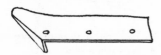 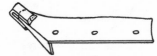

FIG. 255. Plowshare to be repointed.   FIG. 256. The point in place for welding.

9) Remove to anvil and strike a few blows until point is welded. Reheat to welding temperature. Continue to weld on both sides until finished. Cut surplus stock off sides and grind.

10) Reheat as much of share as possible and set to have correct suction and landside, which are 1/8″ and 1/4″, respectively.

11) Reheat to bright red and case-carbonize with potash.

203. Shortening Buggy Tire Without Cutting (Fig. 257).

1) Heat several inches of tire, holding same in vertical position, to light red.

2) Bend heated portion inward over horn of anvil (A).

3) With aid of helper, grasp tire either side of bent portion with flat-jawed tongs over and against rough surface

of horse-shoeing rasp (Fig. 258); place crosswise over surface of anvil, and hammer.

4) Repeat operation No. 3 until stock is upset sufficiently to shorten tire.

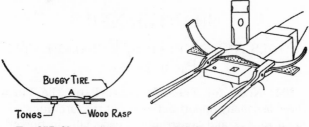

FIG. 257. Shortening buggy tire.

FIG. 258. Details of method of shortening tire.

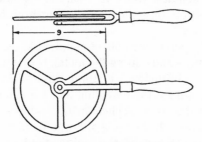

FIG. 259. Tire-measuring tool.

5) Measure outside of felly and inside of tire with tire-measuring tool (Fig. 259). Tire measurement should be about 1/4″ less than felly measurement.

6) Heat tire to red, one-half way around. Slip tire over felly, and shrink on by immediately running in water. If tire is too short, it will "dish" wheel too much. Wheel should be dished (out of true plane) not more than 1″.

PART IV

SHEET-METALWORK

CHAPTER XXII

TOOLS AND SUPPLIES; FUNDAMENTAL PROCESSES

204. Need for Sheet-Metalwork on the Farm.
There are many opportunities about the farm for sheet-metal repairs and construction, especially in tinwork. Kitchen utensils, the equipment of the dairy and creamery, farm machines, water and sanitary systems, and roofs and gutters on buildings, all furnish problems in sheet-metalwork.

The chief operation in sheet-metalwork, aside from calculating sizes and cutting the metal, is that of fastening, which may be divided into three classes, viz., soldering, brazing and riveting. Welding is not included, as it seldom is used in working sheet metal, and, besides, it is considered under the heading of Forge Work.

205. The Process of Soldering. Soldering is the process of joining two pieces of metal by means of a more fusible metal or metallic alloy. The metal, or alloy, called solder, should be selected with the following considerations in mind: (1) Its strength should be as great, or greater, than that of either of the pieces of metal it joins; (2) its color should be as nearly as possible that of the joined metals, and (3) its fusing point should be considerably lower than that of either of them.

206. Classes of Solder. Solder is classed as soft or hard, depending upon the degree of fusibility, and, to some extent, upon the class of metals to be joined by it. Soft solder, sometimes called white or tin solder, is made of soft, readily fused metals or alloys. Such metals as tin, lead-tin and alloys of tin, lead and bismuth are usually used. A good formula for the composition of soft solder is: Lead, 207 parts; tin, 118 parts. To weaken the solder increase the number of parts of tin. Increasing the number of parts of lead will strengthen the solder. The solder may be prepared in a graphite crucible at a low temperature by mixing with an iron rod and then running into iron molds.

207. Soldering Fluxes are substances used to remove the oxide which forms on the surface of a metal. They are melted and run on the metal where the soldered joint is to be formed. The fluxes generally used are powdered rosin or a solution of chloride of zinc, used alone or combined with sal ammoniac.

A soldering fluid is a liquid flux and may be prepared by mixing 27 parts neutral zinc chloride, 11 parts sal ammoniac, and 62 parts of water; or 1 part sugar of milk, 1 part glycerine and 8 parts of water.

A very common liquid is prepared by dissolving in an earthenware vessel small pieces of scrap zinc in commercial muriatic acid. Dissolve one piece at a time to prevent too rapid generation of heat, which might break the jar. Finally secure a saturated solution by adding more zinc than will dissolve. For use in soldering, the solution should be diluted with the addition of its own bulk of water, mixed and filtered. The addition of a few drops of liquid ammonia will increase the activity of the flux, which should be kept in a wide-mouthed bottle and applied to the joint to be soldered, just before the soldering operation begins, by means of a stick or brash.

This flux may be used on almost any metal except aluminum, zinc or galvanized iron.

FIG. 261. Equipment for soldering.

The Soldering Process. Certain metals require special solders and fluxes. For most purposes, however, the solder and fluxes described are serviceable.

The best of tools and materials, however, will not secure good results unless used in the hands of a good workman. To solder successfully the metals to be joined must be fitted accurately and cleaned thoroly, either by some means of mechanical cleaning, such as scraping or grinding, or by removal of dirt and grease with acid.

It is dangerous to use the latter, however, as it may injure the metal surfaces, besides its possible injurious effects upon the workman.

When the metal is clean, apply the flux to all surfaces which will come in contact, join these as planned and run the soldering iron over or against the joint.

208. The Soldering-Iron, which is made of copper, must be "tinned" to serve as a solder carrier. Fig. 261 shows the shape of a soldering-iron. The end is kept filed to form well defined edges and a point. When thoroly clean, heat and rub on solder, then wipe with a cloth, a piece of felt serving the purpose very well.

To use the soldering-iron heat it in a clean fire, using a gasoline torch, a blacksmith's forge, or a tinsmith's gas forge, and place it against a bar of solder, when a little will adhere to the soldering-iron.

Another method of using the soldering-iron is to provide an open-mouthed bottle of chloride of zinc fluxing solution and when the iron is heated, dip the point of it into the solution to clean it. Then place the iron against the bar of solder, and if properly heated a little solder will adhere to it. This is the customary method of tinsmiths. Fig. 261 shows an open-mouthed bottle of the fluxing solution, together with a can of cleaning material, a block of sal ammoniac and a wiping rag. The Bunsen burner shown in this picture is frequently used to heat the tinner's iron when gas is available.

The iron is now run on the joint and the solder which the iron holds will fill the joint, cool, and effect a union of the two pieces of metal. The bar of solder is used to hold the tin in position. In case a long joint is to be made, the iron may be run slowly against the metal with the bar of solder held against the iron. The solder will thus melt, run down and off the iron and fill the joint. Care must be taken not to flood the joint by using too much solder. While an iron may be run over a joint several times, it is advisable to run it over but once. Superfluous solder and the extended use of the soldering-iron are signs of a poor workman. When the soldering-iron is run over the

FIG. 262. A clean joint.

FIG. 262-*a*. A joint where too much solder has been used.

joint many times, the solder will flow out on the surfaces of the metal near the joint, resulting in a "smeared" joint. Fig. 262 shows a soldered joint on which no superfluous solder has been used; Fig. 262-a shows one which has been smeared with too much solder.

CHAPTER XXIII

Projects in Sheet-Metalwork

Problem No. 1:

Making a Lap Joint as Used on Tin Roof.

209. Stock and Tools for Lap Joint. The stock needed is two strips of medium weight, clean new tin, each about 10″ long and 3″ wide.

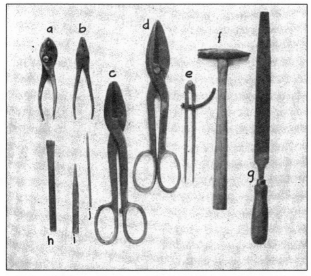

Fig. 263. Tools for sheet-metalwork: *a*, cutting pliers; *b*, flat-jaw pliers; *c*, straight snips; *d*, curved snips; *e*, compass; *f*, tinner's hammer; *g*, flat file; *h*, cold chisel; *i*, punch; *j*, scratch awl.

A limited number of sheet-metalworking tools suitable for ordinary work on the farm is necessary. The equipment may consist of:

1 gasoline soldering torch,

1 soldering iron,

1 pair straight snips,

1 pair curved snips,

1 tinner's hammer,

1 wooden mallet,

1 carpenter's square,

1 pair cutting pliers,

1 pair dividers,

1 punch,

1 scratch awl,

1 bar solder,

1 piece sal ammoniac,

1 bottle cleaning solution,

and tools shown in Fig. 263.

210. Working Instructions for Lap Joint. On one long edge of one piece of tin, scribe a mark 1/2" from the edge with gage of tin made as shown in Fig. 264. This 1/2"

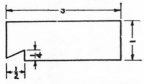

FIG. 264. Gage for making joints.

FIG. 265. Position of pieces for lap joint.

surface will form the joint (Fig. 265). Clean this surface and a corresponding one, not necessarily determined by a scribed line, on the second piece of tin, by wiping clean and applying the flux. Place the two pieces of tin together flat on a board so that the surface of one piece of tin laps over on the surface of the other, the edge of the first coinciding with the scribed line on the second.

The two pieces of tin now lap 1/2″. Grasp a short piece of wood about the size of a screw-driver handle with a square or beveled end in the left hand, and with it press the two pieces of tin together (Fig. 266). This may also be done by using the bar of solder in place of the stick. With the right hand, grasp the handle of the hot, well-

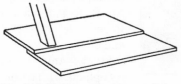

FIG. 266. Holding two pieces of tin for soldering.

tinned soldering-iron, wipe the iron on a cloth or piece of felt conveniently placed on the bench or table on which you are working, touch this iron to a piece of solder and immediately run the end of one of the four "flats" of the iron on the joint (Fig. 267) and near edge of the lap. The holding-stick or bar of solder must be kept near the part of the joint being soldered. It must be moved from point to point as the iron is moved along the joint. The heat of the iron should heat the joint sufficiently to run the solder on the iron between the lapped

FIG. 267. Running solder.

surfaces of the two pieces of tin. As the iron moves from one point to another the heated surfaces will cool, forming a soldered joint. The iron must be touched against the solder frequently to renew the supply of solder on the iron. When the joint has been formed, run the iron slowly the entire length of the joint with one stroke, to make a smooth finish.

This exercise should be repeated, if necessary, until

FIG. 267-*a*. Correct position of soldering iron.

a perfect joint can be made with a few strokes of the soldering-iron. *Problem No. 2:* To Patch a Tin Receptacle (Figs. 268 and 269).

Stock—Any tin receptacle with a hole in it.

Tools—Those used for Problem No. 1.

211. Preparation for Patching. Perhaps one of the most general uses of the soldering-iron in the home is

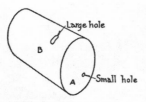

FIG. 268. Patching small hole.   FIG. 269. Patching large hole.

for patching tin utensils. Such work may be listed under two heads, viz., small-hole patching, where an additional piece of tin is unnecessary, and large-hole patching, requiring a piece of tin to cover the hole.

In the first case, the hole is first closed as far as possible by pounding the tin around it with a mallet over a surface as nearly the shape of the tin surrounding the hole as possible. The tin is then cleaned by scraping if very dirty, or by the use of a little muriatic acid, which may be put onto the surface of the tin with a stiff feather. The flux is then applied and solder run into the hole with the soldering-iron used as in soldering a seam (A, Fig. 268).

If the hole is too large to be closed with solder, a patch must be applied and soldered on. *B*, Fig. 268, shows the hole, and Fig. 269 shows it patched.

212. Completing the Patch.

1) Secure a receptacle with a cracked seam or a small hole and with a large hole 1/2″ or more in diameter.

FIG. 270. Shallow watering pan.

Prepare the small hole (A, Fig. 268) for soldering, as described in Sec. 210, and solder, as described there.

Trim the large hole (B, Fig. 268) with a pair of tinner's snips (Fig. 263), either straight or curved, depending upon the shape of the hole and the tin, whether flat or curved.

Cut a piece of tin from an old can or a piece of sheet tin the shape of the hole, but enough larger than the hole to provide for a 1/4″ or 3/8″ lap all around the hole. Clean the tin on the receptacle, and that of the patch also; apply the fluxing material and solder, as described in Sec. 210.

Problem No. 3: To Construct a Shallow Watering Pan for a Chicken Coop (Fig. 270).

Other Projects Suggested for this Group:

Any low, straight-sided tin dish not requiring a wired edge.

Stock—Tin of medium weight cut to size and the same as, or similar to, pattern shown in Fig. 271.

Tools—Those used for Problem No. 1, Sec. 209, and a wooden mallet and ruler, or carpenter's square. It will be necessary, also, to have a sharp-edged piece of hard wood or a straight-edged piece of iron as long or a little longer than the longest edge of the pan.

213. Strengthening the Edge. Ordinarily, it is desirable to strengthen the upper edge of a tin receptacle by

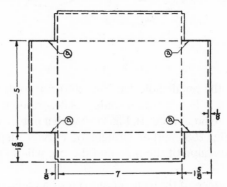

FIG. 271. Tin cut to shape for watering pan.

running a wire around this edge under the tin which is rolled over the wire, as in the case of a tin drinking cup or a funnel (Fig. 278).

This portion of the receptacle may be strengthened, but not so well, by folding a small portion of the upper edge over and pounding it down against the surface of the tin (Fig. 272).

214. Laying Out and Cutting Tin to Shape. With carpenter's square, or with try-square and rule, lay out rectangle, 10-1/4″ x 8-1/4″. Inside of this rectangle, scribe lines with scratch awl and straight-edge (leg of carpenter's square), 1-5/8″ from and parallel to outside edges of this rectangle. Scribe lines in the corners for portion to be cut out. Turn the piece of tin over and scribe lines 1/8″ inside the rectangle and parallel to the outside edges.

With straight snips, cut out the corners, as shown in the drawing (Fig. 271): also cut to the corners of the inside rectangle, formed by the first lines scribed, on the lines marked heavy on the drawing and lettered *a*.

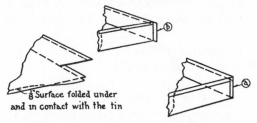

FIG. 272. Details of soldering watering pan.

215. Folding. Over the edge of the piece of hard wood or straight-edged piece of iron, fold with a mallet the 1/8″of tin between the outside edges and the lines scribed 1/8″ from same. These surfaces must be folded toward the surface of the tin on which the lines were scribed 1/8″ from the outside edges. Fig. 272 shows the folding operation. Pound these surfaces down until they are in contact with the sheet of tin to form the strengthened edges of the pan (Fig. 270).

In like manner, but in the opposite direction, fold over the corner of the piece of hard wood or straight-edged piece of iron the 1-5/8″ surfaces to form right angles with the sheet of tin and to make the vertical surfaces on the edges of the pan (Figs. 271 and 272).

Carefully fold the corner laps, lettered *b*, Fig. 272, to come in contact with the long, or 7″, edge of the pan (Fig. 271).

Place each corner of the pan over a square corner of a hard piece of wood and square up and smooth with the mallet.

Solder the inside of each corner of the pan between the end and side edges, and also the edge of the corner lap (*a*, Fig. 272). Apply fluxing material and use soldering-iron, as described in Sec. 209.

Problem No.4: To Construct a Receptacle Requiring the Assembly of Heavy Pieces of Tin or of Galvanized Iron.

Projects Suggested for this Group:

a) Watering trough (Fig. 273).

b) Flower box (Figs. 274, 274-*a*, 274-b).

c) Drip pan (Fig. 275, 275-*a*).

Stock for watering trough: 2 pieces heavy tin, 12″ x 5″; 1 piece heavy tin, 26″ x 12″.

Note: Galvanized iron may be substituted.

Tools—A full set of sheet-metalworker's hand tools (Fig. 263).

216. Constructing Watering Trough. Mark and cut the ends of the piece of metal to form the trough, as shown in Fig. 273. Fold the ends up on lines shown dotted in the figure, and then turn the piece of metal over, laying it along the corners of a square-edged timber on the center line shown as the long dotted line in the drawing (Fig. 276). Bend the metal down over the timber until the surfaces on either side of the line are in contact with the surfaces of the timber, thus forming the trough.

Lay out lines on one surface of each end piece of the trough, to form slits into which the folded ends of the trough piece may be inserted that it may hang on the ends (Fig. 277).

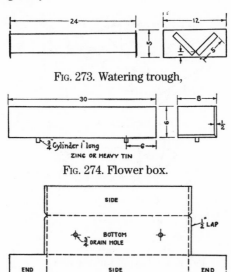

FIG. 273. Watering trough,

FIG. 274. Flower box.

FIG. 274-*a*. Details of flower box.

Lay each end piece of tin with the lined surface up, flat on a smooth, hard board. With a sharp cold chisel and hammer or mallet, cut along each scribed line.

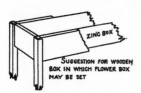

FIG. 274-*b*. Perspective of flower box.

Carefully insert the end laps of the trough into the slits in the end pieces of the trough from the side on which the cold chisel cut, and gently pound into shape with a mallet over the corner of a board. Solder all these joints and run solder in the intersection between end pieces and trough near bottom of trough, where the end laps on trough were not cut, to make trough water-tight.

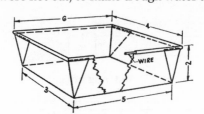

Fig. 275. Drip pan.

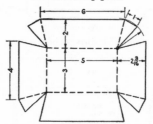

Fig. 275-*a*. Layout for drip pan.

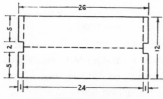

Fig. 276. Layout for watering trough.

Fig. 277. Completed trough.

Problem No. 5: Making a Cylindrical Receptacle with Handle and Reinforced Edge.

Suggested Projects:

   a) Drinking cup (Fig. 278).

b) Small pail (Fig. 279).
c) Cylindrical pan (Fig. 280).

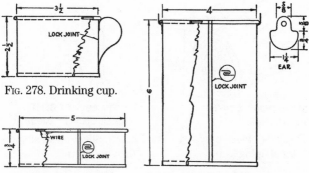

FIG. 278. Drinking cup.

FIG. 280. Cylindrical pan.   FIG. 279. Small pail.

Stock for drinking cup—1 piece circular tin, 3-3/4″ diameter; 1 piece rectangular tin, 11-1/4″ x 2-3/4″; 1 piece rectangular tin, 5″ x 1-1/2″.

Tools—A full set of sheet-metalworker's hand tools.

217. Methods of Inserting Wire. The customary method of strengthening the upper edge of a tin receptacle is to roll the edge of the tin over a piece of wire in what is known as a wiring machine. The wire may be inserted by hand, as described below, altho with less likelikood of securing a perfect job.

218. Shaping Bottom. Pare the end of a round piece of stove wood with a draw-knife to a diameter of 3-1/2″. Sandpaper the surface smooth and saw the end off square (Fig. 280-*a*).

Place the stove wood in a vise with the cylindrical end up. Over this place the circular bottom for the cup so that the 1/8″ surface to be folded projects evenly around the piece of wood (Fig. 281). Hold the tin with the left hand and gently pound the edge of it down around the

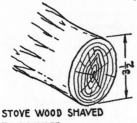

STOVE WOOD SHAVED
TO CYLINDER

Fig. 280-*a*. Piece of wood
for shaping iron.

TIN BOTTOM BENT
OVER END OF WOOD

Fig. 281. Sheet-metal shaped
on wood form.

piece of wood with **a** mallet. It may be necessary to snip the edge of the tin in a few places to prevent it from buckling. The bending must be done carefully. When the edge is finally bent over in contact with the cylindrical surface of the wood, pound the folded portion firmly against the wood until it fits like **a** cap (Fig. 281). The tin may now be pried off.

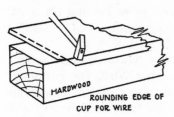

HARDWOOD
ROUNDING EDGE OF
CUP FOR WIRE

Fig. 282. Bending sheet-metal
over piece of wood.

219. Inserting Wire in the Edge. Over a slightly-rounded corner of a piece of hard wood, pound the 1/4″ surface for the wire to strengthen the upper edge of the cup (Fig. 282). When this has been done, place the proper length of 1/16″ wire in the rounded corner turned upward as the tin lies flat on the bench, fasten the bent edge of tin over the wire at each end with a pair of pliers, then carefully pound the remaining portion of the bent edge over the wire until it lies smooth and hugs the wire the entire length (Fig. 283). Fig. 284 shows the process of folding a wire in the edge of a piece of tin.

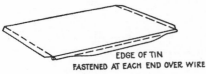

EDGE OF TIN
FASTENED AT EACH END OVER WIRE

Fig. 283. Inserting wire.

Fold the end laps of the pattern in opposite directions to form the lock joint seam for the cup, as shown in insert, Fig. 279. Roll the entire surface over the cylindrical end of the piece of wood used to form the bottom of the cup, having the wire on the outside; lock the joint, pound down with the mallet and, at the same time, slip the cylindrical surface from the wood.

Solder the inside and outside of the lock seam, slip the body of the cup into the bottom, and solder around the bottom edge. The cup is now complete except for the handle.

Fig. 284. Folding metal over wire.

220. Handle for Drinking Cup. Fold the two 1/8″ outside edges of the strip for the handle (Fig. 285) as in the case of the upper edge of the watering pan (Problem 3). With the folded edges on the inside, form the handle, as shown in the drawing for the drinking cup (Fig. 278), and solder both ends to the cup—one against the wire

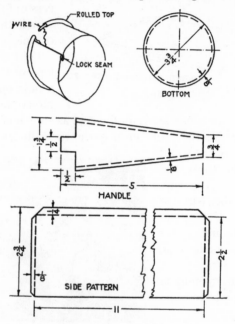

FIG. 285. Details of drinking cup.

and the other against the bottom seam—over the lock seam of the cup. First, gently pound the ends firmly in contact with the cup over the seam. This may be done by putting the cup over the end of a cylindrical stick, such as a tool handle.

Problem No. 6: To Make a Conical Dish.

Suggested Projects:

   a) Funnel (Fig. 286).

b) Flaring pan (Fig. 287).

c) Flaring pail (Fig. 288).

d) Cream dipper (Fig. 289).

Stock for the funnel—1 piece of tin, 12″ x 6″; 1 piece of tin, 5″ x 4″.

Tools—A full set of sheet-metalworker's hand tools.

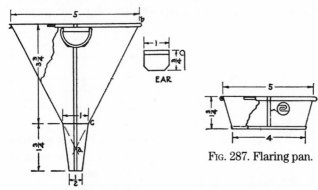

FIG. 286. Funnel.

FIG. 287. Flaring pan.

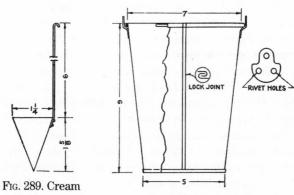

FIG. 289. Cream dipper.

FIG. 288. Flaring pail.

221. Laying Out Conical Shapes. The pattern for a cone or for a frustum of a cone is made by describing an arc of a circle with a compass or a pair of dividers, the distance between the points being the slant height of the cone and the length of the outside arc being circumference of the base of the cone.

222. Construction of Funnel. Lay out the pattern for each of the two parts of the funnel (Fig. 290), produc-

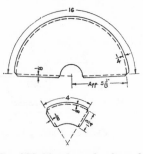

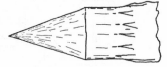

FIG. 291. Piece of wood for shaping funnel

FIG. 290. Metal cut for funnel.

ing the side lines of front view to locate the apex of each cone part, in order to secure the radius to strike the proper arcs (Fig. 290). As in the case of the cylindrical part of the cup, insert a 1/16″ wire in the space marked 1/4″ on the outside of the large pattern, and fold in opposite directions the laps for the lock seam joints. Carefully form each portion of the funnel over a cylindrical piece of stove wood tapered on one end to a cone (Fig. 291). Lock and solder the joint for each part, slip the upper part into the lower, first spreading out the upper opening of the lower part over the surface of the cone-shaped piece of wood, and solder the two parts together. The ear may be made as shown in Fig. 286, and a small piece of wire formed to slip into it to form a hanger. The ear may be soldered on or fastened with rivets.

CHAPTER XXIV

SUPPLEMENTARY PROJECTS IN SHEET-METALWORK

223. Cylindrical Receptacle (Fig. 292).

1) Lay out pattern for bottom, leaving 1/8″ for fold.

2) Lay out pattern for body of receptacle, leaving 1/8″ lap on each end for lock joint.

3) Solder lock joint of body of pattern.

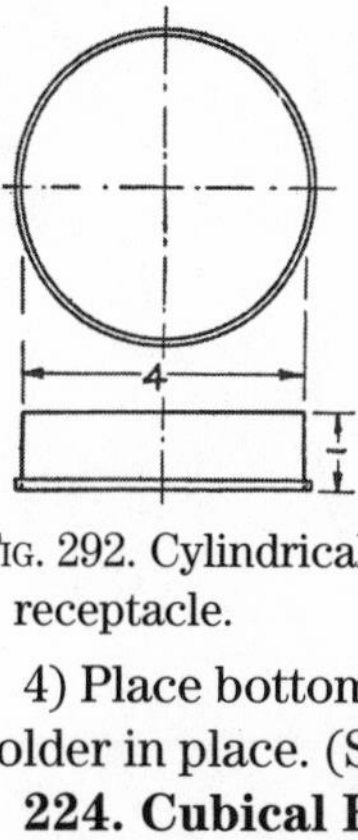

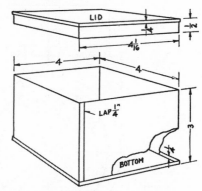

FIG. 292. Cylindrical receptacle.

FIG. 293. Cubical box with lid.

4) Place bottom in position on body of receptacle and solder in place. (See instructions for Problem No. 5.)

224. Cubical Box with Lid (Fig. 293).

1) Lay out pattern for body of box—a rectangle 3″ wide and 12-4/4″ long. The 1/4″ added to the 12″ is to provide a lap which should be formed on one corner.

2) Lay out pattern for bottom of box—a square 4-1/4″ on a side. The 1/4″ added to the 4″ is to provide two 1/8″ laps—one on each side of the square.

3) Solder seam on box after it is folded into shape of square.

4) Fold edges and corner laps on bottom, place in position on box. and solder in place, including corner laps.

5) Construct cover for box by following description for making the body of box.

225. Stovepipe Collar (Fig. 294).

1) Lay out pattern for cylindrical part of collar, allowing 3/8″ for lap to be riveted. Rivet joint.

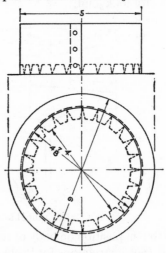

FIG. 294. Stovepipe collar.

2) Lay out pattern for flange of collar—a ring, outside diameter, 6″, and inside diameter, 4″. Scribe a 4-7/8″ circle on this ring. Clip several narrow notches on inside of ring limited by the scribed circle.

3) Fold notched part of ring into cylindrical part of collar and pound in contact with same over cylindrical stock.

4) Solder or rivet two parts of collar together.

226. Conductor Elbows (Fig. 295).

1) Lay out each section of elbow, as shown at A. Livide the end view (circle) into twelve parts, each point to be regarded as the end of a line on the cylindrical section

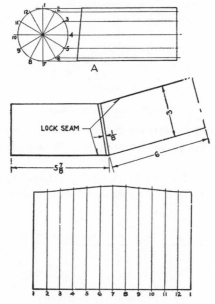

FIG. 295. Conductor elbow.

drawn opposite the point. Space the length of the section of the pattern into twelve parts, and lay off on line thru each point the length of same line in drawing A.

2) Form lock seams, as indicated, and allow for flange for joint between sections.

3) Form each section and solder seams. Place the parts of elbow together and solder. Note that seams of sections are placed on opposite sides of elbow.

227. Roof Ridge Flange (Fig. 296).

1) Lay out cylindrical pattern (Fig. 297). Determine length of lines, as in pattern for conductor (Fig. 295).

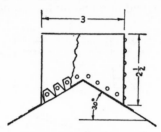

Fig. 296. Roof ridge flange.

2) Lay out pattern for flange, notch and punch holes for rivets, unless solder alone is to be used to fasten it to cylinder.

3) Fit cylinder and flange together, bending flange to proper angle for roof. Rivet or solder cylinder seam and

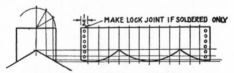

Fig. 297. Details of roof flange.

rivet or solder flange to cylinder. (These joints may be both soldered and riveted.)

228. A Measure (Fig. 298).

1) Lay out pattern for bottom, as in Problem 5. Fold over edge.

2) Lay out pattern for body, as in Problem 6, for upper portion of funnel. Solder lock seam.

3) Lay out pattern for rim of measure, regarding it as a cone with apex (a, Fig. 299). (See pattern, Fig. 290.)

Note radii distances lettered similarly in Figs.298 and 299. Begin at *b* (Fig. 299) and measure the distance *hb* six times in each direction. This will locate points *g* and

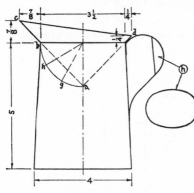

FIG. 298. A measure.

g. Draw lines *ag* and *ag*, and extend both to *d*. Also extend *ab* to c. To secure the arc thru c (Fig. 299), connect points *c* and *d* and erect perpendicular to this line at center point *e* to intersect line *ac* at *f*. Use *f* as a center and draw arc *dcd*. Angles between radii *ad* and *ac*, *ac* and *ad*, reading from left to right in Fig. 299, are equal.

4) Make short and narrow V-cuts with snips in lap surface on lower edge of pattern for rim. Bend this lap to fit into top of body of measure. Bend end laps, form rim and solder to top of measure body after soldering rimseam at *d*.

5) Place bottom in

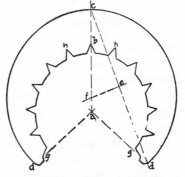

FIG. 299. Details for measure.

position on body, pound firmly in contact with body over end of cylindrical stick and solder seam.

6) Lay out handle, as in Problem 5; form and solder in piece of tin h cut to fit. Solder on handle over seam of body.

229. Three-Piece Elbow (Fig. 300).

1) Lay out pattern for each part of elbow, first making full sized bench drawing. Use methods given in cases of

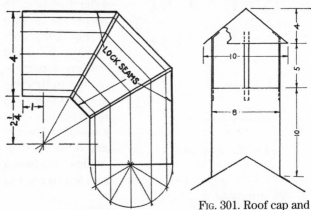

Fig. 300. Three-piece elbow.

Fig. 301. Roof cap and ventilator.

conductor elbow (Sec. 226) and roof ridge flange (Sec. 227). Allow laps for lock lap joint on each section of elbow. Allow 3/32″ lap on each end of central section of pipe to fit over, and solder onto, end sections.

2) Solder lock joint on each section and solder sections together.

230. Roof Cap and Ventilator (Fig. 301).

1) Lay out and construct 8″ cylinder, as in case of roof ridge flange (Sec. 227).

2) Lay out and construct conical cap for ventilator, as for funnel (Problem 6). (Seam should be riveted for ventilator of size given.)

3) Fasten conical and cylindrical parts of ventilator together with four strips of 1/2″ band iron or heavy tin. Ends should be riveted.

231. Gutter Miter (Fig. 302).

1) Lay out pattern for each part of gutter. This will be a rectangle, length the long edge of the gutter and width

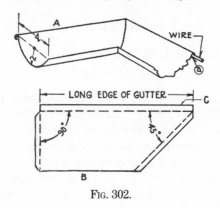

FIG. 302.

one-half the circumference of a 2″ circle, plus 1/2″ to roll over heavy wire on outside edge (A, Fig. 302). Cut one end of pattern square and other edge at 45 degrees (C, Fig. 302). Leave lap on square end to fold, and solder against end of gutter. Leave joint lap on end cut at 45 degrees.

2) Fold edge of gutter over wire. Form gutter (A, Fig. 302.

3) Solder end of gutter in position.

PART V

FARM MACHINERY REPAIR AND ADJUSTMENT

CHAPTER XXV

FARM MACHINERY AS AN ECONOMIC FACTOR

232. Farm Machinery and National Progress. It is not the purpose of this section to furnish information on each type of machine used on the farm, but to present a few general statements, followed by outlined studies of a few machines and their uses, and a few definite problems of repair and adjustment. For a more complete discussion, the reader is referred to the list of books and bulletins given below.

Farm Machinery and Farm Motors. By Davidson and Chase. Orange Judd Co.

Agricultural Engineering. By Davidson. Webb Publishing Co.

Equipment for the Farm and Farmstead. By Ramsower. Ginn & Co.

Farm Machinery. By Wirt. John Wiley & Sons.

Bulletins from the U. S. Department of Agriculture and State Agricultural Experiment Stations.

The greatest growth in agricultural development is marked by the use of modern machinery. We find the plow substituted for the crooked stick; the binder, reaper and mower substituted for the cradle ana scythe; the threshing machine substituted for the flail, and steam and gas power for man and horse power. Every country that is backward in the use of these modern farm machines, is backward also in every other phase of its development. The most striking difference between the American farmer and the Chinese farmer, or the American farmer of today and the American farmer of fifty years ago, is a difference mainly of equipment and the efficient use of that equipment.

The effect of the use of modern machinery on our people is many fold. It has really made possible our high stage of development. In fact, the development of any country is measured by its ability to produce an adequate food supply. It has been only a few years ago that people of this country thot that starvation was staring them in the face. That was in times of peace. It has been estimated that in 1800, 97 per cent of the people of the United States lived on farms, and many of them felt the bite of hunger.

Our farm population decreased slowly until 1850 from 97 per cent to 90 per cent. This was during a period of a half-century, There was no marked development of farm machinery during this period, and our development along other lines was equally retarded. It was the imaginative minds of such men as John Deere, who gave us the first steel plow in 1837; McCormick, who gave us the binder in 1834, and Pitts, who gave us the threshing machine in 1837, that made a start for modern farm machinery. Few of these machines were built before 1850, but after this period, when factories were established and the number of machines built began to increase, the production of food on a much larger scale was made possible, and during the next fifty-year period the population decreased from 90 per cent to about 40 per cent on the farm, or a little over one-third of the total population was on farms.

We can easily imagine the condition that we would be in at the present time if 97 per cent of our people were on the farms without modern equipment. We would be one of the most backward people of the world. We would not have any of the things which go toward making life pleasant and the farm a good place on which to live.

The use of more and better farm equipment has changed the mental attitude of the farmer, it has increased the wages of the farm laborer, it has decreased the necessary labor of women in the field and home, it has increased the production per capita many fold, decreased the cost of production, and improved the quality of products produced.

An abundance of food has made possible our cities, our industries, the arts and sciences, our very civilization. It has made America the greatest nation of the world. These things are made possible because one farmer is capable of producing enough food for three families instead of just his own. Many farmers at the present time are producing even more than this, and doing it with a minimum of labor.

233. Latest Machinery Most Economical. (Fig. 303.) Agricultural production is quite similar to factory production. We find in the factory certain machines for certain particular operations. For example, when we go into a cotton mill, we find a carding machine for making the cotton suitable for use on the spindle. The same thing is true on the farm; we find certain equipment for preparing the soil, special types of seeding machinery for planting, and special equipment for harvesting. The tendency has been too great on the part of many farmers to try to get along without buying the latest improved machines. The farmer can no more get the best results without the latest modern machines than can the manufacturer.

The difference between modern cotton-mill operations and the hand-power method of former days is quite comparable to the modern farmer as compared with the farmer of seventy-five years ago. Production in both cases requires machinery, and without machinery of the right kind and properly taken care of, neither will be successful. The effect of machinery on production

FIG. 303. Motor cultivator, two-row.

per capita is very marked. In those sections where poor equipment is used, the people simply exist and seldom are in a position to improve their living conditions.

The following data collected several years ago illustrates the effect of machinery on the production per capita:

234. Influence of Farm Machinery on Income.

INFLUENCE OF FARM MACHINERY ON INCOME[*]

State	Annual Income of Each Worker	Value of Farm Implements for Each Farm
Florida	$119.72	$ 30.43
Alabama	143.98	33.40
Lowa	611.11	196.55
North Dakota	755.62	238.84

[*]From Circular 21, Bureau of Plant Industry, United States Department of Agriculture.

The use of machinery and modern equipment has not only brought about a greater production per capita, but has also influenced our agricultural conditions along almost every line.

235. The Problem of Farm Power. The farm power problem is one that is being given much more attention at the present time than ever before. To show the tendency toward mechanical power, the census of 1914

Fig. 303-*a*. Two-row cultivator with team.

shows that the power from horses and mules is equal to 14,230,000 H.P., while the power from mechanical sources is equal to 9,675,000 H.P. This vast amount of power is more than that used by all other industries combined. The investment is also much larger than that invested in other forms of power in the United States.

236. Wasting Power and Machinery on the Farm. Some of the greatest losses and wastes on the farm are due to the use of inadequate machines, poor operators, and to lack of care of the machinery. All three of these factors should have the serious attention of every farmer at the present time. Every machine should be adequate for the use for which it is intended. It is very easy to get a machine that is too small or too large to be efficient for

Fig. 304. Checking up machinery for repairs.

a particular use. A great many tractor failures have been due to either the tractor's being too large or too small for a particular farm operation. To use a tractor of 20 to 30 H.P. to drive a pump requiring only 2 H.P. is a mistake often made. It is also as poor economy to operate a single-row corn planter when a two-row planter might be used equally well. In selecting a piece of equipment of any sort, the following points should be kept in mind:

1) It should be the most satisfactory for the particular work at hand.

2) It should be easy to operate with least danger.

3) It should be efficient.

4) It should be capable of easy adjustment.

FIG. 304-*a*. Unprotected machinery.

5) It should be designed so all parts are accessible and easily replaced.

6) It should be well built of good material to resist breakage and wear.

7) It should not cost too much.

Wasting Machinery Thru Ignorance.

The lack of knowledge on the part of the operator has been the cause of many failures with modern machines. This is especially true of power machinery. There have probably been more tractor failures due to this one thing than all other causes combined. Many machines are bought and taken into the field and operated until some trouble develops. It is then found that a wearing part was without lubricant or was not properly adjusted. Every machine should be carefully studied before it is used. An instruction book should be secured with each machine, and it should be studied as a text. With a thoro knowledge of the working parts of a machine, there is little danger of accident, and the best results are assured.

The lack of knowledge of a machine usually results in lack of care and lack of adjustment. It goes without saying that the man who leaves his binder outside to rust and decay does not appreciate its fine points. The same is true of the tractor. If the farm machines were given the attention they deserve, they would be cared for as machinery is cared for in the factory and as the sewing machine is cared for in the home.

Many machines are being run that should be undergoing repairs. The farm machine, as a general rule, is allowed to get in a run-down condition and is not repaired until absolutely necessary, and often such repairs must be made when the machine is in the field and when the work should be in progress. We cannot expect the best results from machines that have been neglected, that have been left in the fields for months, or, if under shelter, are not examined until the day before they are to be used. The farmer would be greatly shocked to see a sewing machine left on the porch for a week at a time where the rain and sun would affect it. Yet, many farmers allow the binders with their delicate tying mechanism to stay out in the weather for months. These machines depreciate in value, become rusty, and are weakened, and there is a loss of time when they fail to give service after they are taken into the field; also, a loss in production.

237. Three Considerations—Housing, Repairing, and Painting. The proper care of machinery might be classified under three heads—(1) housing, (2) repairing, (3) painting. In the housing of farm equipment, we do not have to provide an expensive building. The implements are not affected by cold weather. In sections where the dust is bad, the walls and roofs of the buildings should be made tight enough to prevent its entrance. It has been estimated that the value of machinery on the average farm at the present time is about $1,000. For such an amount of machinery, the farmer can well spend $400 or $500 for a good machinery house. Plans for such a shed can be secured from the U. S. Department of Agriculture or nearly every state agricultural college.

In the repairing of farm equipment, the farmer should be systematic. If the machines are examined on completing a job, and there is not time to repair them at that time, each part should be labeled so that parts can be ordered, and at a later date they can be replaced. The time to repair equipment is not when a machine is needed in the field, but during the time when the machines are in the machinery shed.

In regard to painting, it is well to repaint all wooden parts of farm implements, as it not only increases the life of the implements, but improves their appearance, and where a machine is sold after it has been in use a number of years, the cost of the addition of paint is repaid many fold. Quite often, where the farmer looks after his equipment properly, he will find that discarded machines can be repaired at slight expense and be made to give as good service as a new machine. There are many farmers who discard a machine after it has seen three or four years' service, when it really needs only a few slight repairs. Such machines can often be found standing in fence corners and are used to supply bolts, etc., about the farm.

CHAPTER XXVI

TOOLS AND MATERIALS FOR MACHINERY REPAIR

238. Necessity for Good Tools. Every man who farms will find use for a good kit of tools. In fact, suitable tools will often give an inspiration to do repair jobs that would not be attempted when inadequate tools are provided. Many of the tools described in the sections on woodworking and metal-working are needed

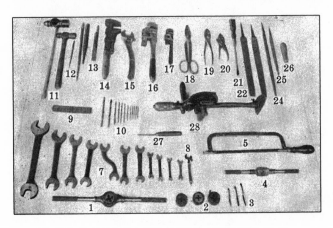

FIG. 305. Principal tools needed in implement repair:

1. Die-stock.	11. Hammers.	20. Cutting pliers.
2. Dies.	12. Punches.	21. Screw-driver.
3. Tap.	13. Cold chisel.	23. Flat files.
4. Stock.	14. Monkey wrench.	24. Round file.
5. Hack-saw.	15. Crescent wrench.	25. Triangular file.
7. End wrenches.	16. Stillson wrench.	26. File handle.
8. Crescent wrench.	17. Trimo wrench.	27. Knife.
9. Rule.	18. Tin snips.	28. Breast drill.
10. Drill bits.	19. Pliers.	

for machine repair and adjustment. There are a few not included in either of these lists that will be mentioned here. All the tools of this group are used without a forge. In fact, the great majority of machinery repair jobs on the farm are "cold jobs" that are made in the field or in the machinery shed. Fig. 305 shows a photograph of the tools which are most likely to be useful in making these repairs.

239. Wrenches. Wrenches for turning nuts and screws are made in various shapes and sizes suited for different uses. They are classed as (1) movable-jaw wrenches adjustable for turning different-sized nuts, the monkey wrench and the crescent wrench being common examples, and (2) fixed-jaw wrenches (the distance between jaws being fixed), the straight- end wrench, the S wrench and the alligator wrench being examples. The alligator type can be used on different-sized nuts, but is not as satisfactory as either the fixed or adjustable type.

Socket wrenches with T-shaped handle are designed for use where the nut cannot be reached with an ordinary wrench. Socket wrenches can be secured in a set of different sizes with a ratchet handle.

Pipe wrenches are made for gripping pipes or cylindrical rods.

In the use of wrenches, one should always be careful to select a wrench that will fit the particular nut snugly. If an adjustable wrench is used, screw the jaw down on the nut tight before attempting to screw it. Always remember to exert force on the handle toward the adjustable jaw.

240. Vise. A bench vise such as described in section on metalwork is well suited for machinery repair.

241. Hammers. A ball-peen machine hammer and a light-weight riveting hammer are needed for many repair jobs.

242. Chisels. The flat chisel, usually referred to as a cold chisel, is useful for cutting rivets or old bolts. Other special-shaped chisels are useful for cutting key ways and oil grooves.

243. Files. There is a number of types of files designed for different uses. Files are used either for smoothing down pieces of work or for sharpening tools such as saws and tools with cutting edges like hoes. Files can be secured of all degrees of coarseness from the rasp used by the horseshoer to the very smooth-cut file used for finishing hard metals. A rasp, one or two flat files, one or two triangular, and several round files should be provided for general repair work.

244. Screw-drivers. Several screw-drivers of different sizes are needed. Keep sides of point of the screw-driver filed parallel to prevent injury to slot in screw.

245. Pliers. Cutting pliers as well as holding pliers are needed. Do not use a pair of pliers where a wrench should be used, or for cutting extremely hard wire when a file will give best results.

246. Hack-Saw. The hack-saw is very useful for cutting pipes, bolts or other pieces of soft metal. It may also be used for cutting slots in screw heads or for similar work.

247. Drills. The most common drills are the breast, post and ratchet drills. The breast and ratchet drills are best suited for general repair work since they can be used at any place without taking a machine apart. The breast drill is de
signed for small holes, while the ratchet can be used for making holes of almost any size.

248. Stock Taps and Dies. Taps and dies are useful for cutting threads on bolts and for threading nuts. Pipe taps and dies are not to be used for bolt work. Machine screw taps can be used for tapping for screws when desired.

249. Materials Needed. For machinery repair, it is essential that there be kept on hand an assorted lot of machine, carriage and stove bolts with nuts and washers; an assorted lot of copper and soft iron rivets; an assorted lot of screws of different kinds and sizes; an assorted lot of cotter keys and pieces of iron rods of different diameters; pipe and pieces of strap iron for general use.

CHAPTER XXVII
How to Study Farm Machinery

250. Three Methods of Approach. Three classes of projects can be worked out to meet the need of the student when studying farm machinery. The first class can hardly be termed projects, but exercises or studies of various types of farm implements and power machines. In taking up these exercises, students will be expected to obtain a general knowledge of all kinds of machinery and make a careful study of those machines used on the home farms. They will be expected to secure booklets describing particular machines under discussion; these booklets may be obtained from manufacturers or from local dealers. The machines are studied on the implement dealer's floor, in farm-machinery sheds, or in the school shop. Most of this work would be done during the time of year when the weather will not allow outdoor work.

The second type of project is the study of the machine while operating under actual farming conditions, the student being given a chance to make adjustment as well as actually operating the machine. A study of the cost of doing the job is carried out in this connection. It may be preparing the the seed bed, planting the grain, or harvesting. Each step is studied, the work is actually done, the time required for it and the cost noted.

The third type of project is a study of the care, adjustment and repair of machinery. Not only can this problem be studied by visiting various farms and studying conditions, but actual repairs can be made. Many machines are left in the shed without checking up repairs at the end of the season's work. Such machines can be inspected, parts ordered and repairs made. Gas engines can be overhauled, tractors gone over and put in first-class shape. The instruction books furnished by manufacturers are an excellent source of information for this work.

A few general exercises and projects such as suggested above will follow, with additional ones briefly outlined. It is suggested that a machinery laboratory manual* be available for student reference for additional subject-matter, it being impossible to cover the subject in this section.

251. Tillage Machinery.

Requirements: To make a careful study and make a complete report on each of the chief tillage machines, including a walking plow, a sulky, a gang and a tractor plow; a peg-tooth, spring, and disc harrow; a disc and a shovel riding cultivator; and a smooth and a corrugated roller.

Tools Needed: Monkey wrench, screw-driver, rule and pair of pliers.

Preliminary Instruction: The importance of a careful study of all types of farm machines is well justified by the part machinery plays in farm production. The lack of knowledge and lack of care of many machines on the farm with the resulting losses should be an example for every boy in his preparation for future farm work. The study of tillage machinery is just as important as the study of the tractor and other power machines, altho it may not be so interesting.

Working Instructions: After being assigned a group of machines, the student will read carefully description

---

*Valuable suggestions can be obtained from *Farm Machinery* by Wirt, John Wiley & Sons, New York.

in text references assigned by instructor. In addition, he should secure catalogs and booklets describing such machines. Next, from a catalog cut out, with a pair of scissors, an illustration of each of the machines being studied. Paste the illustration on a blank sheet of paper. Then, while going over the machine being studied, label all the principal parts. As a report, with the illustration give a statement of the function of each part, its construction and adjustment. These facts may be determined from reference text, from catalogs, from discussion in classroom, by examination, removing parts and taking measurements, or from the instructor in the laboratory.

252. Study of Seeding Machinery.

Requirements: To make a careful study of the different types of seeding machines that the particular type of farming demands, including a study of grain drills, corn planters, cotton planters, broadcast seeders, pea and bean planters and drills. To make a test of the accuracy of planting of the machine studied and calibrate it to plant a definite amount, and make a report.

Tools and Materials Needed: Monkey wrench, screwdriver, rule, pair of pliers, scales for weighing, seed for testing, and paper bags or other containers.

Preliminary Instruction: The accuracy of planting determines to a great extent the final yield of the crop. So every one should know how to test a planter, drill or other seeding device. One should not only know how, but should actually make a test before using the machine in the field.

Working Instructions: Follow instructions under exercise in Sec. 251, and, in addition, the machine may be tested as outlined in Secs. 277 and 279.

253. Study of Fertilizer Drills, Manure and Straw Spreaders.

Requirements: To make a careful study of different types of fertilizer and limestone drills, including the agitator, force-feed and end-gate type. To study manure spreaders and straw spreaders; also straw-spreading attachments for manure spreaders.

Tools Needed: Same as in Sec. 251.

Preliminary Instruction: Keeping up the fertility of the soil is one of the greatest problems of a permanent agriculture. The use of fertilizer drills, manure spreaders and straw-spreaders for distributing materials on the soil greatly facilitates this work.

Working Instruction: Follow instruction in Sec. 251.

254. Study of Haying Machinery.

Requirements: To make a careful study of the various classes of haying machinery—the mower, rakes of different types, tedders, loaders, stackers, presses and other having machinery such as is used in the community.

Tools Needed: Same as in Sec. 251.

Preliminary Instruction: The hay crop is one of the most valuable of the American farmer. By many it is given little consideration; much hay is lost due to lack of care in handling. Modern machinery has made it possible to handle the hay crop with a minimum amount of labor.

Working Instruction: Follow instruction in Sec. 251.

255. Harvesting Machinery.

Requirements: To make a careful study of grain-harvesting machinery, corn binders, grain binders, shocking attachments, push binders, headers, combines and such harvesting machinery as is used in the immediate neighborhood.

Tools Needed: Same as in Sec. 251.

Preliminary Instruction: The modern harvesting machinery on the farm plays a similar part in production to the automatic machines in the factory. They make possible greater production per capita, allowing more people to enter other lines of endeavor. The cost of production where modern harvesting machinery is used is a great deal less than where harvesting is done by hand method.

Working Instruction: Follow instruction as outlined under Sec. 251.

256. Study of Power-Driven Machines.

Requirements: To make a careful study of power-driven machines used on the farm, such as grain separators, silage cutters, feed grinders, corn shellers, limestone grinders, cane mills and other machines in that section.

Preliminary Instruction: With the advent of the stationary engine and the tractor on the farm, power-driven machines in greater numbers will be used each year. Many farmers are already buying small threshing outfits where formerly the grain was threshed by a large threshing outfit; Such practices make for greater efficiency and better products. The grain can be threshed when in the best condition, and the corn cut for silage when at the proper maturity if the farmer has his own equipment. A careful knowledge of such equipment is necessary for its efficient use.

Working Instruction: Same as in Sec. 251.

257. Study of Gas Engines, Tractors and Trucks.

Requirements: To become thoroly familiar with at least one type of gas engines, one type of tractors and one type of trucks.

Preliminary Instruction: There is more power used on the farm than in all other industries combined. The total horse power has been estimated to be more than 25,000,000. More than one-half of this is mechanical power. Altho the farmer is one of the greatest power users, it is only within recent years that he has paid any attention to this phase of his farm problem. Every farmer should become familiar with the construction of an internal-combustion engine.

Working Instruction: Follow Instruction as outlined in Sec. 251. Pay especial attention when studying a stationary engine to its general construction, the ignition system, system of carburation, method of cooling, oiling devices, type of governor, and determine for what type of work the engine is best suited. In addition for tractors and trucks, note how the power goes from engine to drive wheels, the clutch, transmission, differential and drive shaft-and observe the lever control. Note the wheel construction, fenders for protection, seat, arrangement of fuel tanks, etc.

CHAPTER XXVIII
PROJECTS IN FARM MACHINERY OPERATION

258. Conditions for Carrying Out Projects. While this series of projects can be carried on at the same time with the study of machinery, they are better adapted for home projects and can be carried on as outside assignments under actual farm conditions along

FIG. 306. Plowing with horse-drawn riding plow.

with production projects, such as growing five acres of corn, an acre of potatoes, etc. It is not essential that the particular projects that are outlined be followed; the chief thing in mind should be to study the machinery that is being operated on the farm with the idea of, first, becoming familiar with the general method of doing the job; second, determining if the method used is the best or most efficient, and, third, determining how much it costs. For those students who do not live on farms, this work can be done by visiting a farm when a particular operation is being carried on.

259. Preparation of Land for Planting. (Figs. 306 and 306-a.)

Requirements: To operate horse-drawn and tractor-drawn implements in the preparation of a seed bed. To become familiar with all the details of operation,

Fig. 306-a. Disking with tractor power.

determine cost of preparing land for planting by the two methods, and compare the results obtained.

Tools and Equipment Needed: Implements and power available for the particular job.

Preliminary Instruction: A well-prepared seed bed is essential to a good crop, and the work done at the least expense means the greatest income to the operator.

Working Instructions:

1) Each student should harness team and hitch to plow; should lay out a field under supervision of instructor or farmer, and plow at least one acre of ground, noting time required to do the job. The team should then be hitched to harrow, and field harrowed, noting time required. Make all adjustment necessary to make the plow operate effectively and with the least draft.

2) Each student should get tractor ready for field work, make proper hitch to plow and carry out work of plowing as outlined in previous paragraph. Proper adjustment for proper depth and adjustment to avoid side draft should be made. Note time required to plow and harrow one acre.

3) Considering cost and depreciation of the two outfits and all other expense entailed, calculate cost of doing the work by the two methods. In report, compare quality of work done by two outfits.

260. Planting Corn. (Fig. 307.)

Requirements: To operate a corn planter. To select proper plates for particular corn. To make all adjustments necessary to have planter drop and cover effectively. Determine cost of planting corn per acre.

Equipment Needed: Planter complete and team.

Preliminary Instruction: Careful grading and selection of seed corn is as important for good results as proper seed, preparation and careful planting.

Working Instruction:

1) Select proper plates for planting and test them out upon going into the field.

2) Drive in stake, attach check wire and unreel it from the drum the first trip across the field. Place stake at opposite side of field so it will be directly behind the planter tongue after it has been turned into position, and draw check wire up to proper tightness as directed by instructor.

3) Place check wire in trip, set the openers in position, lower marker in place and drive across the field. Note if planter is dropping.

Fig. 307. Planting corn.

4) Observe extreme care to make a straight row the first time across the field. After turning into position, change stake and draw check wire up to proper tightness. Follow marker track with tongue directly above it for second trip across field, and continue as outlined above.

261. Drilling Grain. (Fig. 308.)

Requirements: To operate a drill in drilling grain. To set seeding devices for a definite rate of seeding, and drill a definite area, determining the cost of the operation.

Tools Needed: Drill and team.

Preliminary Instruction: Same general instruction with reference to selecting seed corn also applies to

FIG. 308. Drilling grain.

small grain. In general, it should be remembered to plant *across* the slope instead of *along* the slope. This is to check erosion and avoid starting a small gully by washing at the wheel tracks.

Working Instruction:

1) Adjust feeding device for a definite rate of seeding. It is best to do this by test rather than to be guided by dial.

2) After drill is driven in position, lower the furrow openers into ground.

3) Drive across field, noting that the openers do not clog and that the seed is passing down into the soil.

4) On all following trips, be careful to note where the last track was made in order that no ground will be missed or gone over twice.

262. Harvesting Corn for Silage. (Fig. 309.)

Requirements: To assist in harvesting corn and putting it into the silo. To operate each machine for a period long enough to become familiar with each detail of the work. To determine the cost of each operation in harvesting the corn from the field to putting it into the silo as silage.

FIG. 309. Cutting silage.

Equipment Needed: Corn binder, wagons, silage cutter, teams and engine for power.

Preliminary Instruction: Due to the fact that corn as silage is so highly palatable and nutritious, practical, successful dairymen and cattle feeders have silos. When corn is put into the form of silage, practically none of it is wasted.

Working Instruction:

1) Operate binder in cutting the corn.

2) Note the rate of cutting and estimate the number of acres cut per hour and cost of cutting per ton.
3) Compare cost of cutting by machine and cutting by hand.
4) Haul a load of corn from field to silage cutter.
5) Determine the cost of hauling per ton.
6) Operate silage cutter.
7) Note special safety devices on cutter.

FIG. 310. Harvesting grain with tractor power.

8) Note the rate of cutting in loads and in tons per day.
9) Note the type of engine used to drive cutter.
10) Determine the cost of operating engine.
11) Note the method of elevating the silage into the silo.
12) Assist in packing the silage in the silo.
13) Note the total number in the crew on the various types of work.
14) Determine the total cost of getting the silage into the silo.
15) Determine the capacity of the silo.
16) Determine the cost per ton in getting the corn from the field into the silo as silage.

263. Harvesting Grain. (Fig. 310.)

Requirements: To assist in the various operations of harvesting grain, from cutting with a binder and shocker through threshing. To determine the cost as far as possible for each operation, to be able finally to determine cost of producing a bushel of wheat or a bushel of corn.

Equipment Needed: Binder, teams or tractor, wagon, and threshing outfit.

Preliminary Instruction: It is just as essential that the farmer know how much it costs to grow a bushel of grain as that the manufacturer know how much it costs to mill 100 pounds of flour. The harvest season on the farm is a season when labor is in demand. It is essential that the grain be cut when at the proper stage of ripeness and threshed when properly cured. For these reasons it is important that some study be made of the processes of harvesting grain.

Working Instruction:

1) Get binder ready for cutting, with a satisfactory hitch and properly adjusted and lubricated.
2) Operate binder and note the rate of cutting.
3) Learn to shock the grain properly so it will not fall down or blow over.
4) Note the number of men required behind the binder.
5) Determine the cost of cutting and shocking per acre.
6) Later, when grain is ready for threshing, load grain on rack and haul to threshing outfit.
7) Pitch grain from rack onto threshing feed table.
8) Note each operation that takes place in the threshing machine, from the time the grain bundles are on the feed table until the grain is weighed.

9) Determine amount of grain produced per acre.
10) Make a summary of the cost of each operation in producing an acre of wheat, the total cost per acre, and the cost per bushel.

264. Harvesting Hay Crops. (Fig. 311.)

Requirements: To assist in the various operations of harvesting hay from cutting to baling. To determine the cost of each operation as accurately as possible, and, finally, to determine cost of producing a ton of hay.

Fig. 311. Using a hay loader.

Equipment Needed: Mower, rakes, loaders, balers and power.

Preliminary Instruction: Hay is a crop that has to be made while the sun shines. It must be cut at the right time, and cured to the right degree before it can be stacked, stored or baled. Handling of the hay depends much on the weather. The condition of the crop must also be considered. The proper time to cut alfalfa and other hay crops will be taken up in the study of crops.

Working Instructions:

1) See that mower is properly oiled and that the sickle is sharp. A steady team is essential to the best success in mowing.
2) Lay out land for cutting, size and shape depending on area to be mowed.
3) If ground is rough, adjust cutter bar so it will not cut into the ground.
4) Locate stumps or other obstructions in the field. This is to avoid accident.
5) Drive at a uniform speed. It is the slowing down which causes clogging in heavy grass.
6) Observe care in judging width of swath; cut a full width, but do not leave any uncut.
7) When a side-delivery rake is used, follow in same direction as with mower.
8) Ordinarily, rake after dew is off and before leaves have dried to a point where they shatter.
9) With a dump rake, practice care in dumping so the rick or wind-row of hay will be reasonably straight. This makes loading easier.
10) The loader is best used when the hay has been raked with a side-delivery. Hook the loader on back of wagon and drive straddle of the rick.
11) Keep one man on the wagon to distribute hay on load, and another to drive.
12) If slings and carrier are to be used in unloading, put on three or four slings to the load.
13) If fork and carrier are to be used, put on one sling at bottom of load to clean off the rack in unloading.
14) If hay is baled, carry out each operation in this work, feeding the hay, putting in dividing block, placing wires, tying, etc.

15) Determine as accurately as possible cost of harvesting hay by the ton.

265. Operating Household Equipment. (Fig. 312.)

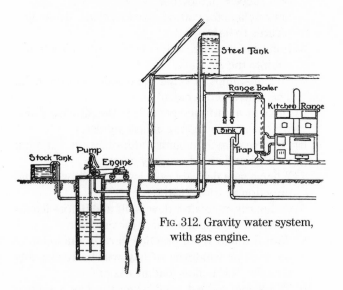

Fig. 312. Gravity water system, with gas engine.

Requirements: To operate each of the various machines about the household under the supervision of some one thoroly familiar with their use. To make a report on value of equipment in the home from the standpoint of time- and labor-saving.

Equipment Needed: The equipment for this project can be found in any modern farmhouse. It is simply a matter of the instructor or students obtaining permission to use equipment in the home as a laboratory.

Preliminary Instruction: So many farm homes are now being equipped with modern lighting, heating and water systems and sewage-disposal plants, that it is essential that every farm boy, and girl as well, become acquainted with the use of this equipment. The best way to become acquainted with its use is to use it. Follow instructions furnished by manufacturers.

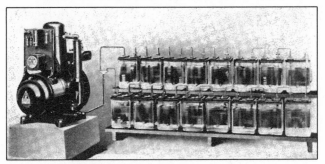

Fig. 312-a. Farm lighting plant with storage battery.

Working Instruction:
1) Lubricate plant, put in fuel and fill radiator.
2) Start electric plant.
3) Turn lights off, and on.
4) Stop plant.
5) Turn lights on so they use power from storage battery.
6) Charge storage battery, note rate of charging.
7) Stop plant; note automatic stopping device.
8) Operate acetylene light plant.
9) Remove water and sludge from plant.
10) Put in a charge of carbide and fresh water.
11) Operate Blau gas plant.
12) Disconnect and replace a container of Blau gas.
13) Operate gasoline plant.

14) Fill tank with gasoline.

15) Crank up pressure motor or pump air into tank.

16) Fire a furnace and clean out all ashes and clinkers.

17) Note use of special devices for controlling draft and temperature.

18) Start and operate different water systems.

19) Note the difference in amount of work required when water is carried in and when it is pumped by machinery.

20) Study washing equipment.

21) Note difference in amount of time required to do the washing when a power washing machine is used and when a scrub board is used.

266. Gas Tractor Operation. (Fig. 313.)

Requirements: To become thoroly familiar with the method of operation of as many types of tractors as possible.

Tools and Materials Needed: Set of tools secured with tractor. Fuel, oil and extra spark plugs.

Preliminary Instructions: In operating and handling a tractor, one should be very careful to avoid breaking any parts. *Always be sure—then go ahead.* Do not attempt to start a tractor for the first time unless under the direction of *some one who knows.* Remember, that there is more danger in starting a tractor than in starting a small stationary engine, on account of danger of personal injury and of damage to the tractor and buildings. Remember, also, that you are handling an expensive machine when operating a tractor.

Working Instructions:

A. Getting tractor ready and starting it.

1) See that the tractor is completely lubricated.

2) See that the clutch works freely.

3) If brakes are provided, see that they are released.

4) Study the manipulation of the various controlling levers.

FIG. 313. Plowing with a tractor.

5) See that the gears are not in mesh.

6) See that the clutch is not engaged.

7) Turn on gasoline.

8) Open needle valve on carburetor.

9) *Retard the spark.*

10) Trip the impulse starter, if any.

11) Prime the motor with gasoline if weather is cold.

12) Crank the motor.

B. Tractor operation.

1) To start the tractor forward or reverse, (a) see that the pulley wheel is not revolving; (*b*) see that clutch is not engaged; (c) shift gears slowly—if they do not mesh, engage the clutch slightly, then disengage it—continue the process until gears mesh; *(d)* engage clutch and the tractor should run.

2) To stop tractor, (*a*) disengage clutch; (*c*) apply brake if necessary; (*c*) shift gears to neutral position.

3) Take tractor outside and practice starting and stopping. (*a*) Run forward a few yards in low, then stop; (*b*) reverse, run backward a few yards, then stop; (*c*) run forward a few yards in high, then stop; *(d)* turn the tractor around, as in plowing, and note the space required to turn it in.

4) Examine the tractor carefully and see that it is in perfect condition. Clean off dust or dirt.

5) Drive tractor back into building under supervision of some one who has had experience.

C. If possible, make study of a tractor while plowing in the field, and obtain the following information:

1) Number of plow bottoms.

2) Size and type of plow.

3) Length of furrows.

4) Width of furrows.

5) Depth of furrows.

6) Time required to plow a furrow.

7) Time required for turning.

8) Kind and condition of soil.

9) Acres plowed per hour.

10) Acres plowed per ten-hour day.

11) Fuel used and cost per ten-hour day.

12) Fuel cost per acre.

13) Lubricant used and cost per ten-hour day.

14) Lubricant cost per acre.

15) Labor cost per ten-hour day.

16) Labor cost per acre.

17) Depreciation cost per acre.

18) Interest on investment per acre.

19) Repair cost per acre.

20) Total cost per acre.

Assume the following condition with reference to a one-man outfit—operator cost, 50 cents per hour; 10 per cent depreciation on original cost of outfit; interest on investment at 6 per cent; cost of repairs, 4 per cent; all three charged to 100 days' service.

D. Write a report on this exercise, giving the information outlined under *A*, *B* and *C* and also:

1) Name of tractor.

2) Where manufactured.

3) Rated brake H.P. and drawbar H.P.

4) Number of cylinders in motor.

5) Arrangement of cylinders.

6) Make and type of carburetor.

7) Make and type of magneto.

8) System of lubrication.

9) Method of cooling.

10) Describe the clutch and transmission system.

CHAPTER XXIX

PROJECTS IN
FARM MACHINERY REPAIR

267. The Proper Time for Checking Up Needed Repairs. The repair and adjustment of machinery is best carried on during the winter months when the weather is not suitable for outdoor work. Especially is this true of the repairs; the final adjustment must often be done after the machine is taken into the field.

It is best to go over a machine carefully when the work is finished for the season and tag all broken or worn parts. By so doing, the work of putting the machine in condition for field use is much easier. One is always more familiar with the condition of the machine just after using it than nearly a year later when it is being taken into the field the first time for the season. When worn and broken parts are not tagged the year before, a careful inspection is very essential. This part of the work should be done some weeks before the actual repair work is to be done and a longer time before the machine is needed in the field. This will give a chance to order parts needed, which often cannot be obtained from the local dealer. This work is best done in the school shop where there are plenty of tools and material. Many students can bring old implements in from the home farm for overhauling. Gas engines can be cleaned up, valves ground, and new piston rings put in place, the cutter bar on the mower can be straightened and the sickle sharpened, and other jobs can be done, a few of which are outlined merely to suggest the possibilities along this line. Such work is of immediate value in putting the machinery in repair, and the practice is of untold value to every student who later is to farm for himself.

The projects in this chapter are arranged in six groups according to the general type of machine. Additional minor groups might be added, but these are the machines in which all farmers are interested: First,

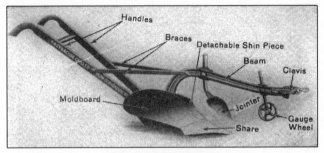

FIG. 314. Walking plow.

tillage machinery; second, planting machinery; third, fertilizer distributors; fourth, harvesting machinery; fifth, belt-driven machinery; sixth, stationary engines and tractors. Projects in the repair of only one or two machines in each group are outlined.

268. Repair and Adjustment of Tillage Machinery. (Fig. 314.)

Requirements: To repair and adjust ready for field use the chief tillage machines, including plows, harrows, rollers and cultivators.

Tools Needed: It is well to have access to a complete set of shop tools. The exact number of tools required will be determined by the repairs needed.

Preliminary Instruction: The plow is the principal implement in the preparation of the seed bed. Because it is simple, it is often neglected and used very inefficiently. Plows not cared for are hard to operate, and a poor job of work is the result. The same principle holds true to a greater or less extent with all other tillage machinery.

269. Repairing a Walking Plow. (Fig. 315.)

1) Share—Badly-worn cast-iron shares must be renewed; steel shares may be sharpened. Provide

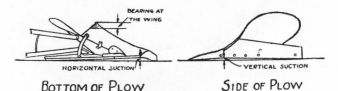

BOTTOM OF PLOW   SIDE OF PLOW

FIG. 315. Detail showing horizontal and vertical suction.

bearing at wing of 3/4″ for 10″ bottom to 1-1/4″ for 16″ bottom, and vertical suction of 1/8″ and horizontal suction of 1/8″ to 1/4″, as shown in figure.

2) Landside—If heel is detachable and worn, renew entire landside.

3) Moldboard—See that moldboard is well bolted to frog. If badly worn, renew.

4) Bracing—Tighten all bolts and brace rods.

5) Handles—See that handles are tight and rigid thruout.

6) Beam—See that beam is bolted tightly to the frog. If a steel beam, be sure it is not sprung.

7) Jointer—Renew or sharpen the jointer. Bolt tightly to beam.

8) Gauge Wheel—Renew bearings if badly worn. Bolt standard rigidly to the beam. Adjust to proper height.

270. Walking Plow Adjustment.

1) Depth of Furrow—Raise or lower clevis hitch vertically. For variable soil conditions, regulate by changing wheel gauge.

2) Width of Furrow—Change the clevis hitch in a horizontal position. Position of beam may be adjusted on some plows. It is usually changed to accommodate a different number of horses.

3) Handles—Change height to suit operator.

4) Jointer—Set so its point is just above the point of the share, slightly to the landside of the shin and 1-1/2″ to *2*″ deep into the soil.

5) Hitch—Plow runs best when hitched to form a straight line from a point on moldboard 2″ from shin thru the hitch at beam clevis to a point midway between the tug rings at harness. A proper hitch means easy operation and less draft for the team.

271. Sulky and Gang Plows. (Fig. 316.)

1) Wheel Bearings—If worn, put in new bearings when possible. Clean thoroly, repack with heavy grease and make adjustments.

2) Frame Beam and Frog—Tighten all bolts. Straighten any part of frame that is twisted.

3) Levers—Tighten all connections, take up lost motion, straighten levers, replace new springs.

4) Share—Sharpen or replace with new share.

5) Landside—Renew entire landside if badly worn. Renew heel when provided.

6) Rolling Coulter—Clean bearings and oil. Tighten standard rigidly to frame.

272. Adjusting Sulky or Gang Plow.

1) Depth—Change depth by lowering bottom in the frame,

2) Width of Cut—Change hitch on frame, change land-
ing of furrow wheel.

3) Jointer—Adjust as on a walking plow.

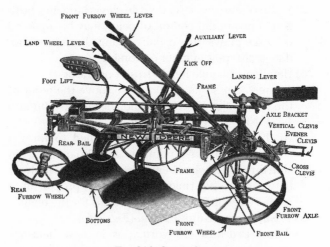

Fig. 316. Gang plow.

4) Rolling Coulter—If used with a jointer, set ahead of
it. If used without jointer, adjust to the position of
jointer when it is used alone and about one-half the
depth of furrow, depending on the soil.

5) Wheels—Adjust land wheel to run straight to the
front. Give the front and rear furrow wheels a slight
lead from the land. Set rear wheel 1″ to 2″ outside
of landside of plow.

6) Hitch—Point of hitch can be changed to take more
or less land; and so the load is carried by wheels.

273. Repair of Peg-Toothed Harrow. (Fig. 317.)

1) Frames—Straighten all bent parts and tighten bolts.

2) Teeth—Adjust to uniform depth. Re-sharpen worn teeth, and renew lost ones.

3) Levers—Straighten all levers, renew worn parts and tighten connections

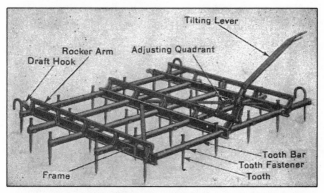

FIG. 317. Peg-toothed harrow.

4) Draft Connections—Renew if badly worn. Straighten if bent.

274. Repairing a Disc Harrow. (Figs. 318 and 318-a.)

1) Frame—See that all bolts are tight and all braces are straight and rigid.

2) Bearings—Clean out bearings by washing with kerosene. Replace if worn, pack grease cups and see that grease gets to bearings.

3) Discs—Sharpen discs on regular sharpener or on emery.

4) Gang Bolts—See that gang bolts are straight and the discs are tight on bolts so they will not wobble.

5) Bumpers—Adjust so they carry end thrust.

6) Scrapers—Replace if badly worn. See that they come in contact with disc without causing undue friction.

7) Snubbing Blocks—Adjust so gangs run level.

8) Levers—Straighten bent levers. Replace worn parts.

9) Draft Connection—If worn, renew.

Adjustment—Change angle of disc to increase or decrease amount of suction. Weighting is sometimes resorted to in hard ground to increase the depth.

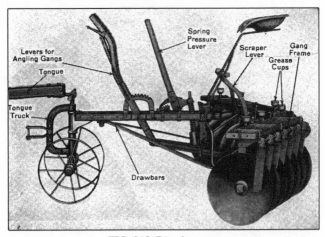

FIG. 318. Disc harrow.

275. Repair and Adjustment of Planting Machinery.

Requirements: To repair and adjust ready for field use planting machinery such as used in the particular locality. A corn planter and drill are outlined.

Preliminary Instruction: Every planting machine should be in first-class repair when taken into the field, to avoid a poor stand due to its poor condition.

276. Repairing a Grain Drill. (Fig. 319.)

1) Grain Feeds—Clean out old grain and dirt. Examine grain feed cup or fluted cylinder, and grain cells. Renew badly-worn or broken parts. Examine method of changing rate of seeding.

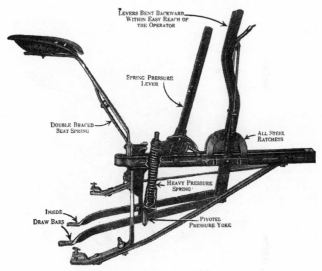

FIG. 318-*a*. Details of construction of disc harrow.

2) Chains, Drive Shaft and Gears—Trace power from wheels thru chains, shaft and gears to feeding device. See that there is no lost motion due to loose, broken or worn parts.

3) Openers—Sharpen opener if dull. If disc opener, examine the bearings and replace if badly worn. See that they are properly lubricated. Adjust springs so enough pressure is on openers.

4) Seed Tubes—Test seed tubes to see that they do not clog easily.

5) Wheels—If wheels are of wood and are dried out so that the tire is loose on the rim, they should be soaked in water until swelled tight. The pawls in the hub are an important part of wheels to give

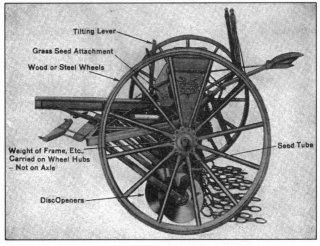

Fig. 319. Grain drill.

positive drive. See that the pawls engage and start the seeding device as soon as the wheels turn.

6) Frame and Levers—See that all bolts are drawn up tight and the frame is rigid. Examine levers and see that they are straight and function properly.

7) Attachments—See that attachments such as surveying device and devices for setting rate of seeding are tight. Check their accuracy if they are to be depended upon.

8) Miscellaneous—See that all covering devices, hitch, braces, etc., are in place and properly adjusted.

277. Adjusting a Grain Drill. Calibration—The principal adjustment on a grain drill is the one for accuracy of planting when the indicator is set at different positions on the scale. The adjustment is accomplished by calibrating the machine. The drill must be calibrated for each kind of grain. The method is as follows:

Set the drill on stands or saw horses so that the wheels clear the ground. Put the grain in the hopper, place the indicator for certain rate of seeding per acre, put paper bags under each of the spouts, throw in the clutch and you are ready to begin. Turn the drive wheel thru 100 revolutions. Weigh the seed caught under each spout. By measuring the circumference of drive wheel in feet and multiplying by 100, the number of turns, the distance traveled is found. Multiply this by the width of seeded strip in feet and the area is obtained. Knowing the area and the total pounds of seed drilled, the rate of drilling is easily obtained. By comparing the rate from test with the actual setting of indicator, the accuracy of the machine is determined. By making several tests at different settings of indicator, the proper adjustment for a certain rate of planting can be established. Unless a drill is carefully tested, the rate of planting is not definitely known, due to the inaccuracy of the indicating device.

278. Repairing Corn Planter. (Fig. 320.)

1) Seed Box and Plates—See that a full set of plates is available and suitable for planting seed at hand. Examine the parts in bottom of seed box and the plates to see that they are not worn. Renew parts as needed.

2) Sprockets, Chains, Gears and Clutch—Trace the power from wheels thru axle, chain, sprockets, drive shaft and clutch to the plate. See that there is no lost motion due to loose, worn or broken parts.

3) Openers—See 3 under Drills.

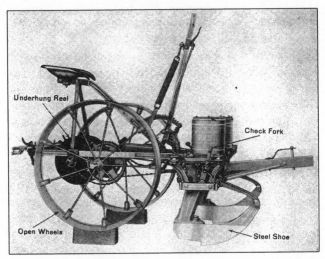

Fig. 320. Corn planter.

4) Valves—Examine valves in feed shank and see that they function properly when the drive wheels are turned.

5) Frame and Levers—See 6 under Drills.

6) Check Wire and Trip—See that the check wire is free from kinks and in good condition. See that the trip works.

7) Miscellaneous—Examine marker, hitch, etc., and see that they are in good condition.

279. Adjustment of Corn Planter.

1) Depth—Adjust for proper depth by setting the tongue; also, by means of the lever, just in front of the seat.

2) Width—The width of rows can be adjusted by shifting the boxes and shoes as a unit on the frame. Shift the position of the wheels on the axle and move the wheel scrapers accordingly.

3) The Drop—the drop is changed by moving the lever to point indicating two, three or four grains per hill.

4) Drilling—Most planters can be adjusted to drill by opening the valves and holding the trip back. A lever is often provided so that the change to drilling can be made from the seat.

5) Accuracy—Proper plates should be selected for the particular seed and the accuracy of drop tested before the planter is taken into the field. If the plates are of the type where one kernel is selected at a time, try out several by taking some kernels of corn and fitting them into the spaces. If they do not fit—are too tight or too loose—try other plates until one is found that fits fairly well. Place this plate in position in box, partially fill it with the corn to be planted, place the planter on a stand or saw horses, and you are ready for test. Set the lever to position of number of grains to be dropped at each hill. Trip clutch and turn drive wheel; catch the grains as they drop out at each hill and count them. Trip for 100 hills; if lever is set for three grains, it should test at least 90 per cent accuracy. Out of 100 hills, if there are 60 threes, 30 twos, 8 ones and 2 fours, a plate should be selected with slightly larger openings. The correct selection of plate is very

important from a standpoint of accuracy in planting. The careful grading of seed and proper selection of plate are big factors in securing a good test.

280. Repair and Adjustment of Fertilizer Distributers.

Requirements: To repair and adjust ready for field use the principal fertilizer distributers, including manure spreader, straw spreader, fertilizer and lime drills (Fig. 321).

Preliminary Instruction: The manure spreader is a machine that is found on most farms where there is stock. Straw spreaders, fertilizers and limestone drills are becoming more common thruout the country.

281. Repairing Manure Spreader. (Fig. 321.)

1) Box and Apron—Tighten all bolts in box so that it is rigid. Examine apron for broken places or damaged chain. Replace broken or worn parts. See that the rollers that carry the apron turn easily and offer little resistance.

2) Frame—Tighten all loose parts on the frame of spreader and renew all broken parts.

3) Beater—See that bearings are in first-class condition. Tighten the bars and see that the teeth are straight and firmly in place. Replace all broken teeth.

4) Driving Mechanism—Examine carefully the drive chains, gears and sprockets that transmit the power from the drive wheel to the beater and to the apron. Weak parts should be replaced. Adjust chains to proper tightness. See that all bolts are drawn up tight.

5) Wheels—Take off drive wheels and examine the pawls. Examine bearings on rear axle and on trucks.

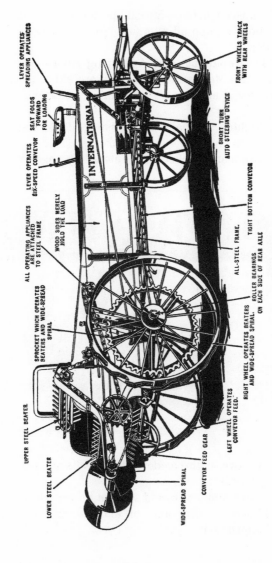

LEVER OPERATES SPREADING APPLIANCES

SEAT FOLDS FORWARD FOR LOADING

LEVER OPERATES SIX-SPEED CONVEYOR

INTERNATIONAL

WOOD SIDES MERELY HOLD THE LOAD

ALL OPERATING APPLIANCES ARE ATTACHED TO STEEL FRAME

SPROCKET WHICH OPERATES BEATERS AND WIDE-SPREAD SPIRAL

UPPER STEEL BEATER

LOWER STEEL BEATER

WIDE-SPREAD SPIRAL

CONVEYOR FEED GEAR

LEFT WHEEL OPERATES CONVEYOR FEED.

RIGHT WHEEL OPERATES BEATERS AND WIDE-SPREAD SPIRAL.

ROLLER BEARINGS ON EACH SIDE OF REAR AXLE

ALL-STEEL FRAME.

TIGHT BOTTOM CONVEYOR

SHORT TURN AUTO STEERING DEVICE

FRONT WHEELS TRACK WITH REAR WHEELS

FIG. 321. Manure spreader.

6) Miscellaneous—Straighten levers and connecting rods. Tighten all nuts and put in new bolts where needed.

282. Repairing and Adjusting Straw Spreader. (Fig. 322.) Most straw spreaders are either an attachment for a manure spreader or an attachment for a wagon.

1) Tighten all chains by adjusting idlers, and renew worn links.

2) See that sprockets are centered on wagon wheel.

Fig. 322. Straw spreader.

3) Go over entire feeder, tighten bolts and renew broken parts.

4) Straighten levers and see that they work easily.

5) Follow instructions of manufacturer in making adjustment.

283. Repairing a Lime and Fertilizer Sower.

1) See that feeding device is free from old lime or fertilizer and rust.

2) Renew badly-worn gears, sprockets or chains.

3) See that adjusting levers work properly.

4) Examine wheels and axles.

5) Repair box or hopper if needed.

6) Renew feeding device if badly worn or broken.

284. Repair and Adjustment of Harvesting Machinery.

Requirements: To repair and adjust a mower, a binder and other harvesting machines such as are used locally. The mower and binder are outlined, as they represent the two most common harvesting machines thruout the country. The tools needed are the same as in previous projects.

Preliminary Instructions: To avoid loss at harvest time, all equipment should be in a first-class condition. Harvest season is a time when delay may mean a great loss. So every farmer should realize the importance of having such equipment ready. The best time to inspect harvest machinery is just at the end of the harvest season rather than the beginning. If the inspection has been properly carried out and parts ordered to take the place of broken and worn ones, the work of repair will be very simple.

285. Repairing a Mower. (Fig. 323.) Place the machine where there is plenty of room and where all sides are accessible.

Working Instructions:

1) Align Cutter Bar—Block tongue to normal position of running with inside shoe just floating. Test alignment by stretching a string from center of

pitman bracing thru center of knife head bracing
to outer side of cutter bar. If properly aligned, the
outside end of knife will lead string by 1" for five-
foot bars and 1-3/8" to 1-1/2" for six-foot bars. If
not properly aligned, examine machine for special provision for alignment and make proper
adjustment.

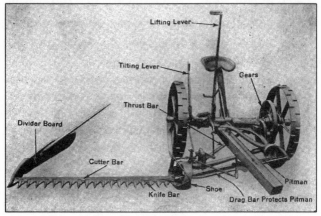

FIG. 323. Mower.

2) Aligning Guards—Remove blade and sight along
   guards, or use a straight-edge to detect the ones
   that are high and the ones that are low. Drive
   guards that are out of alignment back into place
   by a sharp blow with a hammer.

3) Adjusting Cutter-Bar Clips—Examine knife bar
   to see that it is straight; then put it in place. The
   knife should have little play, and the clips should
   fit snugly. Adjust all clips by tapping each with

hammer until it begins to tighten; then loosen it, and begin on the next. When all are adjusted, tighten them.

4) Putting on New Guards—Bolt new guards in place where old ones are damaged. If the new guard brings the ledger plate too high, remedy this by putting pieces of tin between the guard and the bar.

5) Shoes—Examine both the outside and inside shoes on cutter bar. If parts are badly worn, replace them. See that they are adjusted for proper height.

6) Knife Sections—Broken or badly-worn knife sections can be easily removed by placing the vertical edge of bar on an anvil or heavy piece of iron, with a square, straight corner. Strike the back of the section with a hammer, making it cut the rivet off. Use soft steel rivets of proper size for putting on new sections. Test the knife to see that sections center properly. The sections are properly centered if each is directly under a guard when the pitman is at either end of its stroke. Examine to see if a centering device is provided on machine. When steel pitmans are used, they are usually made adjustable for length. This makes centering easy.

7) Pitman—Adjust both the knife head and wrist pin bearing to secure the least amount of lost motion.

8) Gears—If badly worn, make adjustment so they will work properly where possible. If gears are badly worn, replace with new ones.

9) Bearings—Examine all bearings for wear. Free them of all grit, dirt and vegetable matter. Lubricate all parts with new oil.

10) Drive Wheels—Take up all end play by adjustable collar or washers. Examine pawls for wear. Some are reversible, making possible longer use. Renew springs if weak.

11) Miscellaneous—Tighten all nuts, straighten levers, and see that all bolts, cotter keys, etc., are in place. Replace worn parts where needed.

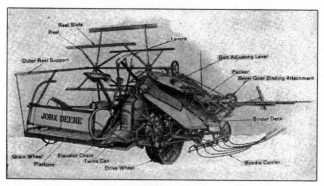

FIG. 324. Grain binder.

286. Repairing and Adjusting a Binder. (Fig. 324.) Practically all the suggestions for the mower also apply to the binder with the following additions:

1) Canvases—See that all rollers work easily, are in good repair, and are properly aligned. Anything wrong with the rollers will cause trouble with the canvases. Test the frame of the machine (either by means of a square or by measuring the diagonals) to see that the canvases are properly squared to it. If canvases are not squared, trouble will result. Replace all broken slats and straps on canvases with new ones.

2) Chains and Sprockets—Replace badly-worn or broken sprockets. See that they are aligned by sighting along the face. Adjust chain tightener so there will not be too much play.

3) Reel—Renew slats if broken. Examine bearings. If they are badly worn, renew them. Take up all lost motion in reel levers.

4) Gears and Bearings—Examine all gears for wear, and if badly worn, replace with new ones. Adjust to mesh where possible. Renew bearings where badly worn.

5) Binder Attachment—The binding attachment is the most complicated device on the binder. Replace all broken or badly-worn parts. See that the tying device is timed to work properly in tying bundles. Use instructions for particular machine furnished by manufacturer.

287. Belt-Driven Machinery.

Requirements: To repair and adjust at least one belt-driven machine. It may be a threshing machine as outlined, or another type of machine. The feed mills, silage cutters and corn shelters, all come in this class. The tools needed are the same as in previous projects.

Preliminary Instruction: A separator that has not been carefully overhauled will cause loss of time and waste of grain. This is a job that should be done some weeks before the threshing season is on, in order that if there is need of any parts, they can be secured and installed without causing a delay. The same general principles as outlined for overhauling a separator apply to other belt-driven machines. It is always a good idea to study carefully the instruction books furnished by the manufacturer before beginning to repair or adjust any part of a belt-driven machine. The points suggested here under working instruction can then be carried out with a much greater degree of intelligence.

288. Repairing a Grain Separator. (Fig. 325.)

Working Instruction:

1) Cylinder—Renew all badly-worn, bent or otherwise damaged teeth. Tighten all loose teeth, and see that the cylinder is firmly keyed to the shaft. If cylinder bearings are worn, they can be made to fit snugly by removing shims. If the bearings are badly worn, they should be re-babbitted or new bushings put in. (See instructions on babbitting at end of this exercise.) Examine the shafts for rough spots. If necessary, smooth them up with a fine file and emery cloth. After the cylinder shaft and bearing are in first-class condition, the cylinder should be carefully balanced before the bearings are adjusted. This is necessary when a number of new teeth have been added. To balance the cylinder, provide two saw horses or other suitable stands to support the ends of the cylinder shaft. Level up the supports and place on them pieces of smooth steel, on which the cylinder is to rest. Place the cylinder on supports and allow it to revolve. Mark the top of cylinder where it came to rest and roll it over again. If it comes to rest in the same position as before, it will indicate that the opposite side is heavy. Provision is made on some cylinders to counterbalance this by driving slugs of lead into the holes in the ends of cylinder. Where no provision is made, new teeth can be put in on the opposite side, or wedges can be driven in under the center band. When cylinder

is put back in place, adjust to the bearings so there is no lost motion. It should make a snug fit, but should not bind. Avoid too much end play. The thickness of wrapping paper at each end of cylinder will be sufficient.

2) Concaves—Replace badly-worn concave teeth. Be careful to avoid breaking the concave bars. Adjust the concaves so the teeth are centered as far as the cylinder teeth are concerned. If the concave and cylinder teeth come closer together than 1/8″, cracking of grain is liable to result. Teeth that are out of line or bent should be brought back into place by the use of a hammer. Inspect the device for raising and lowering the concaves. If badly worn, put in new parts.

3) Separating Grates—See that all bolts are tight and there is no loss motion. Straighten all bent rods or bars. See that all parts work without undue friction.

4) Feeding Attachment—Inspect the frame for looseness, badly-worn or split parts. Tighten all bolts and screws. Tighten the carrier chain; see that slats or canvas is in good repair. Examine band cutter knife; replace it if it is broken or badly worn. See that all bolts are tight and bearings are in good condition on retarder and shaking feed bottom.

5) Beaters and Apron—See that there is no play or lost motion in the beater. If the blades are wood, replace those that are split or badly worn. See that there are no rough surfaces on the blades. See that apron or check board works freely.

6) Racks—Inspect the racks for broken slats. See that the bearings are tight. Replace or adjust worn links and pitmans.

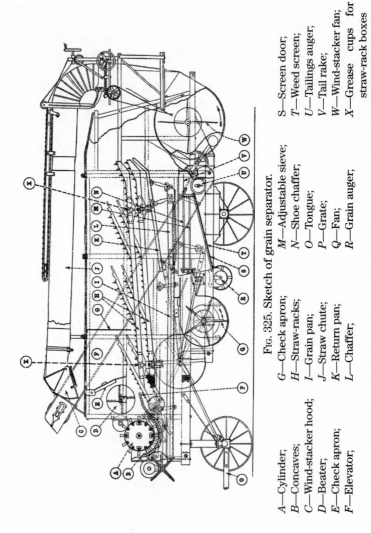

FIG. 325. Sketch of grain separator.

A—Cylinder;
B—Concaves;
C—Wind-stacker hood;
D—Beater;
E—Check apron;
F—Elevator;

G—Check apron;
H—Straw-racks;
I—Grain pan;
J—Straw chute;
K—Return pan;
L—Chaffer;

M—Adjustable sieve;
N—Shoe chaffer;
O—Tongue;
P—Grate;
Q—Fan;
R—Grain auger;

S—Screen door;
T—Weed screen;
U—Tailings auger;
V—Tail rake;
W—Wind-stacker fan;
X—Grease cups for straw-rack boxes

7) Conveyor—See that all parts are tight to avoid wasting grain. Renew metal if it is badly rusted.

8) Screens—See that frames are in good shape. If screens are damaged, renew them. Examine the shoe to see that the castings that carry the screens are in good condition and fastened to the shoe. See that bearings and pitmans are in good shape.

9) Fan—Inspect the fan housing, the bearings and the blades. Replace worn parts where needed.

10) Grain Augers and Elevators—Examine the auger troughs and elevator housing, and all bearings. Replace badly-worn parts. See that the chain is in good condition; also, that the chain tightener is in good working order.

10) Stacker—See that fan, fan housing and bearings are in first-class shape.

11) See that all adjustments are made to insure efficient operation.

Note: Bushing can usually be secured to take care of badly-worn bearings on power-driven machines, but in some cases, babbitting must be resorted to. The following on babbitting will be of interest under such conditions:

289. Babbitting Machine Bearing Boxes. Machine bearings become loose with wear. If the bearings are made in two parts in the form of a split box, adjustments may be made to tighten the bearing until it is practically worn out. If the bearing is in one piece in the form of a solid box, little, if anything, can be done when it is worn to tighten it except to reline or refill it. The process of repairing a bearing by pouring in new metal is called babbitting.

Babbitt is a soft metal consisting of one part of copper, two parts antimony and twenty-two parts tin, melted together. Some of the cheaper grades of babbitt contain some lead and, sometimes, a little zinc.

290. Babbitting a Solid Bearing. Chip or melt out all of the old babbitt and clean out the retaining holes. Warm the box to prevent the babbitt cooling too rapidly when it is poured. This may be done by holding the bearing in the fire or by placing a hot iron against it. Clean the shaft and place it in line in the bearing, first wrapping one thickness of writing paper about the shaft just the length of the bearing, and fastening it by winding twine about it in a spiral shape. The paper will prevent too tight a bearing, and the space occupied by the twine will form oil grooves. Close up the bearing at each end by placing a heavy cardboard over the shaft at each end and puttying up the holes or filling them with soft clay. Reserve the oil hole to pour in the babbitt, or, if it is too small, drive a wooden plug into it clear to the shaft and form a funnel-shaped opening at one end of the bearing with clay.

Heat the babbitt in an iron ladle until it burns or chars a stick, and gently pour it, if necessary, by means of a funnel, thru the hole reserved for the purpose, first making a few vent holes thru the end protections with a wire. When the babbitt is set, and before it thoroly cools, remove the end protections, the plug that fills the oil hole and the shaft. Wipe out the hole formed by the shaft to remove the burned twine and any foreign matter, and the bearing will be ready for use when the babbitt is cold.

291. Babbitting a Split-Box Bearing. Place the shaft in the lower part of the box which forms the bearing and block it in position. Place liners on the box to touch the shaft the full length, first cutting two or three notches on the liner next to the shaft thru which the babbitt can run from the upper half of the box to the lower. Bolt the top part of the box in position, stop the ends and pour the babbitt. When the babbitt is set, drive a cold chisel between the boxes to break the babbitt formed in the notches of the liner, bevel the edges of the babbitt next to the shaft, and cut oil grooves in the babbitt of each half of the box with a diamond point or round nose chisel. These grooves should cross on the oil hole and run to the ends of the box to form carriers for oil.

A split box may also be babbitted by pouring the babbitt on the shaft when placed in the lower half of the box only. When the babbitt reaches the level of the top of the half box, place the liners in position, then the upper half of the box, and pour it full.

292. Scraping a Babbitted Bearing. With the split bearing, it is nearly always necessary to fit the bearing to the shaft by scraping. This is done by coating the surface of the shaft with lampblack and oil, or Prussian blue, and adjusting it in the bearing; then revolve the shaft. Open the bearing and note if it formed a good contact with the shaft; if it only touched the shaft at spots, scraping is necessary. Scrape the high places in bearing with regular bearing scrapers or with a triangular file that has been ground for this purpose, until practically the entire surface of bearing is in contact with shaft.

293. Repair and Adjustment of a Motor. (Fig. 326.) *Requirements:* To repair and adjust a gasoline engine, either a simple type or a tractor, truck or automobile engine. The tools needed are the same as in previous projects.

Preliminary Instructions: The gas engine is the most common type of mechanical motive power on the farm. Every boy needs to know how to make the simple repairs and adjustments, because the gas engine that is not in good adjustment will waste fuel, will develop only a fraction of its power, and will waste the time of the operator. First become thoroly familiar with the engine before trying to repair it. Study it carefully, analyze its troubles before trying to remedy them. In the work of dismantling and putting an engine in shape for operating, the workman must observe extreme care to avoid breaking or marring any part of the machine. Do not use pliers where a wrench should be used, nor a screw-driver where a cold chisel is best suited. Be careful not to tear or destroy the packing. Do not screw the coupling on fuel line too tight, as the threads are liable to be stripped. Clean all parts as they are removed. Place small parts, as nuts and screws, in a box provided for that purpose. Where timing gears are removed, see that they are marked so they will be meshed properly when re-assembled. Secure instruction book on engine as furnished by manufacturer.

294. Overhauling an Engine. The method of procedure will vary slightly with different engines, but the following steps will indicate the general procedure:

1) Disconnect the wiring.
2) Remove the magneto.
3) Remove the igniter block or spark plug.
4) Remove the cylinder head.
5) Scrape the carbon from the face of the cylinder head.
6) Remove the valves and free them from all carbon.

7) Note the valve seats to see if they are free from carbon and not pitted.
8) If valves are in poor condition, they should be ground as follows: (a) Apply a little coarse

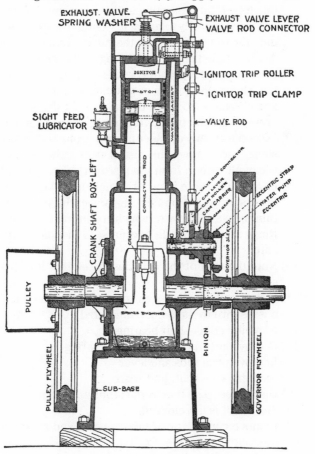

FIG. 326. Gasoline engine.

valve-grinding compound to the valve seat, put the valve in place and grind it by inserting the point of the screw-driver in the slot, or use a valve-grinding tool, and grind by revolving back and forth about one-fourth turn, exerting a little pressure. Lift the valve occasionally to reseat it. Continue the process until the rough surface on the valve is removed. (*b*) Apply a little fine valve-grinding compound to the valve seat and repeat the process as outlined under (*a*). Continue process of grinding until the valves are all seated. (*c*) Clean the valve and valve seat to prevent any compound from entering the cylinder.

9) Disconnect the connecting rod from the crank shaft, and remove the piston from cylinder. (*a*) Clean all carbon from the piston. (*b*) Examine all piston rings; note if any are stuck or broken, (*c*) If necessary to put on new rings, use three or four thin pieces of tin with which to slip on the rings. Rings are very brittle and must be handled with care.

10) Note the wall of the cylinder to see that it is not scored.

11) Remove all oiling devices and see that oil or grease passes thru to the points lubricated.

12) Examine crank-shaft bearings.

13) Remove governor. Examine the spring.

14) Remove push rods and lever.

15) Examine cam shaft and gears.

16) Disconnect pipe line from carburetor to fuel tank. See that it is not clogged.

17) Remove carburetor or mixing valve and examine the following points: (*a*) Type of air valve, if any; (*b*) how the gas is drawn to carburetor; (*c*) how it

is controlled at the carburetor; *(d)* screw out the needle valve and note its condition.

18) Clean out any dirt or other material that may be collected in the cooling system.

19) Reassemble the engine in the reverse order in which it was dismantled.

20) Adjust the engine by timing the valves and the ignition and setting the governor for rated speed.

General Questions to Answer in Report on this Exercise:

1) Does the engine have high- or low-tension ignition?

2) Draw a diagram of the wiring.

3) Why is insulation provided on the wire?

4) Is the fixed or the movable electrode insulated on the igniter block ? Why ?

5) If a spark plug is used, draw a sketch showing its construction.

6) How far apart are the spark plug points?

7) Why are the points on the spark plug separated and those on the igniter block brought together?

8) Why is it necessary to clean the motor cylinder occasionally?

9) Why grind the valves?

10) What causes carbon to collect in the cylinder?

11) How is the carbon best removed?

12) What happens if a piston ring is broken or stuck?

13) What happens if the valves are not seating properly?

14) What happens if the oil line is stopped up?

15) What happens if bearings are too loose?

16) What happens when bearings are too tight?

17) What are shims?

18) What happens if the governor spring gets weak?

19) What happens if the governor sticks?

20) What is the result if the valve stem sticks?

21) What causes the valves to open too late or too early?

22) About when should the valves open and close on a small stationary engine?

23) What happens if the fuel pipe is partially clogged?

24) What is the effect when the air valve is closed?

25) What is the result if the carburetor is not fastened to the intake manifold with an air-tight joint?

26) The feed to carburetors on most tractors and automobiles is controlled by means of a float. What happens if the float becomes soaked full of gasoline? How remedied?

27) What is the effect of using dirty water in the cooling system?

PART VI

BELTS AND BELTING

CHAPTER XXX

KINDS OF BELTS AND BELT LACES

295. Methods of Connecting Machines. There are three common methods of connecting machines— (1) by shaft, known as direct-connected; (2) by gear wheels, the one on the driving machines being known as the driver and the one on the driven machine being known as the follower, and (3) by belts, in which case the names of the machines are those given when gears are used as connectors.

296. Four Kinds of Belts. There are four common forms of belts—chain, canvas, rubber and leather. Chain belts, except for slight wear in link joints, remain constant in length; hence, need no tightening as they grow old. The other three materials named, however, stretch, and, consequently, belts made from them need tightening from time to time to prevent their slipping. The usual method of tightening is to cut the belt, remove a piece and fasten the ends together.

Canvas, rubber and leather belts may be cemented together. However, the result with canvas belts is not very satisfactory. When rubber cement is used, a rubber belt, if not too old, may be cemented successfully. However, the method of fastening the ends of a belt is applicable principally to leather belts.

297. Cement Splice. The most satisfactory splice is one which keeps the belt at the joint the same in shape and general conditions as at any other point. Such a splice is made by squaring the ends (Fig. 327),

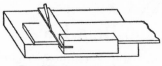

FIG. 327. Squaring a belt.

and then carefully dressing the joining surface, as indicated in Fig. 328, making the thickness at the squared end as thin as possible—a feather edge.

A cement splice can easily be made without removing belt from pulleys. Tighten belt with a belt clamp (Fig. 329), fitting it squarely on the belt.

The length of the splice should be 1″ greater than the width of the belt, up to 12″, which is regarded

FIG. 328. Tapering for glue-joint.

as the maximum length for splicing a belt, no matter how wide it is. When the clamp has pulled the belt to

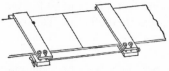

FIG. 329. Belt clamped for gluing.

the desired tension, cut one end to make the lap 1″ longer than the width of the belt. Lay the end of the belt on a board, the end of the two coinciding, and plane the lap joint with a sharp, small plane until it has the shape shown in Fig. 328.

298. Cementing Belt. The surfaces may be joined with any good belt cement procurable at leather and harness shops. Tack the belt at the joint down to a board, and then securely clamp it to the board to dry for at least twenty-four hours (Fig. 330). When the clamps are removed and the tacks withdrawn, the belt is ready for service. The particular advantage of this splice is that it forms a continuous belt with no extensions to interfere with smooth-running.

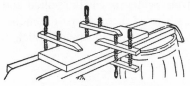

FIG. 330. A glue-joint in clamps.

A canvas belt is usually laced, altho it may be cement-spliced. If so, however, the joint should be cut, as shown in Fig. 331.

299. Laced Joints. These are common for leather belts up to 6″ to 10″ in width. A laced joint is made by calculating the length desired and cutting the belt a little short of this length to allow for stretching.

300. The Process of Lacing. Projects in belt-lacing may be selected from the practical problems of the farm as belts need tightening. It will be well to precede the first lacing of a belt in service by the lacing together

FIG. 331. Joint on canvas belt.

of two scrap pieces of belt. Holes are punched in both ends of the belt. Thru these is drawn a lace, usually a strip of untanned hide known as rawhide, in some manner to fasten the two ends securely together and to permit the lacing to pass over the pulleys with as little thumping and wearing as possible. Laced joints are usually classed as single-cross-laced and double-cross-laced, of which the former is the most used except for heavy belts.

Single-cross lacing gets its name from the fact that a single strand of lacing, or whang, joins the holes punched to receive it, and, also, because these strands cross each other on the side opposite the pulley but once, as shown in Fig. 332.

Patent Belt Fastenings. Many patented belt fastenings are on the market. Some of them are very good, and most of them can be applied in less time than it takes to lace a belt. The pattern which is easily applied and removed consists of a series of metal loops extending thru each end of the belt, thru which a rawhide stick is passed (Fig. 337).

CHAPTER XXXI

Projects in Lacing Belts

301. Single-Cross Lacing; One Row of Holes Punched on Each End. (Fig. 332.)

Tools and Stock: A 6″ leather belt or (for practice) two short pieces of 6″ belting, 56″ of lace, belt punch, square and knife.

Note: A narrower belt can be laced by modifying the following instructions accordingly:

Working Instructions: Square the ends of the belt to make its length 1″ less than that calculated or measured. Square a pencil line across each end of the belt

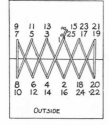 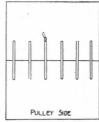

Fig. 332. Single-cross lacing; one row of holes.

l″ from the end and punch 3/16″ holes to dimensions given in *A*, Fig. 333. Point the end of the lace and pass it thru hole 00 (Fig. 332) from the outside, leaving 1/2″ of the end protruding. Pass the lace up thru hole 0 and down thru hole 1, then across to hole 2 and over to hole 3, continuing to pass the lace down thru the odd-numbered holes from the outside of the belt and up thru the even- numbered holes. Continue the lacing, passing thru the holes in rotation, finally returning to hole No. 1, which is also marked 15 and 25. The lacing will now be double. Care must be taken to pass the lacing back to 7 the first time it comes thru 8 in order to get it double at the end. It should be tight and straight. In passing thru a hole the second time or the third, as in case of hole No. 1, use an awl to enlarge

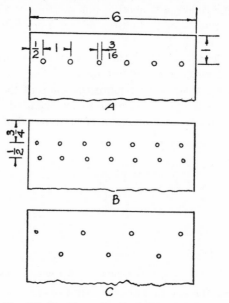

FIG. 333. Position of holes for various laces.

the hole slightly. After passing the end of the lace thru hole No. 1, coax it thru holes 0 and 00, leaving the end extending with the first one. Pull these ends thru and level with the belt, cut half-way thru the lacing at an angle with the lace. This forms a notch in each end of the lace to hold it from slipping thru to the pulley side of the belt.

302. Single-Cross Lacing; Two Rows of Holes Punched on Each End.

Note: The instructions given below are for a 6″ belt. It will be noted by referring to Fig. 334 that the lacings on the pulley side of the belt do not lap one on another. The holes being staggered, cause the lacings to lie singly, which is a decided advantage in overcoming noise in running, and wear.

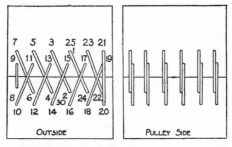

FIG. 334. Single-cross lacing; two rows of holes.

Working Instructions: Punch two rows of holes with a belt punch parallel to the end of the belt. The diameter should be about two-thirds the size of the lace to be used. The first row is placed about 3/4″from the end of the belt, and the second row about 1-1/2″ from the end. In case the belt is old, these distances are increased slightly. The holes are from 3/4″ to 1″ apart with one-half this distance separating the end holes from the edges of the belt. Determine these outside distances first and then divide up the intervening space so that the distances between points will be as nearly as possible 3/4″ (B, Fig. 333). Beginning with the end points on the first row from the end of the belt, punch a hole at every other point. Only one-half the number of holes may be used, as indicated in C, Fig. 333. This will make a less substantial lacing. To lace the belt, place a lace thru the middle holes from the pulley side—holes 1 and 2 (Fig. 334)—allowing the two ends of the lace extending on the side of the belt opposite the pulley to be as nearly as possible the same length. The end which extends thru hole No. 2 is put thru hole No. 3, then thru holes Nos. 4, 5, 6, 7 and 8, passing thru the first row of holes on one part of the belt and thru the second row on the other part; then to No. 9, crossing the belt joint, and back thru holes Nos. 10, 11, 12, 13 and 14; then thru hole No. 15, and, finally, thru a tie hole, No. 16, when the end should be cut off about 3/8″ from the belt. The second end of the lacing should follow a similar course, and, upon its return, should go thru hole No. 2, and, finally, thru the tie hole, No. 30. Note that on the side opposite the pulley, the large crosses or plies are over the short ones. This is desirable to reduce friction and wear. Always pull the lacing taut, but do not buckle the belt.

303. Double-Cross Lacing; One or Two Rows of Holes Punched on Each End. For this problem, two laces rather than one must be used. It is not deemed necessary to give detailed instructions for a double-cross lacing, as the instructions given for Problem 1 and Problem 2 apply, except as indicated below.

Double-cross lacing is similar to single-cross lacing except that two strands of lace are drawn thru each hole and that the holes are spaced twice as far apart across the belt. It is necessary that the two strands be drawn equally tight.

This method of lacing a belt is quicker than the single-cross lacing, but is more bulky and, consequently, is noisier and causes more vibration. It is particularly adaptable to the canvas belt because it does not weaken the material as the single-cross lace does, since there are only half as many holes. These should be punctures rather than cut holes, to still further preserve the strength of the material.

304. The Wire Belt Lacing. Wire lacing is now generally used. It is strong, and the strands are not as large

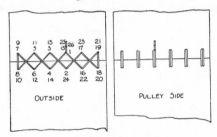

FIG. 335. Wire lacing.

as rawhide lacing. The holes are placed nearer the ends of the belt and nearer together, also. The end holes are about 1/4″ from the edge of the belt, and the remaining holes about 1/4″ apart. The row of holes is about 5/16″ from the end of the belt.

A No. 18 soft copper wire may be used for lacing. If it is hard, it can be annealed by heating it to red and plunging in water.

There are now several good makes of patented wire lacing on the market. These are made up from several metals in a proportion which will give a maximum degree of service. Generally, they will be found superior to the copper wire. When using patented wire lacing, care should be taken to follow the directions which are given on the box in which it comes. The size and length of lacing should be selected according to the width of the belt.

When lacing, start at hole No. 1 and pull one-half the wire thru. Then, using the end extending on the pulley side, lace as indicated by Nos. 1, 2, 3, 4, etc., in Fig. 335, returning thru No. 15. Now, use the other end which is protruding on the outside of the belt thru hole No. 2, and pass it thru 16, 17, 18, etc., returning to 25. The ends are now in the same holes, but in opposite directions. To

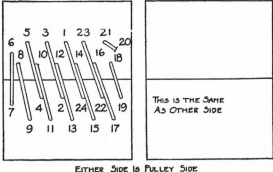

EITHER SIDE IS PULLEY SIDE

FIG. 336. The Annan lacing.

fasten the ends, make a small hole with a nail, as shown by No. 26, and pass both ends thru this. Make another small hole and pull the wire which is now on the pulley side up on the outside. Cut both ends about 1/2″ from the surface of the belt. Make square hooks on the ends of the wires and clinch them thru the belt in a similar manner to that indicated for the hinge lace in Fig. 337.

Pliers are used for pulling the wire thru when lacing. It is better to take hold of the wire at the extreme end so as to avoid nicking it in order to get the maximum durability in the lacing.

305. The Annan Lacing. This lacing (named after the man who designed it) is very satisfactory, and has the advantage of making the belt reversible on the pulley if necessary, as the lacings on both sides are the same. Besides, the lacings do not cross; thus, the disadvantage of a double thickness of lace is avoided. Fig. 336 shows the steps in making the lacing.

Start the lace as for the single-cross lacing, and continue by following thru holes as numbered, fastening the last end of the belt at hole 21.

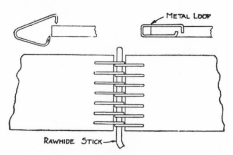

FIG. 337. Hinged belt lacing with wire hooks.

306. The Hinge Belt Lacing. Hinge lacing consists of using practically the same layout of holes as described for the single-cross lacing, but the lace is brought between the edges of the belt where the ends come together and thru the next hole from the opposite side of the belt. In this manner, the plies form a sort of a hinge between the belt ends. They tend to chafe at this point, however, and, consequently, have a short life; therefore, this lacing is no longer popular. Fig. 337 shows method of lacing.

307. Belt-Hook Joint. Belt hooks are obtainable in various sizes and shapes. Some are made to the required shape and are inserted into slits made in the ends of the belt, while others are bent to shape and fitted, as shown in Fig. 338.

FIG. 338. Wire books used in lacing.

PART VII

Farm Home Lighting and Sanitary Equipment

CHAPTER XXXII

Farm Lighting and Farmhouse Heating

308. Necessity for Good Light. During the long winter nights, those on the farm who would spend a part of the evening in reading the current events of the day, studying the various farm problems, and planning for the next year's work, feel the need of a modern lighting system. On farms where there are boys and girls in school and where they are required to prepare lessons at night, there should be the best lights possible. Shortsightedness in school children is a very common defect, which increases with age. It is due principally to poor school room and home lighting.

A good lighting system improves the sanitary condition in the home and makes for better health and higher efficiency. The farmer should give a great deal of throught and attention to the proper lighting of his buildings. The dairy farmer, especially, should have his house and barns well lighted. A well-lighted barn and dairy makes possible the production of a higher quality of products, makes work more pleasant and decreases the danger from fires, thus reducing the insurance rate.

309. The Cheapest Light. Probably the old-fashioned flat-wick kerosene lamp is the cheapest from the standpoint of cost of fuel. This is not true when one considers the cost of operation, however. Again, a consideration of the poor quality of the light produced by this lamp, its effect on the eyes, its danger, and the fact that no workman can do his best work under poor lighting conditions, makes this pioneer means of home-lighting an expensive one.

The kerosene tubular lamp is an improvement over the flat-wick type in the amount of illumination, especially when it is provided with a mantle which improves the quality and increases the amount of light produced.

310. A More Modern Lighting Plant. The farmer who would install a truly modern lighting system in his home has four kinds of plants from which to make his selection, namely, the electric, acetylene, gasoline gas and Blau gas plants.

311. Electric Lighting Plants. (Fig. 339.) There are definite advantages that the electric light has over other forms of lighting that are recognized by every one. It is clean, safe, its cost is not prohibitive, and it does not make the air impure.

Where the power for electric lights can be secured at a reasonable price from power-distribution lines passing the farm, the situation is ideal. Many farmers do not care to be burdened with the chore of looking after a lighting plant.

Until recent years, there were few unit plants on the market; that is, an engine and generator built together. Most of the generators were formerly belt-driven by a small engine that could be easily used for some other purpose. There is a number of unit plants on the market that are arranged with a belt pulley for power purposes. Some farmers use a power windmill to drive the generator.

In the installation of a small low-voltage electric plant, be sure that all wire is of ample size. The mistake is often made of using the same size wire as used in wiring city residences where a higher voltage is used. All wiring should be properly inspected to see that it meets all insurance requirements. The National Board of Fire Underwriters of Chicago will provide rules for this work.

In operating a small electric plant, pay especial attention to the care of the storage batteries. The upkeep and replacement cost of the storage battery is the most expensive item in the cost of operating an electric plant.

FIG. 339. Farm electric plant.

312. Acetylene Lighting Plants. Many farmers purchase the acetylene light plant because it is cheaper to operate than the electric plant and requires less attention. Most farmers like the outdoor type of plant best, because it is safe, easily charged, easily cleaned out, and where a 100-pound capacity plant is secured, it does not require re-charging oftener than three or four times

Fig. 339-a. Gasoline gas generator.

during the year. Any acetylene plant that is constructed or located so that the gas will escape into a closed room is dangerous. Acetylene gas is a little more dangerous than gasoline; both must be handled with great care.

313. Gasoline Gas Lights. Most of the gasoline equipments are either of the small portable-lamp type or the one by which the gas is piped thru small tubes to the individual lamp. These types of gasoline lamps are objectionable from an insurance standpoint. Only where the gas is produced outside of the building (Fig. 339-a) and piped in like ordinary city gas, is the gasoline system really safe. The greatest danger of gasoline lights comes from taking gasoline inside the house.

From a standpoint of economy, the gasoline gas lamps are really cheaper than either acetylene or electric lamps.

314. Blau Gas Lights. Blau gas is an oil gas that is liquified under high pressure. It is freed from all poisonous gases and is practically non-explosive. It is sold in tubes similar to presto-lite—twenty pounds of gas to the tube. The light produced by Blau gas is quite satisfactory and not prohibitive in price.

315. Farmhouse Heating. A well-heated house makes for comfortable living. It has been only during rather recent years that much development has been made in farmhouse heating. Many progressive farmers are now installing systems of heating that will maintain an even temperature thru-out the house, and provide an abundance of fresh air. Heating and ventilation go hand-in-hand.

The modern heating system is located in the basement. It keeps the litter and dirt from the main floors, which are difficult to keep clean when fuel and ashes are handled over them in caring for a stove.

316. The Hot-Air System. There are two types of hot-air systems found on the market. One is the pipeless furnace, which is essentially a special type of stove located in the basement and surrounded by a jacket which carries the heat to the rooms above. A down shaft is provided to keep the air in circulation. This type of furnace can be easily installed in any home already built that is provided with a basement or cellar. The other type of hot-air plant is provided with large pipes that carry the hot air direct to the various rooms. These pipes, or "leaders," as they are sometimes called, must run as direct to the rooms to be heated as possible, and they should be wrapped with asbestos to prevent loss of heat. A house can be heated more quickly with hot air than with water or steam, but it will cool off more quickly when the fires die down. During extremely cold, windy weather, it is difficult with a hot-air system to heat rooms on the side of a house from which the wind is blowing.

317. Steam and Hot-Water Systems. The steam-heating system can be installed as a single-pipe or a two-pipe system. The hot-water heating is a two-pipe system. The two systems are quite similar as far as installation is concerned, and can be installed fairly easily in a house already built. The hot-water system works on the principle of water being lighter when hot than when cold. The heated water rises to the various radiators, the heat is given off in the rooms, and the water at a lower temperature flows back to the boiler. Care must be observed in installing the pipes to get proper circulation.

Most steam systems are for low-pressure steam. The steam is generated in the boiler; it rises thru the pipes to the radiators, where it loses its heat, and is condensed and flows back to the boiler. In the one-pipe system, the condensed steam flows back to the boiler thru the same pipe thru which the live steam flows to the radiator.

A house can be heated much more quickly with steam than with hot water, but in a hot-water system the water will hold the temperature more uniformly and a more even heat is maintained. This is the big advantage of the hot-water system over all other systems.

The installation of most lighting and heating equipment should be left to an experienced man. To install a pipeless furnace, however, is not a very great task, and can be done by a person with little experience.

CHAPTER XXXIII

FARM WATER SUPPLY AND SEWAGE DISPOSAL

318. Importance of Sanitation on the Farm. It is high time that every farmer give serious thought to the sanitation problems of farm life. Water is thought to be cheap and thus little value is put upon it; this is the chief cause of neglect. Many shallow farm wells are contaminated due to poor protection at the top, poor surface drainage, seepage and general neglect. Cistern water is often made unfit to drink by impurities washing in from the roof due to lack of a good filter, or to one improperly cared for. It is sometimes impure because the cistern is not properly built and seepage water gets in.

The first consideration for health on the farm should be a pure and wholesome water supply of capacity to take care of all the needs of the place. A deep well is about the safest source of water supply. Shallow wells and cisterns, however, can be made safe by proper protection at the top, careful surface drainage, and by preventing the entrance of seepage water. For cisterns, the water should be collected only after the roof has been thoroly washed off. A well-built filter, cleaned out and refilled with filtering material at regular intervals, will go a long way toward purifying such water.

319. Simplest Water System. The simplest system of water supply is an ordinary suction, or force, pump attached to a sink in the kitchen. The pipe leads from the pump thru the floor and into the well or cistern. The source of water for a system of this kind must be near the house and not very deep. For satisfactory service, there should not be more than twenty-five feet between the pump cylinder and the lowest level of the water. A drain must be provided to take off all waste water from the sink. Such a system can be easily installed.

320. Gravity System. The simplest gravity system is one that has a small tank located in the attic and is connected by means of a pipe to a force pump in the kitchen. Such a system makes possible the installation of all other plumbing equipment. Fig. 340 shows a system with a sixty-gallon tank in the attic. Water is pumped to the tank by means of a force pump and a small gasoline engine. The overflow from the storage tank runs to the stock tank in the lot. A good feature of this system is that all of the water for stock is pumped thru the house tank, thereby keeping it always fresh and cool. In Fig. 340 is shown also the installation of complete plumbing connections. Where there is a hill or slight elevation near the house, a tank can be placed on the ground. The concrete tank shown in Fig. 341 is a farm storage tank. It is large enough to supply the house, hog house, hog wallows, barns and garage, all of which are provided with faucets. With the tank placed on the ground and provided with a good foundation, there is no danger of supports giving away as with an elevated tank and the danger of the pipes that lead to the tank freezing is eliminated. Where a satisfactory means of elevating the tank is at hand, the gravity system is the most satisfactory for average farm conditions. A tank supported by concrete or masonry walls is a very good arrangement. A room underneath the tank can thus be provided to be used as a milk house.

321. Water Air-Pressure System. This system, shown in Fig. 342, is usually called the hydro-pneumatic

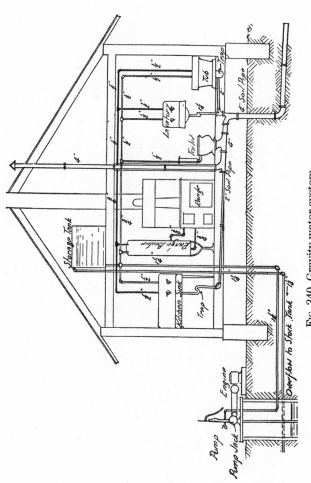

FIG. 340. Gravity water system.

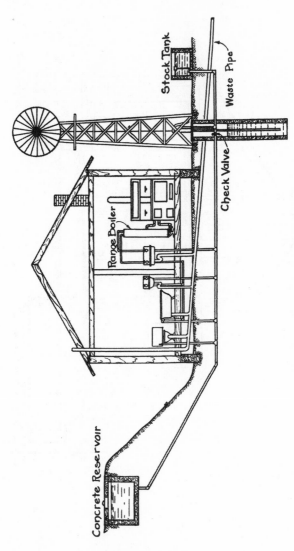

FIG. 341. Concrete reservoir for water storage.

system. In it the water is stored in an air-tight steel tank and is forced thru the pipes to the fixtures by air pressure. As the water is used, the pressure is gradually reduced. In some systems of this type there is both a water and an air pump. The most common type is equipped with only a water pump with air intake. To operate the system, the tank is filled with air, the water is pumped in, and the air pressure increases as the volume of the air decreases. Only about two-thirds to three-fourths of the volume of the tank is effective for water storage. This is one of the principal objections to this system, because to avoid pumping so often, an extremely large tank must be provided if the water requirements are very large. However, with electric power available, an automatic control can be provided and a smaller tank be used. Complete equipment for a system of this kind includes an air-tight tank, a force pump, pressure gauges, and other fittings, and plumbing fixtures.

322. Hydraulic Ram. Where there is a large quantity of water with sufficient fall, a hydraulic ram is the cheapest means of providing water pressure in the home. The first cost is small, there is practically no upkeep, and it will run continuously without any attention. Under ordinary conditions, a ram will elevate about one-seventh of the water that flows to it thru the drive pipe. A rule that can be used to determine the approximate amount of water that will be delivered with a certain flow is: Multiply the number of gallons of flow per minute by the number of feet of vertical fall between the source of water and the ram. Divide this by the height it is desired to elevate the water, and reduce the result by one-third to take care of friction and losses in the pipes. The remainder will be the quantity of water delivered. For example,

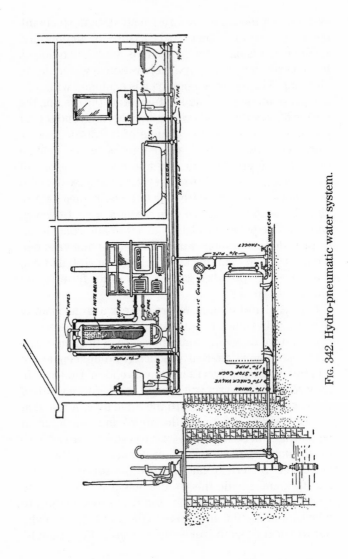

FIG. 342. Hydro-pneumatic water system.

if the flow is 4 gallons per minute, the fall is 9 feet and the water is to be elevated 24 feet, we have four times 9 equals 36; 36 divided by 24 equals 1-1/2; reduce this by 1/3, and we have 1 gallon per minute delivered, or 1,440 gallons per 24 hours.

323. Selecting a System. In selecting a water system, many make the mistake of installing one that does not furnish sufficient water. It is much better to have a cistern or tank with greater capacity than actually needed than to have one too small. The same is true in selecting a pressure tank for the hydro-pneumatic system or an air tank for the fresh-water system. The first cost will be a little greater, but the expense will be less in the end. As a basis for estimate, one must remember that after a modern water system is installed, much more water will be used than before. For each person, one should estimate at least 25 or 30 gallons per day; for each cow, 15 gallons; for horses, 10 gallons, and hogs and sheep, 3 gallons per day, allowing for an additional supply to care for chickens, for watering the garden, washing the car or buggy, sprinkling the lawn, etc.

324. The Septic Tank. No modern water system is complete without proper disposal of the waste water and sewage. Oftentimes the sewer is simply tile that leads down to the field or into a ditch or small stream. This method of sewage disposal is not sanitary, nor is it safe from a standpoint of health. If a large stream is at hand, into which to discharge the sewage, it can be used with safety; a small creek, however, would soon become contaminated.

The septic tank is a means of disposal of sewage from the farm home. The septic tank alone will not purify sewage; it will partially purify it and put it in condition

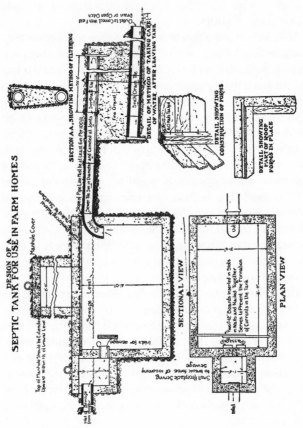

FIG. 343. One-chamber septic tank.

to be completely purified by means of a filter or thru a system of tile. The septic tank illustrated in Fig. 343 is a one-chambered tank. Its action is as follows: In the septic tank the sewage is acted on by bacteria—minute organisms that thrive under conditions where there is neither air nor light. The solids in the sewage are broken down and put into solution. It must be remembered that only one or less than one per cent of the sewage is solids—the rest is all water. Soon a thick leathery scum forms on the surface of the tank; this indicates that it is working properly. The solid part that is not dissolved settles to the bottom of the tank. It is necessary to clean this out every few years.

To completely purify this sewage, it is discharged onto a filter or into a system of tile arranged to allow it to filter away into the soil. In the filter or in the surface soil, there are billions of bacteria that thrive in the presence of air and light. These are called the nitrifying bacteria. They completely purify the sewage. This is nature's method of purification.

325. The Art of Plumbing. Plumbing has been called a sanitary art and defined as the art of placing in buildings, pipes and other apparatus used for introducing water supply and for removing wastes.

The Plumber as a Specialist: In big jobs in large building work, there are special plumbers for doing the heavy roughing-in work, putting in the large pipes and the general network of smaller pipes. Then there are other plumbers to do the finishing work.

There are certain essentials in handling a house-plumbing job. The man in charge should be thoroly competent to see that the connections are properly made. A plumbing job that is poorly finished may be a source of a great deal of danger, and should be thoroly inspected. Simplicity in the laying out of piping and fixtures will tend to eliminate plumbing troubles. The principles of drainage must be ever in mind when installing a plumbing system. All supply pipes, as well as drains, must be installed so they have an outlet and with a gradual slope toward this outlet. There must be no low points or pockets where water will collect when the system is drained. Such a defect would cause stoppage in drain pipes, and the supply pipes, when exposed, would freeze at these points. Main soil pipe made of 4″ pipe should extend 5' from outside of the cellar wall to act as a sewer connection into the house and thru the roof. This pipe should be straight from the cellar to the roof. All fixtures should discharge thru the main soil pipe, and should be provided with traps thoroly ventilated to prevent the escape of sewer gas into the house. In some plumbing jobs, an additional ventilation pipe is carried from each trap into a main 2″ pipe which is independent of the soil pipe and is also carried thru the roof. This prevents leakage of the seal or trap.

Plumbing materials and fixtures should be of good quality, simple in design, with all joints and connections made air- and water-tight. They should be of entirely non-absorbent material

All plumbing should be as nearly accessible as possible. Removable wooden panels over the soil pipe and other main pipes are worth considering. Fixtures near main drain and all bath and kitchen fixtures should be open work. Free access of air and light should also be obtained. Boxed-in sinks and bath tubs are insanitary because dirt and moisture are bound to collect around the base.

326. Materials Used for Plumbing. For sinks, the solid porcelain is the most expensive. The iron enamel is just about as good as the solid porcelain and can be obtained much cheaper. For laun-dry equipment, the slate, reinforced concrete and enameled iron can be used. Slate tubs for laundry are very satisfactory. The most sanitary equipments are those which are in one piece with all parts properly rounded. The general equipment is usually listed as to quality as No.1, No. 2 and No. 3. No. 1 is Fig. 344. Pipe usually guaranteed and is very expensive; No. 2 is very satisfactory. It is usually not advisable to buy the No. 3 quality.

FIG. 344. Pipe vise.

The person who would do the simple plumbing jobs herein described should become familiar with the more common plumbing tools and their uses; also, the various pipe fittings required. The following tools are needed for even the simplest job: Vise, cutter, die-stock and dies, wrenches, reamer or half-round file, and rule.

327. Pipe Vise. The hinged type of vise (Fig. 344) with gravity pawl is about the best to secure. The revers-ible type may be secured. The latter can be thrown open either to the right or to the left, with a clutch on either side to engage the pawl. Such a vise has a distinct advan-tage when cutting a pipe which has fittings that will not pass thru the frame of an ordinary vise.

328. Pipe Cutters. Pipe cutters (Fig. 345) are divided into two general types—the three-cutter wheel and the one-cutter wheel types. The one-cutter wheel can be secured with solid back or with two rollers; the latter type is probably in most general use. The three-cutter wheel type has the advantage of being used in close

Fig. 345. Three-wheel pipe cutter.

quarters. This type of cutter forms a burr on the outside of the pipe which must be removed with a file before the threads can be cut. The pipe does not need to be reamed out, however.

Fig. 346. Stock and die.

329. Die-stocks and Dies. It must be remembered that a different die is used for threading a pipe than for threading a bolt. The pipe thread is a taper thread, making possible a tight joint. The solid type of die is most commonly used (Fig. 346). A number of dies for different-sized pipe can be secured and used in the same stock. The adjustable type of die is used in a special stock. A ratchet stock is sometimes used.

330. Pipe Wrenches. The Trimo and Stillson wrenches are the two types of wrenches in most common use. At least two sizes of wrenches should be provided—one for small pipes and fittings and one for

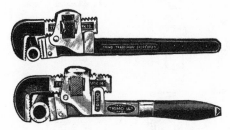

FIG. 347. Pipe wrench.

larger sizes. For extremely large pipe, chain tongs are usually used. (See Fig. 347 for picture of pipe wrenches. Fig. 347-*a* shows many of these tools in a group.)

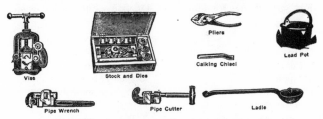

FIG, 347-*a*, Several common plumbing tools.

331. Reamers. The reamer is used to remove the burr formed on the inside of pipe by cutting the pipe. A reamer fitted in a hand wheel is quite satisfactory. A one-half round or a round file can be used.

332. Rule. A folding rule should be provided. For a neat job of pipefitting, careful measuring is necessary.

333. Pipe Fittings. Pipe fittings are used in joining one pipe to another, to change direction, to reduce size, and to branch off. Fittings are made of malleable,

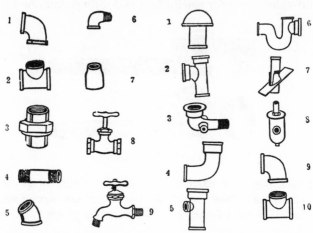

Fig. 348. Fittings for supply pipes:

1. Elbow.
2. Tee.
3. Union.
4. Nipple.
5. 45° elbow.
6. Street elbow.
7. Reducer.
8. Valve.
9. Faucet.

Fig. 348-*a* Fittings for waste pipes:

1. Ventilating cap.
2. Sanitary T-branch.
3. Closed bend.
4. Quarter bend.
5. Tapped T-branch.
6. Trap with hand hole.
7. Roof flange.
8. Drum trap.
9. 90° elbow.
10. Tee

cast and wrought iron; the latter are usually galvanized. There are also brass and nickel fittings for special uses. Figs. 348 and 348-*a* give the names of the principal fittings for supply and waste pipes.

CHAPTER XXXIV
Drainage and Pipe Fitting

334. Fitting Pipe Handle for Lawn Roller.
(Fig. 349.) (See concrete project, Sec. 146.)

Requirements: To cut, thread and assemble pipe and fitting to form a handle of proper dimension for a concrete roller as outlined under Concrete Projects, Sec. 146.

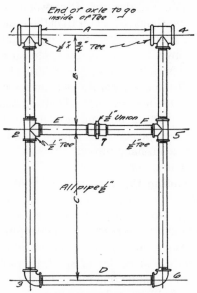

Fig. 349. Pipe handle for garden roller.

Tools and Materials Needed: Pipe cutter, vise, die-stock and die, wrenches, and a rule. Pieces of 1/2″ pipe, lengths depending on requirements of particular handle; two 1/2″ elbows, two 1/2″ tees, one 1/2″ union,

357

and two 3/4″ x 1/2″ tees. The latter is to serve as bearings for axle of roller. The size specified is sufficiently large where a 1/2″ pipe is used for axle.

Preliminary Instruction:

335. Measuring Materials for Handle. Extreme care must be observed in making measurements to have the handle fit smoothly. The distances A and B (Fig. 349) will depend on the length and diameter of roller. The distance A should be made about 1/2″ greater than the length of roller. The distance B between center of fittings should be about 2″ greater than the radius of roller. The distance C should be made a length that will make the roller convenient to operator. Measurements are usually taken from the center of one fitting to the center of the next. To make accurate measurements, each fitting should be made tight before the next piece of pipe is cut. The 1/2″ union can be eliminated if one of the tees in which the cross pipe is threaded has a right-hand thread and the other a left-hand thread.

Working Instructions:

336. Threading Pipe. Place a piece of 1/2″ pipe in the vise. If not threaded, thread it with a right-hand die as follows: Note that proper die is placed in stock; place guide bushing in place; oil end of pipe with lard oil; place bushing end of die-stock on pipe and start die with hands near center of stock by pressing hard on handles and rotating one-fourth turn at a time. After die has taken hold, move hands out to the ends of handle and continue rotating with less pressure. After each complete turn, rotate backward slightly to allow chips to drop. Continue this process until thread of sufficient length is cut. It is often necessary to remove die and try on fitting to get the best results. The fitting should go on at least three threads by hand. Screw fitting No. 1 on end of pipe by means of the pipe wrench.

337. Cutting Pipe. Draw the pipe thru the vise and lay off length B with rule. Place pipe cutter on pipe so that the cutting wheel comes on the mark. Drop a little lard oil on pipe and cutting wheel, screw the handle in until cutting wheel begins to cut, then rotate cutter. At each revolution of the cutter, feed the cutter wheel inward by screwing in on handle; continue until pipe is cut off. Place the pipe B in vise and thread blank end, after which screw on fitting No. 2.

Proceed by cutting and threading pipe length C and screwing into fitting No. 2 and No. 3. Make up other side of handle in the same manner; then cut, thread and assemble pipe lengths D, E and F so that the two sides of handle will be parallel.

After handle is assembled to proper dimension, unscrew union No. 7 and spring handles apart until fittings Nos. 1 and 4 will slip over the ends of axle, after which tighten union.

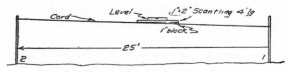

FIG. 350. Establishing a grade for tile.

338. Installing Drain from Kitchen Sink to Sewage Disposal System.

Requirements: To install drain pipe and tile from kitchen sink to outlet or disposal system. (See Figs. 350, 351, 352.)

Tools and Materials Needed: Plumbing tools and tiling spade, hook and scoop, and a carpenter's level are required. Obtain one trap, sufficient 1-1/4″ pipe to carry water from sink to a point 5' outside house, suitable fittings, white lead, 50' of 4″ sewer tile, and sufficient farm drain tile to reach outlet.

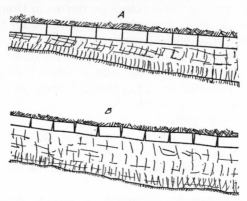

Fig. 351. *A*, tile properly laid; *B*, tile poorly laid.

Preliminary Instruction: The first requirement of every drain is an adequate outlet or point of discharge. This point must be low enough so the tile can be given ample fall to prevent the sewage or water backing up in the drain. It is considered best practice to discharge sewage from sanitary fixtures thru sewer tile direct to septic tank while the kitchen waste water is usually taken care of by ordinary drains. A smooth, uniform grade must be provided for every drain. In farm drainage work, this is usually established by means of a drainage level.

Working Instruction:

339. Establishing Grade Line for Drain. Determine point of outlet and establish a grade line by which to dig the ditch. For a small job such as this, when there is a decided slope of the ground, place the grade parallel to the slope. If the ground is practically level, a grade line can be established by means of an ordinary carpenter's level. Drive in a series of stakes from 4' to 5' long at intervals of 25'; for long drains, stakes are placed every 50'. By using long stakes, a guide line to dig by can be placed as the grade is established. If it is desired to have a fall of 1/4" to the foot, take a straight 1" x 2" scantling 4' long; tack a 1" block under one end, and fasten to lower side of level with block on lower side. Tie a cord to stake at the outlet at a point about 3' above the surface of

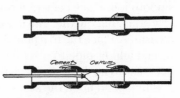

FIG. 352. Sewer tile made tight.

the ground, stretch the cord to the second stake, and test it for grade by placing the level in position so the block will be down grade (Fig. 350). When the bubble of the instrument indicates that it is level, it shows that there is a rise of 1" to every 4' along the cord.

340. Digging Ditch to Grade. Use a gage rod of definite length, and dig ditch so its bottom will be parallel to the cord. If it is desired to have the drain 4' deep at the outlet, the gage rod should be 7' long since the cord was placed 3' above the surface of the ground at the outlet. If the soil is heavy and sticky, an open spade can be used to advantage; use a round-nose spade or a tile scoop for cleaning the bottom of ditch to receive the tile.

341. Laying the Tile. Lay the tile as the ditch is completed, beginning at the outlet. The ordinary farm tile can be laid either by hand or by means of the tile hook. The tile must be made to fit closely together in the ditch (A, Fig. 351); this is best accomplished by rotating the tile until it is in place. The sewer tile (Fig. 352) is provided with bell end or bell mouth, and the joints are made tight. Place the tile in place so the direction of flow will be into the bell end; place the spigot end of each tile into the bell end of the preceding tile as it is laid. A small piece of oakum or tarred rope forced in between the spigot and the bell with a flat stick will make possible a smooth job. Place cement mortar in the joint after properly adjusting the tile. The use of a *swab*, as indicated, is advisable. Place tile to a point 5' outside of building. Cut, thread and fit pipe to discharge into the sewer that has been laid. The depth to place this pipe and slope to give it will depend upon the sewer outlet.

342. Installing Kitchen Sink and Pump.

Requirements: To install a kitchen sink and pump so that water may be pumped directly from a well or cistern to kitchen (Fig. 353). The installation of drain for this sink is outlined under Secs. 338-341.

Tools and Materials Needed: The tools needed for the project are the same as in Sec. 334. The following materials are needed: Pump, sink, trap and sufficient 1-1/4" pipe to reach from pump at sink to cistern or well, as illustrated (Fig. 353). Such elbows and couplings as needed and a check valve for suction pipe. White or red lead for making joints.

Preliminary Instruction:

343. Maximum Depth for Pumping Water.

It must be remembered that the vertical distance from cylinder of pump to low level of water must be 25' or less to give satisfaction. Where there is a likelihood of water in pipe freezing during cold weather, the check valve in suction pipe should be omitted to allow the water to drain back into well. The only difference in the work in this and the preceding project, Sec. 334, is that the pipe joints must all be made absolutely tight.

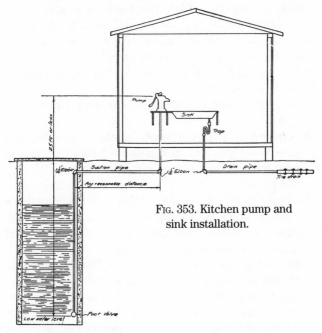

FIG. 353. Kitchen pump and sink installation.

Working Instructions:

344. Location of Kitchen Sink. Locate sink in the kitchen so that it will be convenient and have ample light. Most sinks are fastened to a wall by means of hangers or brackets which can be easily installed. Adjust height of sink to convenience of user.

Excavate for pipe from cistern to point underneath sink. If a basement is under house, excavate from cistern to wall. The pipe should be placed below frost line.

Take measurements from the cistern to the sink, determining the exact length of each piece of pipe needed and the necessary fittings.

345. Connecting Pipe for Pump. Cut and thread pipe as outlined under Secs. 336, 337. In making the various joints, apply a small amount of white lead to the first three threads in fitting or on the pipe. Begin at the pump and screw each fitting and piece of pipe perfectly tight before beginning on the next.

346. Installing Plumbing in Country Home (Fig. 354).

Requirements: To install rough plumbing, including soil pipe, vent pipes and various drains for fixtures in wall partition while house is under construction. (See Fig. 354.)

Tools and Materials Needed: Plumbing tools, plumber's furnace, ladle and caulking tools. Soil pipe and soil-pipe connections, vent pipes, drains for fixtures and traps. Lead and oakum for joints.

Preliminary Instruction: Every plumbing system should be designed with an idea of simplicity in the layout of piping and fixtures. If possible, the bath room should be directly above the kitchen, and with the laundry room below, as shown in Fig. 354. This will make it possible for one soil pipe to take care of the discharge from fixtures on each floor. Fig. 354-*a* shows a system in a three-story house with a bath room on each floor. The soil pipe should extend from a point 5' outside the wall, where it connects with the sewer, up thru the house roof. It should be straight from cellar to roof. Tight joints are an essential requirement of every plumbing system. Provide a trap for every fixture; the best practice provides a

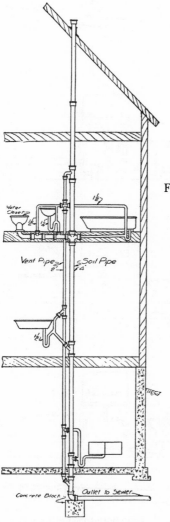

FIG. 354. Waste and
ventilating pipe.

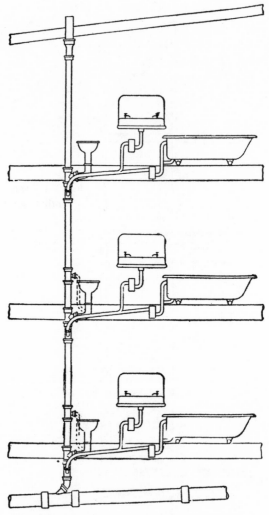

Fig. 354-a. Waste pipe without ventilating pipe.

$2''$ ventilation pipe with connection to each trap. The location of each fixture should be carefully considered with a view to convenience for the user and to make a simple, efficient layout.

Working Instructions:

347. Sewer Tile. Lay a sewer tile from sewer connection, or from septic tank, to a point 5' outside of building. Follow instructions as outlined under Secs. 338-341. Make connection of soil pipe with sewer, and extend it to a point in the basement where it will be most convenient to fixtures and where it will pass thru partition to roof.

Soil Pipe. The joints of soil pipe are similar to joints of sewer tile. Each section of pipe is provided with a bell end into which is placed the spigot end of the next section. The joints must be perfectly tight; to make them so, oakum and lead are used. The pipe is set in place, a roll of oakum is packed into the bottom of joint, after which molten lead is poured into joint caulked with the joint, filling it completely (Fig. 355). To pour the lead where

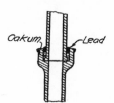

Fig. 355. Soil pipe joint caulked with oakum and lead.

a joint is made in a horizontal pipe, a sort of collar must be provided with opening at the top. If the oakum is not carefully packed into place, the lead will run thru. After the lead has cooled, pack it solidly into the joint with a hammer and caulking tool. Well-caulked joints are absolutely tight.

349. Connecting Fixtures and Vents. Provide suitable Ys and Ts for all fixtures, as illustrated. Connect the vent pipe from a point below the bottom fixture and extend it up, and connect back into the soil pipe at a point above the highest fixture. Give all horizontal soil pipes, whether for drainage or ventilation, a fall of at least 1″ to the foot. To support soil pipe, provide suitable concrete or stone footing at the bottom. Support all horizontal lines with suitable hangers to prevent line from getting out of place.

350. Connecting Cast-Iron and Lead Pipe. To make a connection between a cast-iron pipe and a lead pipe, first connect the lead pipe to a brass ferrule by means of a soldered joint; the ferrule is then caulked into the cast-iron hub or bell end, as outlined above.

CHAPTER XXXV
SUPPLEMENTARY PLUMBING PROJECTS

351. Piping Water to Stock Tank.

Requirements: To construct a pipe line from source of water at well or storage tank to stock tank in barnyard, as shown in Fig. 356.

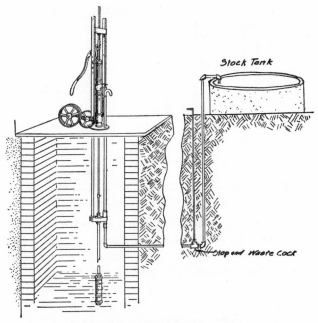

FIG. 356. Piping for stock tank.

Tools and Materials Needed; Plumbing tools, as in Sec. 334. Pipe and fittings determined by particular job.

369

Instructions:

1) Take measurement for pipe.
2) Cut and thread pipe not threaded.
3) Excavate for pipe line.
4) Connect pipe with fittings above ground.
5) Place pipe in ditch.
6) Provide cut-off and means of draining lines to prevent freezing.

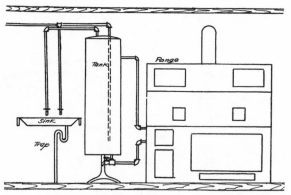

FIG. 357. A hot-water tank installation.

352. Installing Hot-Water Tank with Kitchen Range Having Hot-Water Back.

Requirements: To install a hot-water tank in kitchen with proper connection to water supply, and to hot-water back in kitchen range, as shown in Fig. 357.

Tools and Materials Needed: Tools as in Sec. 334. Materials dependent on particular job. Tank, hot-water back, pipe fittings and white lead or pipe cement.

Instructions:

1) Locate tank so it is out of the way and convenient for connection to mains and to range.

2) Take measurements for pipe. Cut and thread pipe not threaded.

3) Tap main water line with a tee.

4) Make all connections.

Note: Cold water must enter at the bottom of the tank, and hot water is drawn off at the top. Remember, also, that the bottom connection from water back must enter tank at the bottom, and the top connection must enter the tank several feet above the bottom and at a point above the back so the water will rise on being heated and will have proper circulation.

353. To Make a Stock Water Heater.

Requirements: To make a stock water heater when steam pressure and a supply of water under pressure is available. (See Fig. 358.)

Tools Needed: Plumbing tools as in Sec. 334. One breast drill and 1/8″ bit.

Materials Needed: 3' of 1-1/2″ galvanized pipe, 4-1/2' of 1/2″ galvanized

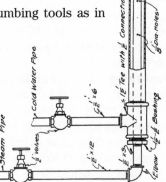

FIG. 358. Stock water heater.

pipe, three 1/2″ cut-off valves, two 1/2″ elbows, one 1/2″ cap, one 1-1/2″ to 1/2″ bushing, one 14/2″ x 1/2″ tee, one 1-1/2″ to 1/2″ coupling reducer, one 1-1/2″ x 3″ nipple.

Instructions:

1) Cut and thread all pipe to dimension Indicated on plan.

2) Drill 1/8″ holes at 8″ intervals on opposite sides of 1/2″ pipe.

3) Assemble 1/2″ steam pipe in following order: Screw 12″ pipe into valve on steam line, elbow onto 12″ pipe, 3″ nipple into elbow, 1-1/2″ x 1/2″ bushing onto nipple, 33″ pipe into opposite side of bushing, screw cap on end of pipe.

4) Assemble water jacket as follows: Screw tee on 1-1/2″ bushing, connect 6″ nipple into tee, screw valve onto nipple, connect valve to water main. Screw 1-1/2″ pipe into tee, on opposite end screw 1-1/2″ x 1/2″ reducing coupling, connect 3″ nipple, elbow, another 3″ nipple and valve to control flow of warm water.

Note: The temperature and flow of water can be controlled by regulating the flow of steam and cold water. Where a boiler is used in connection with dairy room, this is a good way to heat the water for the cows.

354. Installing a Hydraulic Ram.

Requirements: To install a hydraulic ram for elevating water from a lower to a higher elevation for household consumption, as shown in Fig. 359.

Tools Needed: Plumbing tools as in Sec. 334.

Tiling tools as in Project No. 3, Sec. 338.

Material Needed: Sufficient drive and discharge pipe of proper size and length, with necessary fittings; this depending on the individual installation.

Instructions:

1) Locate position of ram.
2) Make measurements and lay out position of drive pipe and discharge pipe.
3) Excavate for drive and discharge pipes.

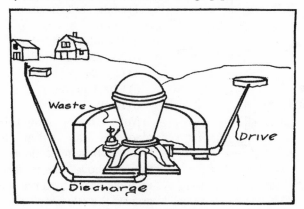

FIG.359. Hydraulic ram.

4) Proper length and slope of drive pipe depends on particular ram. Secure proper information from manufacturer.
5) Connect drive pipe from ram to source of water.
6) Connect discharge pipe from ram to storage tank.
7) Provide drain for waste water at ram.
8) Cover drive and discharge pipes.
9) Protect ram from high water.

Note: A hydraulic ram is practical only where there is a large quantity of water flowing with several feet fall.

355. Installing Drain Tile at Foundation of House
(Fig. 360).

Requirements: To install a drain tile at foundation of house to intercept any seepage water that flows into basement.

Tools and Materials Needed: Tiling tools as in Sec. 338. Sufficient drain tile to extend along side of house and to outlet. Actual amount depending on local conditions.

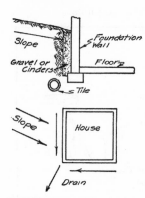

Fig. 360. Location of tile to drain house foundation.

Instructions:

1) Stake out location of drain.
2) Establish grade line.
3) Excavate to grade.
4) Lay tile.
5) Check grade.
6) Back-fill.

Note: Drains should be placed a little below the level of basement floor. If trench above tile is filled with a porous material like gravel, the tile more effective in intercepting the water. This tile should be placed in addition to drains for down-spouting and for basement floor.

356. Additional Jobs on Farm.

a) Install drain to septic tank.

b) Install farm drains.

c) Install tile for down-spouting on barnyard buildings.

d) Install an automatic waterer for stock.

e) Re-charge an acetylene light plant.

f) Put a new pump in a well or cistern.

g) Repair a farm pump.

h) Construct and install a filter.

PART VIII

ROPE AND HARNESS WORK ON THE FARM

CHAPTER XXXVI

CONSTRUCTION AND USE OF ROPE

357. The Need for Rope Work. A working knowledge in the use of rope is of value to every one on the farm. Rope is used in a great many ways, and often much time may be saved by knowing how to make a simple splice, or tie a satisfactory knot or hitch for a particular purpose. Accidents are often averted by knowing how to tie the right knot for the right place. To become expert in tying and splicing rope requires a great deal of practice. One can learn this kind of work only by actually doing it. The work outlined under this head is to give the reader an idea of the principal knots and splices and their applications. Practice work is grouped into several projects. The student should not expect to make progress in rope work without carrying thru these projects.

358. Materials of Which Rope Is Made. The greater part of rope is made from either manilla or sisal fiber. Manilla fiber, a product of the Philippine Islands, is obtained from a plant similar to the banana. The sisal fiber, from which most binder twine is made, a product of Yucatan, is secured from a plant similar to the American aloe. The two kinds of rope are ordinarily known as hemp rope. The sisal is neither as strong nor as durable as manilla fiber. A distinguishing characteristic of the best quality manilla fiber is its glossy appearance. The poorer quality of manilla is of a brownish color, and its glossy characteristic is only slight. Sisal has a dead, lifeless color. The difference between the two might be compared with enamel paint and flat paint. Cotton rope is little used at present, altho, at one time, it was used almost exclusively in some localities.

359. How Rope Is Made. In the actual process of making a rope, the fibers are twisted right-handed into yarns; several yarns are twisted right-handed into a strand, and the strands are twisted left-handed into a rope.

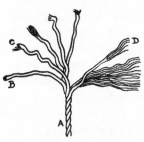

FIG. 361. Parts of rope: *A*, rope; *B* and C, strands; *D*, fiber twisted into yarn.

360. Rope Terms.

Fiber—material as obtained from plant.

Yarn—twisted fiber.

Thread—two or more small yarns twisted together left-handed.

(Usually cotton, wool and silk.)

String or *Twine*—same as thread, but made of a little larger yarns, (Jute and hemp also used.)

Strand—same as string, but with larger yarns, for making rope.

Cord—two or more threads or strings twisted together.

Rope—two or more strands twisted together right-handed.

Hawser—a rope of three strands.

Shroud-laid—a rope of four strands.

Cable—three hawsers twisted together left-handed.

Standing part—long end of rope not used.

Bight—is formed when the rope is turned back on itself, forming the letter *U*.

End—part used in leading.

Loop—is formed by crossing the sides of a bight.

Lay—to twist the strands of a rope together.

Unlay—to untwist the strands of a rope.

Relay—to twist strands together that have become untwisted.

Whip—to bind the end of the rope to prevent raveling.

Splice—to join two ends of a rope by interweaving the strands.

Crown splice—to interweave the strands at the end of a rope.

Pay—to paint, tar or grease a rope to resist moisture.

Haul—to pull on a rope.

Taut—drawn tight or strained.

361. Care and Treatment of Rope. A new rope that is kinky when unwound can best be straightened out by drawing it across the floor or over a sod-covered field. If it is very stiff, it should be immersed in raw linseed oil, tallow or lard, and boiled. This treatment not only makes the rope more pliable, but serves as a lubricant, preventing internal wear. The wear inside a rope is the result of the fibers slipping back and forth over each other, frequently caused by using a pulley that is too small. This wear in a rope can be easily seen by pulling the strands apart. Often a rope is greatly weakened before the wear is noticed. External wear is the result of drawing the rope over rough surfaces which tears the fibers. This source of wear can be easily detected and removed. Where it is desired to protect the rope from dampness, as well as to prevent external wear, the application of an exterior coating such as tallow, graphite, beeswax, or black lead and tallow, will lengthen the life of a rope. Always keep rope in a dry place. If it does get wet, stretch to dry it. Do not allow the end of the rope to unravel.

362. Requirements of a Good Knot. The three requirements of a good knot have been stated as follows: (*a*) Rapidity with which it can be tied, (*b*) its ability to hold fast when pulled tight, (*c*) the readiness with which it can be untied.

363. Theory of Knots and Splices. Method of making various types of knots can be acquired only by practice. The method of making many good knots is obtained by close observation. There are no very definite rules that one can follow. The following principles should be kept in mind:

"The principle of a knot is that no two parts which move in the same direction, if the rope were to slip, should lie alongside of and touch each other."* . . . "A knot or hitch must be so devised that the tight part of the rope must bear on the free end in such a manner as to pinch and hold it, in a knot against another tight part of the rope, or in a hitch against the object to which the rope is attached." †

The student should try to apply these two principles until they are thoroly mastered.

*Wm. Kent, *Mechanical Engineers' Hand Book.*
†Howard W. Riley, *Cornell Reading Course.*

CHAPTER XXXVII

WHIPPING AND MAKING END KNOTS; END SPLICES

364. Tools and Materials Needed for Rope Work.

Tools and Materials Needed: A knife and a large nail or marlin spike (Fig. 362) which can be whittled out of a piece of hard wood, are the only tools needed for this work. A few pieces of 3/8″ rope and some pieces of cord will complete the equipments.

FIG. 362. Marlin spike.

365. Treatment of Raveled Ropes.

In ropes that are raveled, the strands should be twisted and carefully relaid to the point where the knot is to be formed. In unlaying the end of a new rope in preparation for making

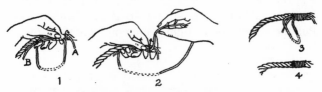

FIG. 363. Whipping end of rope.

a knot, care must be taken not to untwist the strands. Neither whipping nor down crowns can be called knots, but they serve the purpose of a knot and can be used to advantage where it is desirable to have a knot on the end of a rope.

366. Whipping.

Place the piece of cord on the rope, allowing one end to hang loosely over the end of the rope about 2″ (*A*, Fig. 363). Make a loop by passing the other

(B) end of the string along the rope to make a loose end of about 2″. Hold the rope and cord with left hand, as shown in 2, Fig. 363. Grasp the loop of cord with the right hand and wrap it tightly down the rope over itself, as shown in third sketch. When wrapped as much as desired, draw up the loop by pulling on the ends *A* and *B*. This will complete the job of whipping.

367. Crown Knot. The crown knot (Fig. 364) in itself is of little value, but it is the first step in making a crown

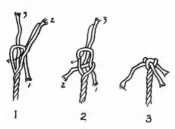

FIG. 364. Crown knot.

or end splice. First unlay several inches of rope, then bring strand No. 1 between strands Nos. 2 and 3, forming a loop, as shown in sketch 1. Pass strand No. 2 across the loop, as shown in sketch 2. Pass strand No. 3 over strand No. 2 and thru the loop. Pull the strands down tightly and complete the crown.

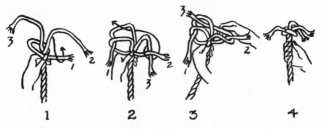

FIG. 365. Wall knot.

368. Wall Knot. (Fig. 365). Unlay several inches of rope as in previous case. Hold rope with left hand and with right hand bring strand No. 1 around, forming a loop as in 1. Strand No. 2 is passed around No. 1, as indicated by arrow in 1. Strand No. 3 is passed around No. 2 and up thru loop formed by No. 1, as indicated in 2 and 3. The loose ends are then drawn up, as shown in 4.

369. Wall and Crown Knot (Fig. 366). As the name would imply, the knot is a combination of the two previous knots. The wall knot is made and then the crown knot, as shown in 1 and 2, Fig, 366.

FIG. 366. Wall and crown knot.

370. Manrope Knot. (Fig. 367.) This knot is also a combination of the wall and crown knot, but is made just the reverse of the wall and crown knot. The crown knot is first made and the wall knot drawn down over it.

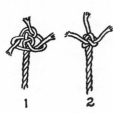

FIG. 367. Manrope knot.

371. Matthew Walker Knot (Fig. 368). This is a very permanent end knots It is made by first making a loosely-constructed wall knot, then by passing *A* thru the loop with *B*, *B* thru the loop with *C*, and *C* thru the loop with *A*, as shown in 1, Fig. 368. When drawn up tight, we have knot, as shown in 2, Fig. 368.

372. End or Crown Splice (Fig. 369). This is one of the best end fastenings for lead ropes. It is made by making a crown knot and then splicing back the loose ends.

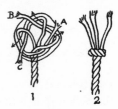

FIG. 368. Matthew Walker knot.

A large nail or marlin spike is best for weaving the loose ends back. Each loose strand is passed over the adjacent strand in a diagonal direction and under the next one, as shown in 1, 2 and 3, Fig. 369.

373. Overhand Knot (Fig. 370). The overhand knot is one of the most common and the simplest of end knots.

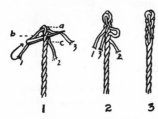

FIG, 370. Overhand knot.

FIG. 369. Crown splice.

It forms a part of many other knots and hitches. It is made by making a loop in the rope and passing one end thru it. Either a right- or left-hand knot may be made.

374. Blood Knot (Fig. 371). This knot is larger than the overhand knot, but made in the same way, except

FIG. 371. Blood knot.

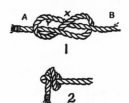

FIG. 372. Figure 8 knot.

that the end of the rope is passed thru the loop several times before it is pulled tight. A similar knot is made by the seamstress by rolling the end of the thread between the finger and thumb.

375. Figure 8 Knot (Fig. 372). This knot is a good one to use on the ends of ropes to prevent them from being pulled thru a pulley or a hole. It is made by forming a loop, then passing the short end *A* of the loop over

Fɪɢ. 373. Stevedore knot.

the standing part of the rope *B* at *X* and bringing it back thru the loop at *Y*

376. Stevedore Knot (Fig. 373). This knot is the same as the figure 8 knot, but instead of one turn around the standing part of rope, three turns are made, as shown in 1 and 2, Fig. 373.

CHAPTER XXXVIII
TYING KNOTS AND HITCHES

377. Practice in Tying Knots. Whip the ends of each piece of rope. Study each knot carefully, and make the same knot several times for practice with and without sketch. Be sure the knot is correctly tied before attempting a new one.

378. Binder Knot. This knot (Fig. 374) is one of the

FIG. 374. Binder knot.

simplest for fastening two pieces of rope together. It is made by taking the two rope ends, placing them side by side, and tying an overhand knot.

379. Square Knot (Fig. 375). This is a smooth knot that is easily tied and easily untied. It is used a great deal for tying packages; also, for fastening the ends of binder

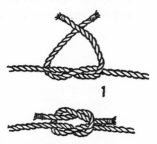

FIG. 375. Square knot.

twine when threading the binder. To make this knot, first make a right-hand overhand knot, then cross the strand and tie the left-hand overhand knot. This knot will not hold if the two ropes are of different sizes.

380. Granny Knot (Fig. 376). The granny knot slips easily and is therefore a very poor knot. The difference between the granny knot and square knot can be easily noted by comparing Fig. 375 and Fig. 376. A great many make the granny knot when attempting to make the square knot.

FIG. 377. Surgeon's knot.   FIG. 376. Granny knot.

381. Surgeon's Knot (Fig. 377). This knot is practically the same as the square knot, except that when making the right-hand overhand knot, the rope is twice wrapped instead of only once. The second part of the knot is completed by making a left-hand overhand as in completing the square knot.

FIG. 378. Weaver's knot.

382. Weaver's Knot (Fig. 378). This knot is one of the best, due to the fact that it holds well, is easily tied and easily untied. To tie this knot, grasp the ends of the rope with left hand, as shown in 1. With end *A* under *B*, grasp rope at *X* and pass it around end A, forming a loop as in 2; complete the knot by passing end *B* thru loop as in 3. Draw it up tight.

383. Carrick's Bend (Fig. 379). This knot is used as a fancy knot in braids. It is also a very satisfactory knot to fasten ropes together. In tying this knot, form a loop with the end *Y* under the standing part *A*, as shown in

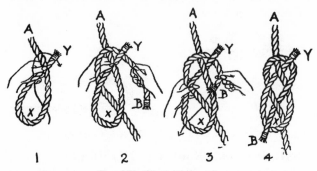

Fig. 379. Carrick's bend.

1. Pass the other end of rope under the loop X, over the standing part *A*, under end *Y*, again over *A*, under standing part *B* and over *A*, the final knot being completed as in 4. When drawn tight, it will assume the shape of a double bowline.

Knots for Fastening Cattle, Tying Hay Ropes, Etc.:

A point to be considered in use of rope is the correct selection of knot for right place; this is especially true where a knot is to be loosened often, or where it is desired to have a knot that will slip.

384. Bowline Knot (Fig. 380). This is one of the best knots for fastening the end of a rope as in hitching. There are several kinds, but the overhand is probably the easiest and quickest to make. To make the knot, form a small loop (C) in 1 near the end of the rope, as in Fig. 380. Hold the loop with the left hand, grasp the end

A with the right hand, pass it thru the loop *C* and around the standing part *B*, and back thru the *loop*, as in 2 and 3.

385. Double-Rope Bowline Knot (Fig. 381). This knot is quite similar to the knot just described, but is

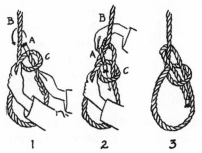

FIG. 380. Bowline knot.

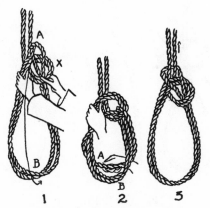

FIG. 381. Double rope bowline.

used when made in the middle of a long rope or at the end when doubled. A loop (X) is formed and part *A* passed thru as in previous case. Part *A* is drawn thru far enough so that the double loop *B* can be drawn thru it, as shown in 2 and 3. This knot is especially useful in throwing horses and cattle.

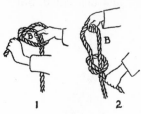

FIG. 382. Slip knot.

386. Slip Knot (Fig. 382). The slip knot is a very common one and often used when a different type of knot would be much more satisfactory. To tie this knot, form a loop, grasp rope *B* and draw it thru, as shown in 1 and 2 in Fig. 382.

387. Manger Knot (Fig. 383). This knot is quite similar to the ordinary slip knot, but much better on account

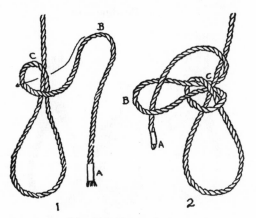

FIG. 383. Manger knot.

of being easier to untie. To tie this knot, form a loop *C*, grasp the bight *B* and pass it around the standing part of the rope and thru loop *C*; then complete the knot by bringing end *A* around the standing part and thru *B*.

388. Lariat Knot (Fig. 384). As the name would indicate, this knot is used in forming a lariat. It is tied by first forming an overhand knot near the end of the rope, as

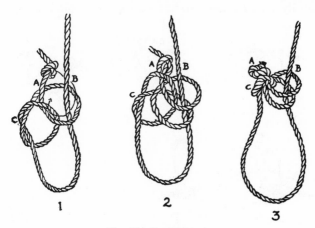

FIG. 384. Lariat knot.

at *C* in 1, Fig. 384. The end *A* is then passed around the standing part *B* and thru the loop twice. The overhand knot is then drawn tight and the knot is complete.

389. Hangman's Noose (Fig. 385). This is another knot with a slip loop. It is a knot that is easy to tie and

FIG. 385. Hangman's knot.

holds well. Make a double loop, as in 1; then wind the end of rope back the number of rounds desired, passing it thru loop *Y*, 2. By drawing on the noose, the knot is completed, as in 3.

390. Farmer's Loop (Fig. 386). If it is desired to tie a loop in the middle of a rope when both ends are fastened, the farmer's loop is suitable. It is easily tied and easily untied. Make two turns in rope and hold it, as in 1,

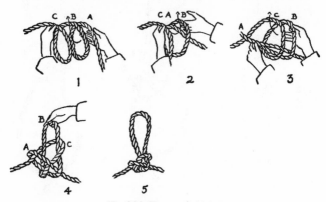

FIG 386. Farmer's loop.

Fig. 386. Pass loop *A* under loop *B* between *B* and *C* in 2. Next pass loop *C* under loop *A*, as in 3. Now, pass *B* under loop *C* and up between *A* and *C* in 4. The knot is completed by drawing the standing part tight.

Temporary Hitches.

Note: A hitch should be selected for a particular use. One should be very careful in making a scaffold hitch where life is in danger. It must be kept in mind that the hitches here outlined are for temporary use.

391. Half Hitch (Fig. 387). The half hitch is one step in making other hitches and knots. It is useful, however, when the standing part of the rope is drawn tight and pinches the end against object tied, as in Fig. 387.

392. Timber Hitch (Figs. 388 and 389). This hitch is one step in advance over the half hitch. The end of the rope is wrapped several times instead of simply drawn

FIG. 387. Half hitch.

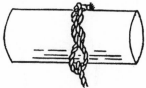

FIG. 388. Timber hitch.

under once as in the half hitch. A combination of the timber and half hitch is much more secure. (See Fig. 389.)

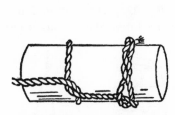

FIG. 389. Timber and half hitch.

FIG. 390. Rolling hitch.

393. Rolling Hitch (Fig. 390). This hitch is very easily and quickly made, and is a suitable fastening for most any purpose. Wrap rope three times about the object to which it is to be fastened, then make two half hitches about the standing part.

394 Clove Hitch (Fig. 391). This is one of the simplest and yet one of the most secure methods of fastening tent ropes, guy ropes or any rope when there is to be a direct pull against it. There are several methods of making the clove hitch, but probably the farmer's method is best. Cross the arms, the left in front of the right; grasp the rope, as in 1, Fig. 391; then bring the hands to position.

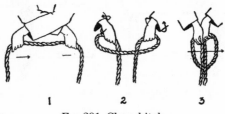

FIG. 391. Clove hitch.

shown in 2; then complete the hitch by turning both hands to the right, as in 3.

395. Scaffold Hitch (Fig. 392). The scaffold hitch is a modified form of the clove hitch. Make a rather loose

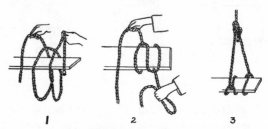

FIG. 392. Scaffold hitch.

clove hitch and place over the end of scaffold, as in 1, Fig. 392. Draw the ropes tight in opposite direction, turn the plank over and fasten short end to the standing part by means of a bowline knot, as in 3.

396. Blackwall Hitch (Fig. 393). This hitch can be used only when the pull on rope is continuous and a hook is provided. Make a bight in the rope and pass around the hook; the free end is then passed thru the hook, and the standing part passed over it from the opposite side.

397. Sheepshank (Fig. 394). The sheepshank is not a hitch in the same sense as the other hitches described.

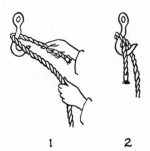

1   2

FIG. 393. Blackwall hitch.

It is used as a means of shortening ropes. To tie this hitch, a loop is formed that is large enough to reduce the rope to desired length (see 1, Fig. 394) and held in the left hand; a half hitch is formed of the standing part

FIG. 394. Sheepshank.

of the rope and passed over each end of the loop, as in 2. To make this hitch permanent, the standing part is drawn thru the bight at each end of the loop.

Splices:

398. End or Crown Splice. This type of splice has been described under head of means of preventing rope from raveling (Fig. 369).

399. Loop Splice (Fig. 395). This splice is used when a permanent loop is to be constructed at any point of the rope other than the end. The size and location of the loop is first determined, then two strands are raised on

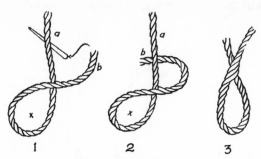

FIG. 395. Loop splice.

the short end and the lead rope passed under them. To complete the splice, two strands in the long part of the rope are raised, as in a, 1, Fig. 395; and the short end *b* is passed thru and drawn up, as in 3.

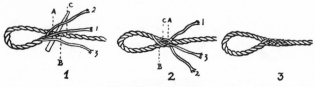

FIG. 396. Eye splice.

400. Eye Splice or Side Splice (Fig. 396). The eye splice is used when it is desired to form a loop at the end of a rope or as a side splice where it is desired to fasten one rope to another at any point other than the end. Unlay the end of the rope for several inches, determine the size of loop to form, then place the two outside strands to straddle the main rope and the center strand to run along the top of the rope, as in 1, Fig. 396. Now, by means of the marlin spike or large nail, raise one of the strands A and pass the center strand No. 1 under it. Pass strand No. 2 over A and under B, and pass strand No. 3 thru from the opposite side so that it comes out where No. 1 enters. Draw all ends up snug and weave in the strands, as described for the end splice.

401. Short Splice (Fig. 397). The short splice is used for joining the two ends of rope together when it

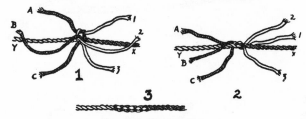

FIG. 397. Short splice.

is not desired to draw it thru pulleys. This splice is not as smooth as the long splice, but it is strong and quite easily made. To make the splice, unlay the ends of the two ropes for a sufficient distance, depending on size of rope and load—for a 3/8″ rope, at least 6″. Bring the ends of the rope together so that the strands of one pass alternately between those of the other, as in 1, Fig. 397. Take each pair of strands from opposite sides and tie a right-hand overhand knot, draw the knots tightly and pass each strand diagonally to the left, then weave it in as in making the end splice (1, 2 and 3, Fig. 397).

402. Long Splice (Fig. 398). This type of splice is so nearly the same size as the other part of the rope, that it can be used thru pulleys without hindrance. Every user

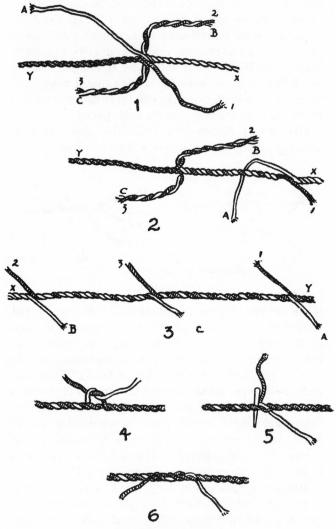

FIG. 398. Long splice.

of rope should know how to make the long splice. Unlay the end of the rope, as in making a short splice. A 1/4″ rope should be unlaid about 12″, a 3/8″ rope 16″, a 1/2″ rope 24″, and a 1″ rope 36″, to obtain best results. Lock the strands as in the beginning of the short splice, pair the strands from each end, as in 1, Fig. 398, twisting two of the pairs together. As for the remaining pair, unlay one strand and relay the other strand in its place. Continue until within a few inches of the end of the relaid strand No. 1, as in 2. Repeat the process with either pair of the other strands. Untwist the last pair; the rope should appear, as in 3, with each strand coming from the left and passing in front of the strands from the right. To complete the splice, tie each pair of strands with a right-hand overhand knot, as in 4. Weave the loose ends into the rope by passing one end over the adjacent strand and under the next, as in 5. Cut the ends of strands off and pound down the uneven ends to make finished splice, as in 6.

CHAPTER XXXIX
PROJECTS IN ROPE WORK

403. Making a Halter.

Preliminary Instructions: There will be 12' or more of
rope needed for this project. Temporary halters
are much more satisfactory for leading an animal
than is a rope placed about the animal's neck. To

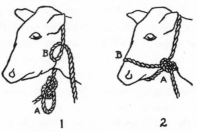

FIG. 399. Temporary halter.

make a temporary halter, it is necessary to know
how to make a few of the various knots and hitches
described in previous pages and referred to in this
project.

Halter No. 1 (Fig. 399). To construct this halter, first
make a loop in the end of the rope A, tying it with a simple overhand bowline, as described in Sec. 389. Pass
the end of rope with loop about animal's neck and form
a second loop *B* in the standing part of the rope thru
which draw loop *A* and place around the animal's nose.
The slack is drawn out with the free end of the rope, as
in 2.

Halter No. 2 (Fig. 400). This type of temporary halter is usually called the Hackamore. It is used for leading either cattle or horses, and is made by passing one

end of the rope about the animal's neck and tying with
a bowline knot. A half hitch is thrown in the standing
part of the rope and passed over the animal's nose, as

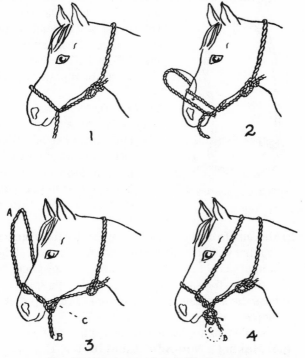

Fig. 400. Temporary halter (Hackamore).

in 1, Fig. 400; a second half hitch is made below the first
and passed over the nose, as in 2. The first half hitch is
wrapped about the second and passed over the aninal's
head, as in 3. To complete the halter, the standing part of
the rope is passed thru the loop *C* below the half hitch,
as indicated in 4.

404. An Adjustable Halter (Fig. 401).

Preliminary Instructions: To make a satisfactory adjustable halter, it is necessary to be familiar with the method of making the eye splice, loop splice and end splice. The size of rope to use will depend on the use of halter. Most halters are made from 1/2″ to 3/4″ rope. The length of rope needed is 12'.

Fig. 401. Adjustable halter.

Working Instructions:

1) Make an eye splice in end of rope as at *a*, Fig. 401. This splice should be only large enough to allow the standing part of rope to pass thru it freely.

2) Measure from the loop of the eye splice the distance *(d)* that will be required to reach nearly around the animal's nose. At this point make a loop splice (b) with loop the same size as that of the eye splice.

3) Pass the standing end of the rope thru loop *a* and loop *b*.

4) Complete halter by making end splice on end *c*.

405. Making a Non-Adjustable Halter (Fig. 402).

Note: The only difference between this halter and the one described in Sec. 404 is that the head piece and nose piece are made of definite length, depending on the head dimensions of the particular animal for which the halter is made.

1) Determine the necessary length of head piece and nose piece by measuring animal's head.

2) Make loop splice (b, Fig. 402), leaving *c* long enough to form nose piece.

3) Side splice end of *c* into standing part of rope at *a*, making head piece *d* of suitable size.

4) Thread end *c* thru loop *b*.

5) Make end splice *e* in end of standing part of rope to complete the halter.

406. The Trip Rope (Fig. 403).

Materials Needed: Thirty feet of 1/2″ rope, three 2″ rings, and two heavy straps with buckles to go around ankles.

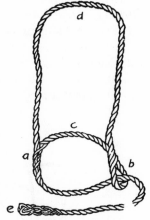

FIG. 402. Non-adjustable halter.

Preliminary Instructions: In handling young horses, it is sometimes very essential to have some means

FIG. 403. Trip rope.

of tripping them when the horse does not obey the

command of the Trainer. Knee pads should be provided when the trip rope is used.

Working Instructions:

1) Place ankle straps on front ankles with a ring on each strap.

2) Place surcingle with ring at bottom around horse just back of shoulders, or tie around the body at this point a piece of rope, using a single bowline knot.

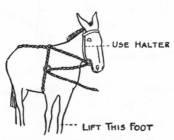

Fig. 404. Throwing rope.

3) Take long rope provided, pass thru ring on ankle strap of near foot, up thru ring at bottom of surcingle, and down to other ankle ring where it is tied. The trip is then ready to use by pulling on standing part of long rope.

407. Throwing or Casting Rope (Fig. 404).

Materials Needed: Thirty feet of 1/2″ rope and straps.

Preliminary Instructions: In handling horses, it is sometimes necessary to throw the animal for the purpose of an operation or otherwise. To avoid chafing or burning the animal with a rope, straps should be provided for those places where a rope would rub.

Working Instructions:

1) Tie a double rope bowline knot, as described in Sec. 385, in middle of rope to serve as crupper.

2) Adjust crupper in place, run to withers and tie a square knot (Sec. 379).

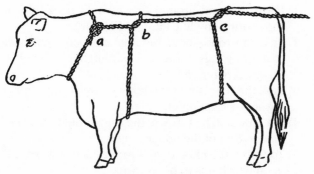

FIG. 405. Casting rope.

3) Pass rope about body at withers just back of front legs; tie with another square knot, forming a surcingle and crupper properly adjusted to the animal. Provide ring or girth, as shown in Fig. 404.

4) Run the free end of the rope from top of surcingle thru the ring in halter back thru ring on girth. The rope is then ready for use.

5) Instead of rope under No. 3, a regular crupper and surcingle may be used.

6) To use rope, lift the front foot of the animal on the side opposite that on which the rope is passed and pull on the free end of the rope.

408. Rope for Casting Cattle (Fig. 405).

Material Needed: Thirty-five feet of 1″ rope.

Preliminary Instructions: The instruction for throwing a horse should be kept in mind in throwing a cow or steer. Care must be observed to avoid hurting the animal. One need be acquainted only with the bowline knot and the simplest half hitches to adjust a rope for throwing cattle.

Working Instructions:

1) Place one end around the animal's neck and tie rope with a bowline knot (a, Fig. 405).
2) Pass the rope about the animal's body just back of the fore legs, forming a half hitch at withers, as shown at *b*, Fig. 405.
3) Pass the rope about the body (c) at the hips, forming another half hitch.
4) If a cow is to be thrown, the rope should be placed just in front of the udder.
5) To throw the animal, pull to rear and toward side upon which it is to be thrown.

CHAPTER XL
HARNESS REPAIR

409. The Importance of Good Harness. Nothing adds more to the appearance of a well-groomed horse than a neat, clean and properly-fitting harness. A good set of harness not only adds to the appearance of a team, but makes the team more efficient in its operation. A first-class teamster will take pride in keeping his team properly fitted. Such negligence as allowing the harness to be bound up with pieces of baling wire and with binder twine is inexcusable. Often the hame straps are allowed to loosen, the breeching to hang too low, resulting in sore shoulders and chafed sides and back.

The farmer cannot afford to neglect the care and upkeep of his harness. In fact, each farmer should be provided with a simple harness repair outfit and keep on hand a few supplies for adding a strap by sewing or riveting where one is worn.

The life of a harness can be greatly increased by systematic care. The practice of oiling the harness at least once a year should not be overlooked. Take the harness apart and wash it thoroly in warm, soft water and soap; allow it time to dry; then apply a coat of good quality harness oil. Allow the oil to soak in before it is rubbed off. Before the harness is reassembled, each part should be gone over carefully and needed repairs made. This work can easily be done on the farm on rainy days.

410. The Harness Room. A conveniently-located harness room is of great value in caring for the harness. It is very objectionable to store the harness in most stables due to the effect of the moisture and the ammonia from the manure. When the stable is kept thoroly cleaned and is well ventilated, harness can be kept with little damage and are thereby much more conveniently located for use.

411. Harness Oil. Be careful not to use a mineral oil for the harness or leather belts. Mineral oils will cause the leather to dry out and crack. Buy only standard brands found on the market. A good oil can be made by melting three pounds of tallow without letting it boil, and gently adding one pound of neat's-foot oil. Stir continuously until cold so that it will be perfectly mixed. Color by adding a little lampblack.

412. Repair Leather. Leather for repairing can be bought from any harness shop. It is best to buy a fairly large piece, as it can be secured much more cheaply that way. Some men buy a half hide, and, thereby, secure some of both the best quality leather from the back of the hide and the poorer, cheaper belly leather. The latter can be used where there is little strain.

413. Equipment for Harness Work. A clamp is needed for holding the work. This can easily be made at home. Some men prefer a vise to a clamp. A common type of clamp is illustrated and described under woodwork. (Fig. 86.) In addition to the clamp, the repair outfit should consist of the following: One dozen sewing needles, different sizes; a sharp knife, half dozen awls, ball of shoe thread, two awl handles, one revolving punch, one small riveter with rivets. The entire repair outfit can be purchased for less than $2. Instead of shoe thread and wax, the prepared thread can be secured. An advantage of the shoe thread and wax is that it can be prepared to meet the requirement of the special job.

414. Splicing Worn Harness Strap.

Requirements: To make a satisfactory splice that will be smooth and not chafe, and if used thru a ring, will not catch or bind. It must also be strong enough to resist the force applied to it.

Materials Needed: Suitable leather strip for repair, thread, wax.

Tools Needed: Clamp (such as shown in Fig. 86), one selected awl, two selected needles, one sharp knife.

415. Preparing Strap for Sewing. Prepare the worn strap for splice by cutting away the worn part. Thin the ends down with a sharp knife to a gradual taper for about 3″. If the strap is one that can be shortened, it is then ready for splicing; otherwise, an insert will have to be prepared and a double splice made. Small wire tacks are useful in holding the straps together while the stitching is being done. Prepare thread for sewing by waxing it. To do this, the thread must first be broken with a ragged end. Pull the thread out of the center of the ball, hold it on the knee, and roll it to take out the twist. When the twist is out, give the string a pull and it should break with long ragged ends. Give the end a twist around the first finger of the left hand and draw it thru the right hand. When about 6' have been drawn out, throw the center over a hook in the wall and pull until the ends are about even and each about 3' long. Keep the string tight with the left hand, and with the right hand rub it on the knee as before and break it. Repeat this until the required number of strands have been secured, depending on the work to be done. Make the ends of the strands slightly uneven in length to provide a long tapering point for threading the needle. Wax the free ends before twisting. Twist the thread carefully and wax it thoroly. Put the two needles on the thread ready for sewing.

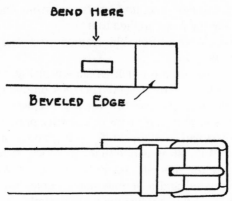

FIG. 406. Sewing buckle or strap.

416. Sewing the Splice. Put the splice in the clamp, using extreme care to keep edges perfectly even. Mark off holes a definite distance apart. Make hole with the awl, insert needle and draw the thread half way thru, leaving one needle on each side. Make another hole with the awl, insert the needle thru and draw the thread thru a few inches; then put the other needle thru the same hole from the other side and pull both threads up tight, being careful to avoid knots. Continue this process along both sides and across the ends of the splice. To do a good job, keep the stitches straight and of uniform length. To complete the job, draw the ends of the thread out between the splice and tie.

417. Sewing Buckle and Ring on Harness Strap. Suitable buckle, ring and strap for the work intended are the required materials for this job. The proper selection of strap and buckle for the particular job is very essential. The buckle should be slightly wider than the strap to insure ease in buckling and to reduce the amount of wear on the strap. The strap should be prepared for sewing as in preceding exercise. Fig. 406 illustrates this operation.

418. Instructions for Sewing Buckle. If strap is wider than buckle, trim it down until it is a very little

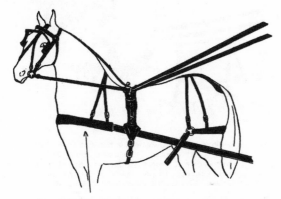

FIG. 407. Single harness, breast collar type.

narrower than the buckle. Double end of strap back thru buckle at least 2″ for a 1″ strap; cut a slot for the tongue of buckle long enough to move the tongue thru 180 degrees. Next shave the inner surface of the end of the strap to a beveled edge to make a smooth joint when it is sewed. Cut a narrow strap of leather to pass around the strap as shown (Fig. 406), to hold the opposite end of the strap when buckled. Put buckle in place, fold strap back, and clamp tightly in sewing clamp. Proceed to sew, as in preceding problem. Riveting and sewing can often be employed together on such work.

419. Overhauling a Set of Harness (Figs. 407 and 408).

Preliminary Instructions: Soap and water, oil, harness dressing and metal polish must be provided for this work. Harness in poor repair means a loss of time during the busy season. Inspect and repair

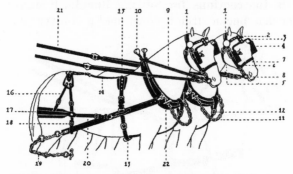

1-BRIDLE CROWNPIECE, 2-BROWBAND, 3-BLINDSTAY, 4-BLINDS,
5-THROATLATCH, 6-CHEEKPIECE, 7-NOSEBAND, 8-BIT, 9-REIN,
10-HAMES, 11-NECKYOKE CHAINS, 12-BREAST STRAP, 13-BELLYBAND,
14-BACKSTRAP, 15-SADDLE, 16-HIPSTRAPS, 17-BREECHING, 18-LAZY-
STRAP, 19-HEELCHAIN, 20-HOLDING BACK STRAP, 21-LINES, 22-COLLAR

FIG. 408. Double harness.

all harness before the spring season work begins. The best time to do this is when the weather is bad and outside work cannot be done to advantage. To keep the harness in best condition, they should be gone over at least twice each year.

Working Instructions: First take the harness apart so that each strap, buckle and ring can be carefully inspected. Carefully clean with a little warm soft water. If the harness is very dirty, soak for a few minutes in warm water; then scrub with a brush, using soap freely; wipe and hang up to dry. When dried, apply oil, prepared as outlined in Sec. 411, or a special harness oil. Make several applications and rub the oil into the leather to get the best results. To give the harness a good, glossy, black finish, it is necessary to apply some good standard harness dressing as recommended by the harness-maker. Ordinary black shoe polish may be used, but would probably be a little more expensive than the material prepared for the purpose. After application, rub vigorously with a polishing cloth to get the best results. To clean the metal mountings, use some form of metal polish or cleansing compound, like Old Dutch Cleanser or Bon Ami. Careful polishing is a big factor in giving the harness a good appearance. Lastly, put harness back together, making all necessary repairs, adding new straps, buckles or rings where needed, following instructions outlined in Secs. 415–418.

420. Adjusting Harness to Horse. Every one who handles a team should realize the importance of a well-fitted harness. A poorly-fitted harness not only hinders the horse in working, but is liable to make a balker out of a good worker, and, in addition, is liable to damage the horse by causing a sore mouth, shoulders or back. Well-fitted harness insures more work done during the busy season.

421. The Bridle. The fitting of the bridle will depend on the individual animal. Adjust the check pieces so that the bit will not hang too low in the mouth or so high that it will raise the corners of the mouth, thereby causing soreness. Each part of the bridle should fit snugly, but not so tight as to cause pinching. The blinds should fit snugly up to the head. Do not adjust the throat latch too tight.

422. The Collar. Pay especial attention to the collar, as it must bear the load. Test the fittings of the collar by pressing it back against the shoulder when the horse is holding its head in working position. The collar should have an even contact against all parts of the shoulder and have ample space at the wind pipe for the place of one's hand. Collars often need to be readjusted after the animal has been worked a while in the spring, due to its losing flesh. Adjust breast collar to a height where it will neither hinder movement nor interfere with breathing.

423. Hames. After the collar is adjusted, adjust the hames at the top to fit the collar and then buckle or tie as tightly as possible at the bottom.

424. Other Adjustments. All other parts of the harness should be adjusted to make them fit snugly, neither too tight nor too loose. Adjust the breeching the proper height. Fit the saddle to the back at the low place just back of the withers. Adjust the crupper strap, back straps, hip straps, holding back straps and traces to proper length in the order mentioned.

Note: Avoid accidents in hitching the team to implement or vehicle by taking down the lines and adjusting them first.

INDEX

413

FARM ENGINES AND HOW TO RUN THEM

by James H. Stephenson

PREFACE

This book makes no pretensions to originality. It has taken the best from every source. The author believes the matter has been arranged in a more simple and effective manner, and that more information has been crowded into these pages than will be found within the pages of any similar book.

The professional engineer, in writing a book for young engineers, is likely to forget that the novice is unfamiliar with many terms which are like daily bread to him. The present writers have tried to avoid that pitfall, and to define each term as it naturally needs definition. Moreover, the description of parts and the definitions of terms have preceded any suggestions on operation, the authors believing that the young engineer should become thoroughly familiar with his engine and its manner of working, before he is told what is best to do and not to do. If he is forced on too fast he is likely to get mixed. The test questions at the end of Chapter III. will show how perfectly the preceding pages have been mastered, and the student is not ready to go on till he can answer all these questions readily.

The system of questions and answers has its uses and its limitations. The authors have tried to use that system where it would do most good, and employ the straight narrative discussion method where questions could not help and would only interrupt the progress of thought. Little technical matter has been introduced, and that only for practical purposes. The authors have had traction engines in mind for the most part, but the directions will apply equally well to any kind of steam engine.

The thanks of the publishers are due to the various traction engine and threshing machine manufacturers for cuts and information, and especially to the *Threshermen's Review* for ideas contained in its "Farm Engine Economy," to the J. I. Case Threshing Machine Co. for the use of copyrighted matter in their "The Science of Successful Threshing," and to the manager of the Columbus Machine Co. for valuable personal information furnished the authors on gasoline engines and how to run them. The proof has been read and corrected by Mr. T. R. Butman, known in Chicago for 25 years as one of the leading experts on engines and boilers, especially boilers.

CONTENTS

CHAPTER I

BUYING AN ENGINE

There are a great many makes of good engines on the market to-day, and the competition is so keen that no engine maker can afford to turn out a very poor engine. This is especially true of traction engines. The different styles and types all have their advantages, and are good in their way. For all that, one good engine may be valueless for you, and there are many ways in which you may make a great mistake in purchasing an engine. The following points will help you to choose wisely:

1. Consider what you want an engine for. If it is a stationary engine, consider the work to be done, the space it is to occupy, and what conveniences will save your time. Remember, TIME IS MONEY, and that means that SPACE IS ALSO MONEY. Choose the kind of engine that will be most convenient for the position in which you wish to place it and the purpose or purposes for which you wish to use it. If buying a traction engine, consider also the roads and an engine's pulling qualities.

2. If you are buying a traction engine for threshing, the first thing to consider is FUEL. Which will be cheapest for you, wood, coal or straw? Is economy of fuel much of an object with you—one that will justify you in greater care and more scientific study of your engine? Other things being equal, the direct flue, firebox, locomotive boiler and simple engine will be the best, since they are the easiest to operate. They are not the most economical under favorable conditions, but a return flue boiler and a compound engine will cost you far more than the possible saving of fuel unless you manage them in a scientific

429

way. Indeed, if not rightly managed they will waste more fuel than the direct flue locomotive boiler and the simple engine.

3. Do not try to economize on the size of your boiler, and at the same time never get too large an engine. If a 6-horse power boiler will just do your work, an 8-horse power will do it better and more economically, because you won't be overworking it all the time. Engines should seldom be crowded. At the same time you never know when you may want a higher capacity than you have, or how much you may lose by not having it. Of course you don't want an engine and boiler that are too big, but you should always allow a fair margin above your anticipated requirements.

4. Do not try to economize on appliances. You should have a good pump, a good injector, a good heater, an extra steam gauge, an extra fusible plug ready to put in, a flue expander and a header. You should also certainly have a good force pump and hose to clean the boiler, and the best oil and grease you can get. Never believe the man who tells you that something not quite the best is just as good. You will find it the most expensive thing you ever tried—if you have wit enough to find out how expensive it is.

5. If you want my personal advice on the proper engine to select for various purposes, I should say by all means get a gasoline engine for small powers about the farm, such as pumping, etc. It is the quickest to start, by far the most economical to operate, and the simplest to manage. The day of the small steam engine is past and will never return, and ten gasoline engines of this kind are sold for every steam engine put out. If you want a traction engine for threshing, etc., stick to steam.

Gasoline engines are not very good hill climbers because the application of power is not steady enough; they are not very good to get out of mud holes with for the same reason, and as yet they are not perfected for such purposes. You might use a portable gasoline engine, however, though the application of power is not as steady as with steam and the flywheels are heavy. In choosing a traction steam engine, the direct flue locomotive boiler and simple engine, though theoretically not so economical as the return flue boiler and compound engine, will in many cases prove so practically because they are so much simpler and there is not the chance to go wrong with them that there is with the others. If for any reason you want a very quick steamer, buy an upright. If economy of fuel is very important and you are prepared to make the necessary effort to secure it, a return flue boiler will be a good investment, and a really good compound engine may be. Where a large plant is to be operated and a high power constant and steady energy is demanded, stick to steam, since the gasoline engines of the larger size have not proved so successful, and are certainly by no means so steady; and in such a case the exhaust steam can be used for heating and for various other purposes that will work the greatest economy. For such a plant choose a horizontal tubular boiler, set in masonry, and a compound engine (the latter if you have a scientific engineer).

In general, in the traction engine, look to the convenience of arrangement of the throttle, reverse lever, steering wheel, friction clutch, independent pump and injector, all of which should be within easy reach of the footboard, as such an arrangement will save annoyance and often damage when quick action is required.

The boiler should be well set; the firebox large, with large grate surface if a locomotive type of boiler is used, and the number of flues should be sufficient to allow good combustion without forced draft. A return flue boiler should have a large main flue, material of the required 5-16-inch thickness, a mud drum, and four to six hand-holes suitably situated for cleaning the boiler. There should be a rather high average boiler pressure, as high pressure is more economical than low. For a simple engine, 80 pounds and for a compound 125 pounds should be minimum.

A stationary engine should have a solid foundation built by a mason who understands the business, and should be in a light, dry room—never in a dark cellar or a damp place.

Every farm traction engine should have a friction clutch.

CHAPTER II

BOILERS

The first boilers were made as a single cylinder of wrought iron set in brick work, with provision for a fire under one end. This was used for many years, but it produced steam very slowly and with great waste of fuel.

The first improvement to be made in this was a fire flue running the whole length of the interior of the boiler, with the fire in one end of the flue. This fire flue was entirely surrounded by water.

Then a boiler was made with two flues that came together at the smoke-box end. First one flue was fired and then the other, alternately, the clear heat of one burning the smoke of the other when it came into the common passage.

The next step was to introduce conical tubes by which the water could circulate through the main fire flue (Galloway boiler).

Fig. 1. Orr & sembower's standard horizontal boiler, with full-arch front setting.

433

The object of all these improvements was to get larger heating surface. To make steam rapidly and economically, the heating surface must be as, large as possible. But there is a limit in that the boiler must not be cumbersome, it must carry enough water, and have sufficient space for steam.

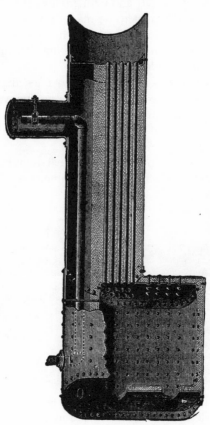

FIG. 2. Gaar, Scott & Co.'s locomotive boiler.

The stationary boiler now most commonly used is cylindrical, the fire is built in a brick furnace under the sheet and returns through fire tubes running the length of the boiler. (Fig. 1.)

LOCOMOTIVE FIRE TUBE TYPE OF BOILER

The earliest of the modern steam boilers to come into use was the locomotive fire tube type, with a special firebox. By reference to the illustration (Fig. 2) you will see that the boiler cylinder is perforated with a number of tubes from 2 to 4 inches in diameter running from the large firebox on the left, through the boiler cylinder filled with water, to the smoke-box on the right, above which the smokestack rises.

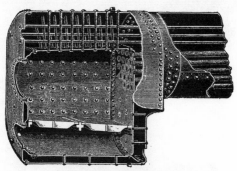

FIG. 3. The huber fire box.

It will be noticed that the walls of the firebox are double, and that the water circulates freely all about the firebox as well as all about the fire tubes. The inner walls of the firebox are held firmly in position by stay bolts, as will be seen in Fig. 3, which also shows the position of the grate.

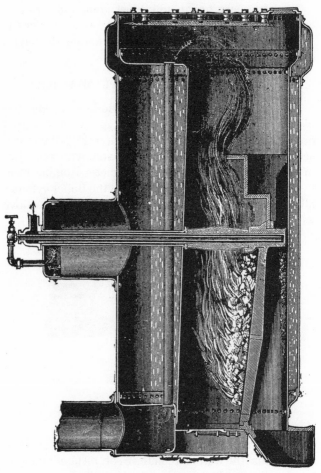

FIG. 4. Hubber return flue boiler.

RETURN FLUE TYPE OF BOILER

The return flue type of boiler consists of a large central fire flue running through the boiler cylinder to the smoke box at the front end, which is entirely closed. The smoke passes back through a number of small tubes, and the smokestack is directly over the fire at the rear of the boiler, though there is no communication between the fire at the rear of the boiler and it except through the main flue to the front and back through the small return flues. Fig. 4 illustrates this type of boiler, though it shows but one return flue. The actual number may be seen by the sectional view in Fig. 5.

The fire is built in one end of the main flue, and is entirely surrounded by water, as will be seen in the illustration. The long passage for the flame and heated gases enables the water to absorb a maximum amount of the heat of combustion. There is also an element of safety in this boiler in that the small flues will be exposed first should the water become low, and less damage will be done than if the large crown sheet of the firebox boiler is exposed, and this large crown sheet is the first thing to be exposed in that type of boiler.

Fig. 5. Section view of huber return flue boiler.

WATER TUBE TYPE OF BOILER

The special difference between the fire tube boiler and the water tube boiler is that in the former the fire passes through the tubes, while in the latter the water is in the tubes and the fire passes around them.

In this type of boiler there is an upper cylinder (or more than one) filled with water; a series of small tubes

FIG. 6. Freeman vertical boiler.

running at an angle from the front or fire door end of the upper cylinder to a point below and back of the grates, where they meet in another cylinder or pipe, which is connected with the other end of the upper cylinder. The portions of the tubes directly over the fire will be hottest, and the water here will become heated and rise to the front end of the upper cylinder, while to fill the space left, colder water is drawn in from the back pipe, from the rear end of the upper cylinder, down to the lower ends of the water tubes, to pass along up through them to the front end again.

This type of boiler gives great heating surface, and since the tubes are small they will have ample strength with much thinner walls. Great freedom of circulation is important in this type of boiler, there being no contracted cells in the passage. This is not adapted for a portable engine.

UPRIGHT OR VERTICAL TYPE OF BOILER

In the upright type of boiler the boiler cylinder is placed on end, the fire is built at the lower end, which is a firebox surrounded by a water jacket, and the smoke and gases of combustion rise straight up through vertical fire flues. The amount of water carried is relatively small, and the steam space is also small, while the heating surface is relatively large if the boiler is sufficiently tall. You can get up steam in this type of boiler quicker than in any other, and in case of the stationary engine, the space occupied is a minimum. The majority of small stationary engines have this type of boiler, and there is a traction engine with upright boiler which has been widely used but it is open to the objection that the upper or steam ends of the tubes easily get overheated and so become leaky. There is also often trouble from mud and scale deposits in the water leg, the bottom area of which is very small.

DEFINITION OF TERMS USED IN CONNECTION WITH BOILERS

Shell—The main cylindrical steel sheets which form the principal part of the boiler.

Boiler-heads—The ends of the boiler cylinder.

Tube Sheets—The sheets in which the fire tubes are inserted at each end of the boiler.

Fire-box—A nearly square space at one end of a boiler, in which the fire is placed. Properly it is surrounded on all sides by a double wall, the space between the two shells of these walls being filled with water. All flat surfaces are securely fastened by stay bolts and crown bars, but cylindrical surfaces are self-bracing.

Water-leg—The space at sides of fire-box and below it in which water passes.

Crown-sheet—The sheet of steel at the top of the fire-box, just under the water in the boiler. This crown sheet is exposed to severe heat, but so long as it is covered with water, the water will conduct the heat away, and the metal can never become any hotter than the water in the boiler. If, however, it is not covered with water, but only by steam, it quickly becomes overheated, since the steam does not conduct the heat away as the water does. It may become so hot it will soften and sag, but the great danger is that the thin layer of water near this overheated crown sheet will be suddenly turned into a great volume of steam and cause an explosion. If some of the pressure is taken off, this overheated water may suddenly burst into steam and cause an explosion, as the safety valve blows off, for example (since the safety valve relieves some of the pressure).

Smoke-box—The space at the end of the boiler opposite to that of the fire, in which the smoke may accumulate before passing up the stack in the locomotive type, or through the small flues in the return type of boiler.

Steam-dome—A drum or projection at the top of the boiler cylinder, forming the highest point which the steam can reach. The steam is taken from the boiler through piping leading from the top of this dome, since at this point it is least likely to be mixed with water, either through foaming or shaking up of the boiler. Even under normal conditions the steam at the top of the dome is drier than anywhere else.

Mud-drum—A cylindrical-shaped receptacle at the bottom of the boiler similar to the steam-dome at the top, but not so deep. Impurities in the water accumulate here, and it is of great value on a return flue boiler. In a locomotive boiler the mud accumulates in the water leg, below the firebox.

Man-holes—Are large openings into the interior of a boiler, through which a man may pass to clean out the inside.

Hand-holes—Are smaller holes at various points in the boiler into which the nozzle of a hose may be introduced for cleaning out the interior. All these openings must be securely covered with steam-tight plates, called man-hole and hand-hole plates.

A boiler jacket—A non-conducting covering of wood, plaster, hair, rags, felt, paper, asbestos or the like, which prevents the boiler shell from cooling too rapidly through radiation of heat from the steel. These materials are usually held in place against the boiler by sheet iron. An intervening air-space between the jacket and the boiler shell will add to the efficiency of the jacket.

A steam-jacket—A space around an engine cylinder or the like which may be filled with live steam so as to keep the interior from cooling rapidly.

Ash-pit—The space directly under the grates, where the ashes accumulate.

Dead-plates—Solid sheets of steel on which the fire lies the same as on the grates, but with no openings through to the ash-pit. Dead-plates are sometimes used to prevent cold air passing through the fire into the flues, and are common on straw-burning boilers. They should seldom if ever be used on coal or wood firing boilers.

Grate Surface—The whole space occupied by the grate-bars, usually measured in square feet.

Forced Draft—A draft produced by any means other than the natural tendency of the heated gases of combustion to rise. For example, a draft caused by letting steam escape into the stack.

Heating Surface—The entire surface of the boiler exposed to the heat of the fire, or the area of steel or iron sheeting or tubing, on one side of which is water and on the other heated air or gases.

Steam-space—The cubical contents of the space which may be occupied by steam above the water.

Water-space—The cubical contents of the space occupied by water below the steam.

Diaphragm-plate—A perforated plate used in the domes of locomotive boilers to prevent water dashing into the steam supply pipe. A dry-pipe is a pipe with small perforations, used for taking steam from the steam-space, instead of from a dome with diaphragm-plate.

THE ATTACHMENTS OF A BOILER*

Before proceeding to a consideration of the care and management of a boiler, let us briefly indicate the chief working attachments of a boiler. Unless the nature and uses of these attachments are fully understood, it will be impossible to handle the boiler in a thoroughly safe and scientific fashion, though some engineers do handle boilers without knowing all about these attachments. Their ignorance in many cases costs them their lives and the lives of others.

Two-rod water gauge.

The first duty of the engineer is to see that the boiler is filled with water. This he usually does by looking at the glass water-gauge.

THE WATER GAUGE AND COCKS

There is a cock at each end of the glass tube. When these cocks are open the water will pass through the lower into the glass tube, while steam comes through the other. The level of the water in the gauge will then be the same as the level of the water in the boiler, and the water should never fall out of sight below the lower end of the glass, nor rise above the upper end.

Below the lower gauge cock there is another cock used for draining the gauge and blowing it off when

---

*Unless otherwise indicated, cuts of fittings show those manufactured by the Lunkenheimer Co., Cincinnati, Ohio.

there is a pressure of steam on. By occasionally opening this cock, allowing the heated water or steam to blow through it, the engineer may always be sure that the passages into the water gauge are not stopped up by any means. By closing the upper cock and opening the lower, the passage into the lower may be cleared by blowing off the drain cock; by closing the lower gauge cock and opening the upper the passage from the steam space may be cleared and tested in the same way when the drain cock is opened. If the glass breaks, both upper and lower gauge cocks should be closed instantly.

In addition to the glass water gauge, there are the try-cocks for ascertaining the level of the water in the boiler. There should be two to four of these. They open directly ut of the boiler sheet, and by opening them in turn it is possible to tell approximately where the water

Gauge or try cock.

stands. There should be one cock near the level of the crown sheet, or slightly above it, another about the level of the lower gauge cock, another about the middle of the gauge, another about the level of the upper gauge, and still another, perhaps, a little higher. But one above and one below the water line will be sufficient. If water stands above the level of the cock, it will blow off white mist when opened; if the cock opens from steam-space, it will blow off blue steam when opened.

The try-cocks should be opened from time to time in order to be sure the water stands at the

Try cock.

proper level in the boiler, for various things may interfere with the working of the glass gauge. Try-cocks are often called gauge cocks.

THE STEAM GAUGE

The steam gauge is a delicate instrument arranged so as to indicate by a pointer the pounds of pressure which the steam is exerting within the boiler. It is extremely important, and a defect in it may cause much damage.

The steam gauge was invented in 1849 by Eugene Bourdon, of France. He discovered that a flat tube bent in a simple curve, held fast at one end, would expand and contract if made of proper spring material, through the

Pressure gauge.

pressure of the water within the tube. The free end operates a clock-work that moves the pointer.

Steam gauge siphon.

It is important that the steam gauge be attached to the boiler by a siphon, or with a knot in the tube, so that the steam may operate on water contained in the tube, and the water cannot become displaced by steam, since steam might interfere with the correct working of the gauge by expanding the gauge tube through its excessive heat.

Steam gauges frequently get out of order, and should be tested occasionally. This may conveniently be done by attaching them to a boiler which has a correct gauge

already on it. If both register
alike, it is probable that both are
accurate.

Front cylinder cock.

There are also self-testing steam
gauges. With all pressure off, the
pointer will return to O. Then a
series of weights are arranged
which may be hung on the gauge and cause the pointer
to indicate corresponding numbers. The chief source of
variation is in the loosening of the indicator needle. This
shows itself usually when the pressure is off and the

pointer does not return exactly
to zero.

The safety valve is a valve
held in place by a weighted
lever* or by a spiral spring (on
traction engines) or some similar device, and is adjustable by
a screw or the like so that it
can be set
to blow off
at a given
pressure
of steam,
usually the
rated pres-
sure of the

Sectional view of kunkle
pop valve.

boiler, which on traction engines is
from 110 to 130 pounds. The valve
is supplied with a handle by which
it can be opened, and it should be
opened occasionally to make sure it

Safety valve.

is working all right. When it blows off the steam gauge should be noted to see that it agrees with the pressure for which the safety valve was set. If they do not agree, something is wrong; either the safety valve does not work freely, or the steam gauge does not register accurately.

The cut shows the Kunkle safety valve. To set it, unscrew the jam nut and apply the key to the pressure screw. For more pressure, screw down; for less, unscrew. After having the desired pressure, screw the jam nut down tight on the pressure screw. To regulate the opening and closing of the valve, take the pointed end of a file and apply it to the teeth of the regulator. If valve closes with too much boiler pressure, move the regulator to the left. If with too little, move the regulator to the right.

This can be done when the valve is at the point of blowing off.

Other types of valves are managed in a similar way, and exact directions will always be furnished by the manufacturers.

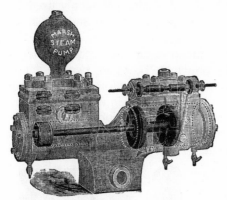

Phantom view of marsh independent steam pump.

---

*This kind of safety valve is now being entirely discarded as much more dangerous than the spring or pop valve.

FILLING THE BOILER WITH WATER

There are three ways in which a boiler is commonly filled with water.

First, before starting a boiler it must be filled with water by hand, or with a hand force-pump. There is usually a filler plug, which must be taken out, and a funnel can be attached in its place. Open one of the gauge cocks to let out the air as the water goes in.

When the boiler has a sufficient amount of water, as may be seen by the glass water gauge, replace the filler plug. After steam is up the boiler should be supplied with water by a pump or injector.

THE BOILER PUMP

There are two kinds of pumps commonly used on traction engines, the Independent pump, and the Crosshead pump.

The Independent pump is virtually an independent engine with pump attached. There are two cylinders, one receiving steam and conveying force to the piston; the other a water cylinder, in which a plunger works, drawing the water into itself by suction and forcing it out through the connection pipe into the boiler by force of steam pressure in the steam cylinder.

It is to be noted that all suction pumps receive their water by reason of the pressure of the atmosphere on the surface of the

Straight globe valve.

water in the supply tank or well. This atmospheric pressure is about 15 pounds to the square inch, and is sufficient to support a column of water 28 to 33 feet high, 33 feet being the height of a column of water which the atmosphere will support theoretically at about sea level. At greater altitudes the pressure of the atmosphere decreases. Pumps do not work very well when drawing water from a depth over 20 or 22 feet.

Water can be forced to almost any height by pressure of steam on the plunger, and it is taken from deep wells

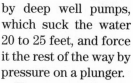

Angle globe valve.

by deep well pumps, which suck the water 20 to 25 feet, and force it the rest of the way by pressure on a plunger.

The amount of water pumped is regulated by a cock or globe valve in the suction pipe.

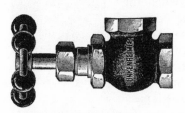

A Cross-head boiler pump is a pump attached to the cross-head of an engine. The force of the engine piston is transmitted to the plunger of the pump.

The pump portion works exactly the same, whether of the independent or cross-head kind.

The cut represents an independent pump that uses the exhaust steam to heat the water as it is pumped (Marsh pump).

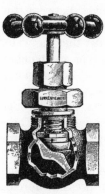

Valve with internal screw.

Every boiler feed-pump must have at least two check valves.

A check valve is a small swinging gate valve (usually) contained in a pipe, and so arranged that when water is flowing in one direction the valve will automatically open to let the water pass, while if water should be forced in the other direction, the valve will automatically close tight and prevent the water from passing.

There is one check valve in the supply pipe which conducts the water from the tank or well to the pump cylinder. When the plunger is drawn back or raised, a vacuum is created in the pump cylinder and the outside atmospheric pressure forces water through the supply pipe into the cylinder, and the check valve opens to let it pass.

Sectional view of swing check valve.

When the plunger returns, the check valve closes, and the water is forced into the feed-pipe to the boiler.

There are usually two check valves between the pump cylinder and the boiler, both swinging away from the pump or toward the boiler. In order that the water may flow steadily into the boiler there is an air chamber, which may be partly filled with water at each stroke of the plunger. As the water comes in, the air must be compressed, and as it expands it forces the water through the feed pipe into the boiler in a steady stream. There is one check valve between the pump cylinder and the air chamber, to prevent the water from coming back into the cylinder, and another between the air chamber and the boiler, to prevent the steam pressure forcing itself or the water from the boiler or water heater back into the air chamber.

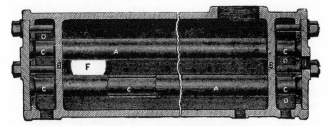

Sectional view of case heater.

All three of these check valves must work easily and fit tight if the pump is to be serviceable. They usually close with rubber facings which in time will

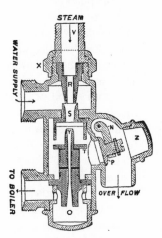

Sectional view of penberthy injector.

U.S. automatic injector.
(american injector co.)

get worn, and dirt is liable to work into the hinge and otherwise prevent tight and easy closing. They can always be opened for inspection, and new ones can be put in when the old are too much worn.

Only cold water can be pumped successfully, as steam from hot water will expand, and so prevent a vacuum being formed. Thus no suction will take place to draw the water from the supply source.

There should always be a globe valve or cock in the feed pipe near the boiler to make it possible to cut out the check valves when the boiler is under pressure. *It is never to be closed except* when required for this purpose.

Automatic injector.

Before passing into the boiler the water from the pump goes through the *heater*. This is a small cylinder, with a coil of pipe inside. The feed pipe from the pump is connected with one end of this inner coil of pipe, while the other end of the coil leads into the boiler itself. The exhaust steam from the engine cylinder is admitted into the cylinder and passes around the coil of pipe, afterwards coming out of the smoke stack to help increase the draft. As the feed water passes through this heater it becomes heated nearly to boiling before it enters the boiler, and has no tendency to cool the boiler off. Heating the feed water results in an economy of about 10 per cent.

The Injector is another means of forcing water from a supply tank or well into the boiler, and at the same

time heating it, by use of steam from the boiler. It is a
necessity when a cross-head pump is used, since such a
pump will not work when the engine is shut down. It is
useful in any case to heat the water before it goes into
the boiler when the engine is not working and there is
no exhaust steam for the heater.

There are various types of injectors, but they all work
on practically the same principle. The steam from the
boiler is led through a tapering nozzle to a small chamber into which there is an opening from a water supply
pipe. This steam nozzle throws out its spray with great
force and creates a partial vacuum in the chamber, causing the water to flow in. As the pressure of the steam
has been reduced when it passes into the injector, it
cannot, of course, force its way back into the boiler at
first, and finds an outlet at the overflow. When the water
comes in, however, the steam jet strikes the water and
is condensed by it. At the same time it carries the water
and the condensed steam along toward
the boiler with such force that the back
pressure of the boiler is overcome and
a stream of heated water is passed into
it. In order that the injector may work,
its parts must be nicely adjusted, and
with varying steam pressures it takes
some ingenuity to get it started. Usually
the full steam pressure is turned on and
the cock admitting the water supply is
opened a varying amount according to
the pressure.

Plain whistle.

First the valve between the check
valve and the boiler should be opened, so that the feed
water may enter freely; then open wide the valve next

the steam dome, and any other valve between the steam supply pipe and the injector; lastly open the water supply valve. If water appears at the overflow, close the supply valve and open it again, giving it just the proper amount of turn. The injector is regulated by the amount of water admitted.

In setting up an injector of any type, the following rules should be observed:

All connecting pipes as straight and short as possible.

The internal diameter of all connecting pipes should be the same or greater than the diameter of the hole in the corresponding part of the injector.

When there is dirt or particles of wood or other material in the source of water supply, the end of the water supply pipe should be provided with a strainer. Indeed, invariably a strainer should be used. The holes in this strainer must be as small as the smallest opening in the delivery tube, and the total area of the openings in the strainer must be much greater than the area of the water supply (cross-section).

The steam should be taken from the highest part of the dome, to avoid carrying any water from the boiler over with it. Wet steam cuts and grooves the steam nozzle. The steam should not be taken from the pipe leading to the engine unless the pipe is quite large.

Before using new injectors, after they are fitted to the boiler it is advisable to disconnect them and clean them out well by letting steam blow through them or forcing water through. This will prevent lead or loose scale getting into the injector when in use.

Set the injector as low as possible, as it works best with smallest possible lift.

Ejectors and jet pumps are used for lifting and forcing water by steam pressure, and are employed in filling tanks, etc.

BLAST AND BLOW-OFF DEVICES

In traction engines there is small pipe with a valve, leading into the smoke stack from the boiler. When the valve is opened, the steam allowed to blow off into the smoke stack will create a vacuum and so increase the draft. Blast or blow pipes are used only in starting the fire, and are of little value before the steam pressure reaches 15 pounds or so.

The exhaust nozzle from the engine cylinder also leads into the smoke stack, and when the engine is running the exhaust steam is sufficient to keep up the draft without using the blower.

Blow-off cocks are used for blowing sediment out of the bottom of a boiler, or blowing scum off the top of the water to prevent foaming. A boiler should never be blown out at high pressure, as there is great danger of injuring it. Better let the boiler cool off somewhat before blowing off.

Diamond spark arrester.

SPARK ARRESTER

Traction engines are supplied as a usual thing with spark arresters if they burn wood or straw. Coal sparks are heavy and have little life, and with some engines no spark arrester is needed. But there is great danger of setting a fire if an engine is run with wood or straw without the spark arrester.

Spark arresters are of different types. The most usual form is a large screen dome placed over the top of the stack. This screen must be kept well cleaned by brushing, or the draft of the engine will be impaired by it.

In another form of spark arrester, the smoke is made to pass through water, which effectually kills every possible spark.

The Diamond Spark Arrester does not interfere with the draft and is so constructed that all sparks are carried by a. counter current through a tube into a pail where water is kept. The inverted cone, as shown in cut, is made of steel wire cloth, which permits smoke and gas to escape, but no sparks. There is no possible chance to set fire to anything by sparks. It is adapted to any steam engine that exhausts into the smoke stack.

CHAPTER III

THE SIMPLE ENGINE

The engine is the part of a power plant which converts steam pressure into power in such form that it can do work. Properly speaking, the engine has nothing to do with generating steam. That is done exclusively in the boiler, which has already been described.

The steam engine was invented by James Watt, in England, between 1765 and 1790, and he understood

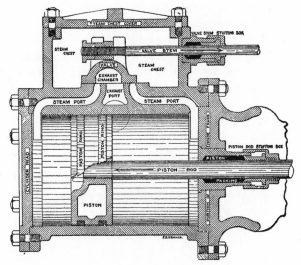

View of simple cylinder.
(J. I. Case Threshing Machine Co.)

all the essential parts of the engine as now built. It was improved, however, by Seguin, Ericsson, Stephenson, Fulton, and many others.

457

Let us first consider:

THE STEAM CYLINDER, ITS PARTS AND CONNECTIONS

The cylinder proper is constructed of a single piece of cast iron bored out smooth.

The *cylinder heads* are the flat discs or caps bolted to the ends of the cylinder itself. Sometimes one cylinder head is cast in the same piece with the engine frame.

The *piston* is a circular disc working back and forth in the cylinder. It is usually a hollow casting, and to make it fit the cylinder steam tight, it is supplied on its circumference with *piston rings*. These are made of slightly larger diameter than the piston, and serve as springs against the sides of the cylinder. The *follower plate* and bolts cover the piston rings on the piston head and hold them in place.

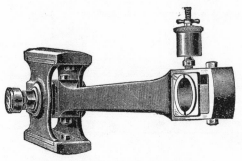

Connecting rod and cross-head.
(J. I. Case Threshing Machine Co.)

The *piston rod* is of wrought iron or steel, and is fitted firmly and rigidly into the piston at one end. It runs from the piston through one head of the cylinder, passing through a steam-tight "stuffing box." One end of the piston rod is attached to the cross-head.

The *cross-head* works between *guides*, and has *shoes* above and below. It is practically a joint, necessary in converting straight back and forth motion into rotary. The cross-head itself works straight back and forth, just as the piston does, which is fastened firmly to one end. At the other end is attached the *connecting rod*, which works on a bearing in the cross-head, called the *wrist pin*, or cross-head pin.

The *connecting rod* is wrought iron or steel, working at one end on the bearing known as the wrist pin, and on the other on a bearing called the *crank pin*.

The *crank* is a short lever which transmits the power from the connecting rod to the *crank shaft*. It may also be a disc, called the *crank disc*.

Let us now return to the steam cylinder itself.

The steam leaves the boiler through a pipe leading from the top of the steam dome, and is let on or cut off by the *throttle* valve, which is usually opened and closed by some sort of lever handle. It passes on to the

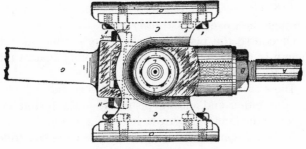

Cross-head.
(J. I. Case Threshing Machine Co.)

Steam-chest, usually a part of the same casting as the cylinder. It has a cover called the *steam-chest cover,* which is securely bolted in place.

The *steam valve,* usually spoken of simply as the *valve,* serves to admit the steam alternately to each end of the cylinder in such a manner that it works the piston back and forth.

There are many kinds of valves, the simplest (shown in the diagram) being the D-valve. It slides back and forth on the bottom of the steam-chest, which is called the *valve seat,* and alternately opens and closes the two *steam ports,* which are long, narrow passages through which the steam enters the cylinder, first through one port to one end, then through the other port to the other end. The exhaust steam also passes out at these same ports.

The *exhaust chamber in* the type of engine now under consideration is an opening on the lower side of the valve, and is always open into the *exhaust port,* which connects with the exhaust pipe, which finally discharges itself through the *exhaust nozzle* into the smoke stack of a locomotive or traction engine, or in other types of engines, into the *condenser.*

The valve is worked by the *valve stem,* which works through the valve stem *stuffing box.*

Of course the piston does not work quite the full length of the cylinder, else it would pound against the cylinder heads.

The *clearance* is the distance between the cylinder head at either end and the piston when the piston has reached the limit of its stroke in that direction.

In most engines the valve is so set that it opens a trifle just before the piston reaches the limit of its movement in either direction, thus letting some steam in before the piston is ready to move back. This opening, which usually amounts to 1/32 to 3/16 of an inch, is called the *lead*. The steam thus let in before the piston reaches the limit of its stroke forms *cushion*, and helps the piston to reverse its motion without any jar, in an easy and silent manner. Of course the cushion must be as slight as possible and serve its purpose, else it will tend to stop the engine, and result in loss of energy. Some engines have no lead.

Setting a valve is adjusting it on its seat so that the lead will be equal at both ends and sufficient for the needs of the engine. By shortening the movement of the valve back and forth, the lead can be increased or diminished. This is usually effected by changing the eccentric or valve gear.

The *lap* of a slide valve is the distance it extends over the edges of the ports when it is at the middle of its travel

Lap on the steam side is called outside lap; lap on the exhaust side is called inside lap. The object of lap is to secure the benefit of working steam expansively. Having lap, the valve closes one steam port before the other is opened, and before the piston has reached the end of its stroke; also of course before the exhaust is opened. Thus for a short time the steam that has been let into the cylinder to drive the piston is shut up with neither inlet nor outlet, and it drives the piston by its own expansive, force. When it passes out at the exhaust it has a considerably reduced pressure, and less of its force is wasted.

Let us now consider the

VALVE GEAR

The mechanism by which the valve is opened and closed is somewhat complicated, as various things are accomplished by it besides simply opening and closing the valve. If an engine has a *reverse lever*, it works through the valve gear; and the *governor* which regulates the speed of the engine may also operate through the valve gear. It is therefore very important.

The simplest valve gear depends for its action on a fixed eccentric.

An *eccentric* consists of a central disc called the *sheave*, keyed to the main shaft at a point to one side of its true center, and a grooved ring or *strap* surrounding it and sliding loosely around it. The strap is usually made of brass or some anti-friction metal. It is in two parts, which are bolted together so that they can be tightened up as the strap wears.

The *eccentric rod* is either bolted to the strap or forms a single piece with it, and this rod transmits its motion to the valve.

It will be seen, therefore, that the eccentric is nothing more than a sort of disc crank, which, however, does not need to be attached to the end of a shaft in the manner of an ordinary crank.

The distance between the center of the eccentric sheave and the center of the shaft is called the *throw* of the eccentric or the *eccentricity*.

The eccentric usually conveys its force through a connecting rod to the valve stem, which moves the valve.

The first modification of the simple eccentric valve gear is

THE REVERSING GEAR

It is very desirable to control the movement of the steam valve, so that if desired the engine may be run in the opposite direction; or the steam force may be brought to bear to stop the engine quickly; or the travel of the valve regulated so that it will let into the cylinder only as much steam as is needed to run the engine when the load is light and the steam pressure in the boiler high.

There is a great variety of reversing gears; but we will consider one of the commonest and simplest first.

Huber single eccentric reverse.

If the eccentric sheave could be slipped around on the shaft to a position opposite to that in which it was keyed to shaft in its ordinary motion, the motion of the valve would be reversed, and it would let steam in front of the advancing end of the piston, which would check its movement, and start it in the opposite direction.

The *link gear*, invented by Stephenson, accomplishes this in a natural and easy manner. There are two eccentrics placed just opposite to each other on the crank shaft, their connecting rods terminating in what is called a *link*, through which motion is communicated to the valve stem. The link is a curved slide, one eccentric being connected to one end, the other eccentric to the other end, and the *link-block*, through which motion is conveyed to the valve, slides freely from one end to the other. Lower the link so that the block is opposite the end of the first rod, and the valve will be moved by the corresponding eccentric; raise the link, so that the block is opposite the end of the other rod, and the valve will be moved by the other eccentric. In the middle there would be a dead center, and if the block stopped here, the valve would not move at all. At any intermediate point, the travel of the valve would be correspondingly shortened.

Valve and link reverse.

Such is he theoretical effect of a perfect link; but the dead center is not absolute, and the motion of the link is varied by the point at which the rod is attached which lifts and lowers it, and also by the length of this rod. In full gear the block is not allowed to come quite to the end of the link, and this surplus distance is called the *clearance*. The *radius* of a link is the distance from the center of the driving shaft to the center of the link, and the curve of the link is that of a circle with that radius. The length of the radius may vary considerably, but the point of suspension is important. If a link is suspended by its center, it will certainly cut off steam sooner in the front stroke than in the back. Usually it is suspended from that point which is most used in running the engine.

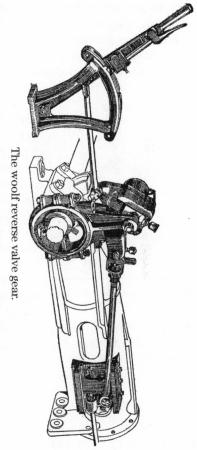

The woolf reverse valve gear.

The *Woolf reversing gear* employs but one eccentric, to the strap of which is cast an arm having a block pivoted at its end. This block slides in a pivoted guide, the angle of which is controlled by the reverse lever. To the eccentric arm is attached the eccentric rod, which transmits the motion to the valve rod through a rocker arm on simple engines and through a slide, as shown in cut, on compound engines.

The Meyer valve gear does not actually reverse an engine, but controls the admission of steam by means

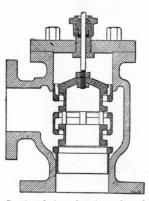

Sectional view showing valve of waters governor.

of an additional valve riding on the back of the main valve and controlling the cut-off. The main valve is like an ordinary D-valve, except that the steam is not admitted around the ends, but through ports running through the valve, these ports being partially opened or closed by the motion of the riding valve, which is controlled by a separate eccentric. If this riding valve is connected with a governor, it will regulate the speed of an engine; and by the addition of a link the gear may be made reversible. The chief objection to it is the excessive friction of the valves on their seats.

GOVERNORS

A governor is a mechanism by which the supply of steam to the cylinder is regulated by revolving balls, or the like, which runs faster or slower as the speed of the engine increases or diminishes. Thus the speed of an engine is regulated to varying loads and conditions.

The simplest type of governor, and the one commonly used on traction engines, is that which is only a modification of the one invented by Watt. Two balls revolve around a spindle in such a way as to rise when the speed of the engine is high, and fall when it is low, and in rising and falling they open and close a valve similar to the throttle valve. The amount that the governor valve is opened or closed by the rise and fall of the governor balls is usually regulated by a thumb screw at the top or side, or by what is called a handle nut, which is usually held firm by a check nut directly over it, which should be screwed firm against the handle nut. Motion is conveyed to the governor balls by a belt and a band wheel working on a mechanism of metred cogs.

Pickering horizontal governor.

There is considerable friction about a governor of this type and much energy is wasted in keeping it going. The valve stem or spindle passes through a steam-tight stuffing box, where it is liable to stick if the packing is too tight; and if this stuffing box leaks steam, there will be immediate loss of power.

Such a governor as has just been described is called a throttle valve governor. On high grade engines the difficulties inherent in this type of governor are overcome by making the governor control, not a valve in the steam supply pipe, but the admission of steam to the steam cylinder through the steam valve and its gear. Such engines are described as having an "automatic cut-off." Sometimes the governor is attached to the link, sometimes to a separate valve, as in the Meyer gear already described. Usually the governor is attached to the fly-wheel, and consequently governors of this type are called fly-wheel governors. An automatic cut-off governor is from 15 per cent to 20 per cent more effective than a throttle valve governor.

CRANK, SHAFT AND JOURNALS

We have already seen how the piston conveys its power through the piston rod, the cross-head, and the connecting rod to the crank pin and crank, and hence to the shaft.

The key, gib, and strap are the effective means by which the connecting rod is attached, first to the wrist pin in the cross-head, and secondly to the crank pin on the crank.

The *strap* is usually made of two or three pieces of wrought iron or steel bolted together *so* as to hold the *brasses*, which are in two parts and loosely surround the pin. The brasses do not quite meet, and as they wear may be tightened up. This is effected by the *gib*, back of which is the *key*, which is commonly a wedge which may be driven in, or a screw, which presses on the back of the gib, which in turn forces together the brasses; and thus the length of the piston gear is kept uniform in spite of the wear, becoming neither shorter nor longer. When the brasses are so worn that they have been forced together, they must be taken out and filed equally on all

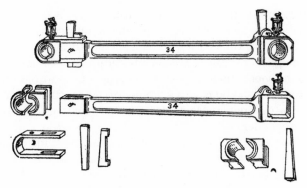

Connecting rod and boxes.
(A. W. Stevens Co.)

four of the meeting ends, and shims, or thin pieces of sheet iron or the like placed back of them to equalize the wear, and prevent the piston gear from being shortened or otherwise altered.

The *crank* is a simple lever attached to the shaft by which the shaft is rotated. There are two types of crank in common use, the side crank, which works by what is virtually a bend in the shaft. There is also what is called the disc crank, a variation of the side crank, in which the power is applied to the circumference of a disc instead of to the end of a lever arm.

The *boss* of a crank is that part which surrounds the shaft and butts against the main bearing, and is usually about twice the diameter of the crank shaft journal. The *web* of the crank is the portion between the shaft and the pin.

To secure noiseless running, the crank pin should be turned with great exactness, and should be set exactly parallel with the direction of the shaft. When the pressure on the pin or any bearing is over 800 pounds per square inch, oil is no longer able to lubricate it properly. Hence the bearing surface should always be large enough to prevent a greater pressure than 800 pounds to the square inch. To secure the proper proportions the crank pin should have a diameter of one-fourth the bore of the cylinder, and its length should be one-third that of the cylinder.

The *shaft* is made of wrought iron or steel, and must not only be able to withstand the twisting motion of the crank, but the bending force of the engine stroke. To prevent bending, the shaft should have a bearing as near the crank as possible.

The *journals* are those portions of the shaft which work in bearings. The main bearings are also called *pedestals*, *pillow blocks*, and *journal boxes*. They usually consist of boxes made of brass or some other anti-friction material carried in iron pedestals. The pillow blocks are usually adjustable.

THE FLY-WHEEL

This is a heavy wheel attached to the shaft. Its object is to regulate the variable action of the piston, and to make the motion uniform even when the load is variable. By its inertia it stores energy, which would keep the engine running for some time after the piston ceased to apply any force or power.

LUBRICATORS

All bearings must be steadily and effectively lubricated, in order to remove friction as far as possible, or the working power of the engine will be greatly reduced.

Besides, without complete and effective lubrication, the bearings will "cut," or wear in irregular grooves, etc., quickly ruining the engine.

Bearings are lubricated through automatic lubricator cups, which hold oil or grease and discharge it uniformly upon the bearing through a suitable hole.

A sight feed ordinary cup permits the drops of oil to be seen as they pass downward through a glass tube, and also the engineer may see how much oil there is in the cup. Such a cup is suitable for all parts of an engine

DESCRIPTION.

C 1—Body or Oil Reservoir.
C 3—Filler Plug.
C 4—Water Valve.
C 5—Plug for inserting Sight-Feed Glass.
C 6—Sight-Feed Drain Stem.
C 7—Regulating Valve.
C 8—Drain Valve.
C 9—Steam Valve.
C 10—Union Nut.
C 11—Tail Piece.
　 H—Sight-Feed Glass.

The "detroit" zero double connection lubricator.

except the crank pin, cross-head, and, of course, the cylinder.

The crank pin oiler is an oil cup so arranged as to force oil into the bearing only when the engine is working, and more rapidly as the engine works more rapidly. In one form, which uses liquid oil, the oil stands below a disc; from which is the opening through the shank to the bearing. As the engine speeds up, the centrifugal force tends to force the oil to the top of the cup and so on to the bearing, and the higher the speed the greater the amount of oil thrown into the crank pin.

Hard oil or grease has of late been coming into extensive use. It is placed in a compression cup, at the top of which a disc is pressed down by a spring, and also by some kind of a screw. From time to time the screw

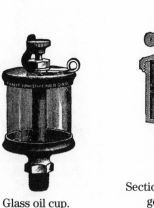

Glass oil cup.

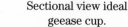

Sectional view ideal geease cup.

is tightened up by hand, and the spring automatically forces down the grease.

The Cylinder Lubricator is constructed on a different principle, and uses an entirely different kind of oil, called "cylinder oil." A sight-feed automatic oiler is so arranged that the oil passes through water drop by drop, so that each drop can be seen behind glass before it passes into the steam pipe leading from the boiler to the cylinder. The oil mingles with the steam and passes into the steam chest, and thence into the cylinder, lubricating the valve and piston.

The discharge of the oil may not only be watched, but regulated, and some judgment is necessary to make sure that enough oil is passing into the cylinder to prevent it from cutting.

Acorn oil pump.

The oil is forced into the steam by the weight of the column of water, since the steam pressure is the same at both ends. There is a small cock by which this water of condensation may be drained off when the engine is shut down in cold weather. Oilers are also injured by straining from heating caused by the steam acting on cold oil when all the cocks are closed. There is a relief cock to prevent this strain, and it should be slightly opened, except when oiler is being filled.

There are a number of different types of oilers, with their cocks arranged in different ways; but the manufacturer always gives diagrams and instructions fully explaining the working of the oiler. Oil pumps serving the same purpose are now often used.

DIFFERENTIAL GEAR

The gearing by which the traction wheels of a traction engine are made to drive the engine is an important item. Of course, it is desirable to apply the power of the engine to both traction wheels; yet if both hind wheels were geared stiff, the engine could not turn from a straight line, since in turning one wheel must move faster than the other. The differential or compensating gear is a device to leave both wheels free to move one ahead of the other if occasion requires. The principle is much the same as in case of a ratchet on a geared wheel, if power were applied to the ratchet to make the wheel turn; if for any reason the wheel had a tendency of its own to turn faster than the ratchet forced it, it would be free to do so. When corners are turned the power is applied to one wheel only, and the other wheel is permitted to move faster or slower than the wheel to which the gearing applies the power.

There are several forms of differential gears, differing largely as to combination of spur or bevel cogs. One of the best known uses four little beveled pinions, which are placed in the main driving wheel as shown in the cut. Beveled cogs work into these on either side of the main wheel. If one traction wheel moves faster than the other these pinions move around and adjust the gears on either side.

FRICTION CLUTCH

The power of an engine is usually applied to the traction wheel by a friction clutch working on the inside of the fly-wheel. (See plan of Frick Engine.) The traction wheels are the two large, broad-rimmed hind wheels, and are provided with projections to give them a firm footing on the road. Traction engines are also provided with mud shoes and wheel cleaning devices for mud and snow.

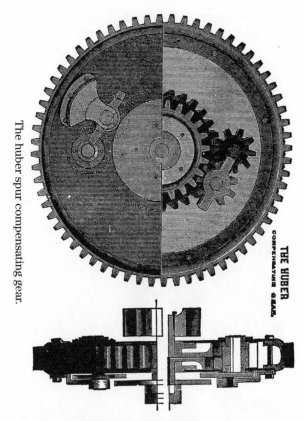

The huber spur compensating gear.

THE HUBER COMPENSATING GEAR.

THE FUSIBLE PLUG

The fusible plug is a simple screw plug, the center of which is bored out and subsequently filled with some other metal that will melt at a lower temperature than steel or iron. This plug is placed in the crown sheet of a locomotive boiler as a precaution for safety. Should the crown sheet become free of water when the fire is very hot, the soft metal in the fusible plug would melt and run out, and the overheated steam would escape into the firebox, putting out the fire and giving the boiler relief so that an explosion would be avoided. In some states a fusible plug is required by law, and one is found in nearly every boiler which has a crown sheet. Return flue

Aultman & Taylor bevel compensating gear.

Differential gear, showing cushion springs and bevel
pinion.

boilers and others which do not have crown sheets (as
for example the vertical) do not have fusible plugs. To
be of value a fusible plug should be renewed or changed
once a month.

STUFFING BOXES

Any arrangement to make a steam-tight joint about a moving rod, such as a piston rod or steam valve rod, would be called a stuffing box. Usually the stuffing box gives free play to a piston rod or valve rod, without allowing any steam to escape. A stuffing box is also used on a pump piston sometimes, or a compressed air piston. In all these cases it consists of an annular space around the moving rod which can be partly filled by some pliable elastic material such as hemp, cotton, rubber, or the like; and this filling is held in place and made tighter or looser by what is called a gland, which is forced into the partly filled box by screwing up a cap on the outside of the cylinder. Stuffing boxes must be repacked occasionally, since the packing material will get hard and dead, and will either leak steam or cut the rod.

CYLINDER COCKS

These cocks are for the purpose of drawing the water formed by condensation of steam out of the cylinder. They should be opened whenever the engine is stopped or started, and should be left open when the engine is shut down, especially in cold weather to prevent freezing of water and consequent damage. Attention to these cocks is very important.

These are small cocks arranged about the pump and at other places for the purpose of testing the inside action. By them it is possible to see if the pump is working properly, etc.

STEAM INDICATOR

The steam indicator is an instrument that can be attached to either end of a steam cylinder, and will indicate the character of the steam pressure during the entire stroke of the piston. It shows clearly whether the lead is right, how much cushion there is, etc. It is very important in studying the economical use and distribution of steam, expansive force of steam, etc.

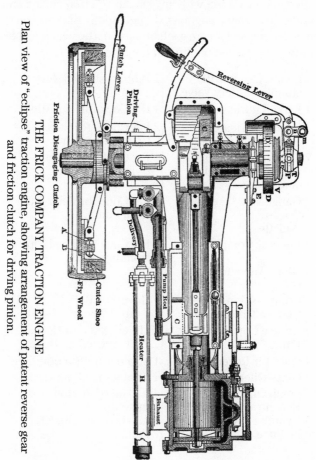

THE FRICK COMPANY TRACTION ENGINE

Plan view of "eclipse" traction engine, showing arrangement of patent reverse gear and friction clutch for driving pinion.

LIST OF ATTACHMENTS FOR TRACTION ENGINE
AND BOILER

The following list of brasses, etc., which are packed with the Case traction engine will be useful for reference in connection with any similar traction engine and boiler. The young engineer should rapidly run over every new engine and locate each of these parts, which will be differently placed on different engines:

1 Steam Gauge with siphon. I Safety Valve.

1 Large Lubricator

1 Small Lubricator for Pump.

7. Glass Water Gauge complete with glass and rods.

2 Gauge Cocks.

1 Whistle.

1 Injector Complete.

1 Globe Valve for Blow-off.

1 Compression Grease Cup for Cross Head.

1 Grease Cup for Crank Pin.

1 Oiler for Reverse Block.

1 Glass Oiler for Guides.

1 Small Oiler for Eccentric Rod.

1 Cylinder Cock (1 is left in place.

1 Bibb Nose Cock for Pump.

1 Pet Cock for Throttle.

2 Pet Cocks for Steam Cylinder of Pump.

1 Pet Cock for Water Cylinder of Pump.

1 Pet Cock for Feed Pipe from Pump.

1 Pet Cock for Feed Pipe from Injector.

1 Governor Belt.

1 Flue Cleaner.

15 ft. lin. Suction Hose.

5 ft. Sprinkling Hose.

1 Strainer for Suction Hose.

1 Strainer for Funnel.

4 ft. 6 in. of in. Hose for Injector.

5 ft. 6 in. of in. Hose for Pump.

2 Stop Cocks to drain Heater.

1 Stop Cock for Hose Coupling on Pump.

2 Nipples 3/4 x 2 1/2 in. for Hose.

2 3/4 in. Hose Clamps.

2 1/2 in. Hose Strainers.

TEST QUESTIONS ON BOILER AND ENGINE

Q. How is the modern stationary fire-flue boiler arranged?

Q. How does the locomotive type of boiler differ?

Q. What is a return flue boiler?

Q. What is a water-tube boiler and how does it differ from a fire-flue tubular boiler?

Q. What is a vertical boiler and what are its advantages?

Q. What is the shell?

Q. What are the boiler heads?

Q. What are the tube sheets?

Q. What is the firebox?

Q. What is the water leg?

Q. What is the crown-sheet?

Q. Where is the smoke-box located?

Q. What is the steam dome intended for?

Q. What is the mud-drum for?

Q. What are man-holes and hand-holes for?

Q. What is a boiler jacket?

Q. What is a steam jacket?

Q. Where is the ash-pit?

Q. What are dead-plates?

Q. How is grate surface measured?

Q. What is forced draft?

Q. How is heating surface measured?

Q. What is steam space?

Q. What is water space?

Q. What is a diaphragm plate?

Q. What is the first duty of an engineer in taking charge of a new boiler?

Q. What are the water gauge and try cocks for, and how are they placed?

Q. What is the steam gauge and how may it be tested?

Q. What is a safety valve? Should it be touched by the engineer? How may he test it with the steam gauge?

Q. How is a boiler first filled with water?

Q. How is it filled when under pressure?

Q. What is an independent pump? What is a cross-head pump?

Q. What is a check valve, and what is its use, and where located?

Q. What is a heater and how does it work?

Q. What is an injector, and what is the principle of its operation?

Q. Where are the blow-off cocks located? How should they be used?

Q. In what cases should spark arrester be used?

Q. Who invented the steam engine, and when?

Q. What are the essential parts of a steam engine?

Q. What is the cylinder, and how is it used?

Q. What is the piston, and how does it work? The piston-rings?

Q. What is the piston rod and how must it be fastened?

Q. What is the crosshead, and how does it move? What are guides or ways? Shoes?

Q. What is the connecting rod? Wrist pin? Crank pin?

Q. What is the crank? Crank shaft?

Q. Where is the throttle valve located, and what does opening and closing it do?

Q. What is the steam chest for, and where is it placed?

Q. What is a steam valve? Valve seats? Ports?

Q. What is the exhaust? Exhaust chamber? Exhaust port? Exhaust nozzle? What is a condenser?

Q. How is the valve worked, and what duties does it perform, and how?

Q. What is clearance?

Q. What is lead?

Q. What is cushion?

Q. How would you set a valve? What is lap?

Q. How is a steam valve moved back and forth in its seat?

Q. How may an engine be reversed?

Q. What is a governor, and how does it work?

Q. What is an eccentric? Eccentric sheave? Strap? Rod?

Q. What is the throw of an eccentric?

Q. How does the link reversing gear work?

Q. How does the Woolf reverse gear work?

Q. How does the Meyer valve gear work? Will it reverse an engine?

Q. What are the chief difficulties in the working of a governor?

Q. What are key, gib, and strap? Brasses?

Q. What is the boss of a crank? Web?

Q. How may noiseless running of a crank be secured?

Q What are journals? Pedestals? Pillow blocks? Journal boxes?

Q. What is the object in having a fly wheel?

Q. What different kinds of lubricators are there? Where may hard oil or grease be used? Is the oil used for lubricating the cylinder the same as that used for rest of engine?

Q. How does a cylinder lubricator work?

Q. What is differential gear, and what is it for?

Q. What is the use of a fusible plug, and how is it arranged?

Q. What are stuffing-boxes, and how are they constructed?

Q. What are cylinder cocks, and what are they used for?

Q. What are pet cocks?

Q. What is a steam indicator?

CHAPTER IV

HOW TO MANAGE A TRACTION ENGINE BOILER

We will suppose that the young engineer fully understands all parts of the boiler and engine, as explained in the preceding chapters. It is well to run over the questions several times, to make sure that every point has been fully covered and is well understood.

We will suppose that you have an engine in good running order. If you have a new engine and it starts off nice and easy (the lone engine without load) with twenty pounds steam pressure in the boiler, you may make up your mind that you have a good engine to handle and one that will give but little trouble. But if it requires fifty or sixty pounds to start it, you want to keep your eyes open, for something is tight. But don't begin taking the engine to pieces, for you might get more pieces than you know what to do with. Oil every bearing fully, and then start your engine and let it run for a while. Then notice whether you find anything getting warm. If you do, stop and loosen up a very little and start again. If the heating still continues, loosen again as before. But remember, loosen but little at a time, for a box or journal will heat from being too loose as quickly as from being too tight, and if you have found a warm box, don't let that box take all your attention, but keep your eye on the other bearings.

In the case of a new engine, the cylinder rings may be a little tight, and so more steam pressure will be required to start the engine; but this is no fault, for in a day or two they will be working all right if kept well oiled.

In starting a new engine trouble sometimes comes from the presence of a coal cinder in some of the boxes, which has worked in during shipment. Before starting a new engine, the boxes and oil holes should therefore be thoroughly cleaned out. For this purpose the engineer should always have some cotton waste or an oiled rag ready for constant use.

A new engine should be run slowly and carefully until it is found to be in perfect running order.

If you are beginning on an old engine in good running order, the above instructions will not be needed; but it is well to take note of them.

Now if your engine is all right, you may run the pressure up to the point of blowing off, which is 100 to 130 pounds, at which most safety valves are set at the factory. It is not uncommon for a new pop to stick, and as the steam runs up it is well to try it by pulling the relief lever. If on letting it go it stops the escaping steam at once, it is all right. If, however, the steam continues to escape the valve sticks in the chamber. Usually a slight tap with a wrench or hammer will stop it at once; but don't get excited if the steam continues to escape. As long as you have plenty of water in the boiler, and know that you have it, you are all right.

STARTING UP A BOILER

Almost the only danger from explosion of a boiler is from not having sufficient water in the boiler. The boiler is filled in the first place, as has already been explained, by hand through a funnel at the filler plug, or by a force pump. The water should stand an inch and a half in the glass of the water gauge before the fire is started. It should be heated up slowly so as not to strain the boiler or connections. When the steam pressure as shown by the steam gauge is ten or fifteen pounds, the blower may be used to increase the draft.

If you let the water get above the top of the glass, you are liable to knock out a cylinder head; and if you let the water get below the bottom of the glass, you are likely to explode your boiler.

The glass gauge is not to be depended upon, however, for a number of things may happen to interfere with its working. Some one may inadvertently turn off the gauge cocks, and though the water stands at the proper height in the glass, the water in the boiler will be very different.

A properly made boiler is supplied with two to four try-cocks, one below the proper water line, and one above it. If there are more than two they will be distributed at suitable points between.

When the boiler *is* under pressure, turn on the lower try-cock and you should get water. You will know it because it will appear as white mist. Then try the upper try-cock, and you will get steam, which will appear blue.

NEVER FAIL TO USE THE TRY-COCKS FREQUENTLY. This is necessary not only because you never know when the glass is deceiving you; but if you fail to use them they will get stopped up with lime or mud, and when you need to use them they will not work.

In order also to keep the water gauge in proper condition, it should be frequently blown out in the following manner: Shut off the top gauge cock and open the drain cock at the bottom of the gauge. This allows the water and steam to blow through the lower cock of the water gauge, and you know that it is open. Any lime or mud that has begun to accumulate will also be carried off.

After allowing the steam to escape a few seconds, shut off the lower gauge cock, and open the upper one, and allow it to blow off about the same time. Then shut the drain cock and open both gauge cocks, when you will see the water seek its level, and you can feel assured that it is reliable and in good working condition. This little operation you should perform, every day you run your engine. If you do you will not *think* you have sufficient water in the boiler, but will *know*. The engineer who always *knows* he has water in the boiler will not be likely to have an explosion. Especially should you never start your fire in the morning simply because you see water in the gauge. You should *know* that there is water in the boiler.

Now if your pump and boiler are in good working condition, and you leave the globe valve in the supply pipe to the pump open, with the hose in the tank, you will probably come to your engine in the morning and find the boiler nearly full of water, and you will think some one has been tampering with the engine. The truth is, however, that as the steam condensed, a vacuum was formed, and the water flowed in on account of atmospheric pressure, just as it flows into a suction pump when the plunger rises and creates a vacuum in the pump. Check valves are arranged to prevent anything passing out of the boiler, but there is nothing to prevent water passing in.

The only other cause of an explosion, beside poor material in the manufacture of the boiler, is too high steam pressure, due to a defective safety valve or imperfect team gauge. The steam gauge is likely to get out of order in a number of ways, and so is the safety valve. To make sure that both are all right, the one should frequently be tested by the other. The lever of the safety valve should frequently be tried from time to time, to make sure the valve opens and closes easily, and whenever the safety valve blows off, the steam gauge should be noted to see if it indicates the pressure at which the safety has been set.

WHEN YOUR ENGINE IS ALL RIGHT, LET IT ALONE

Some engineers are always loosening a nut here, tightning up a box there, adjusting this, altering that. When an engine is all right they keep at it till it is all wrong. As a result they are in trouble most of the time. When an engine is running all right, LET IT ALONE. Don't think you are not earning your salary because you are nerely sitting still and looking on. If you must be at work, keep at it with an oily rag, cleaning and polishing up. That is the way to find out if anything is really the matter. As the practised hand of the skilled engineer goes over an engine, his ears wide open for any peculiarity of sound, anything that is not as it should be will make itself decidedly apparent. On the other hand, an engineer who does not keep his engine clean and bright by constantly passing his hand over it with an oily rag, is certain to overlook something, which perhaps in the end will cost the owner a good many dollars to put right.

Says an old engineer* we know, "When I see an engineer watching his engine closely while running, I am most certain to see another commendable feature in a

---

*J. H. Maggard, author of "Rough and Tumble Engineering," to whom we are indebted for a number of valuable suggestions in this chapter.

good engineer, and that is, when he stops his engine he will pick up a greasy rag and go over his engine carefully, wiping every working part, watching or looking carefully at every point that he touches. If a nut is working loose, he finds it; if a bearing is hot, he finds it; if any part of his engine has been cutting, he finds it. He picks up a greasy rag instead of a wrench, for the engineer that understands his business and attends to it never picks up a wrench unless he has something to do with it."

This same engineer goes on with some more most excellent advice. Says he:

"Now, if your engine runs irregularly, that is, if it runs up to a higher speed than you want, and then runs down, you are likely to say at once, 'Oh, I know what the trouble is, it is the governor'. Well, suppose it is. What are you going to do about it? Are you going to shut down at once and go to tinkering with it? No, don't do that. Stay close to the throttle valve and watch the governor closely. Keep your eye on the governor stem, and when the engine starts off on one of its speed tilts, you will see the stem go down through the stuffing box and then stop and stick in one place until the engine slows down below its regular speed, and it then lets loose and goes up quickly and your engine lopes off again. You have now located the trouble. It is in the stuffing box around the little brass rod or governor stem. The packing has become dry and by loosening it up and applying oil you may remedy the trouble until such time as you can repack it with fresh packing. Candle wick is as good for this purpose as anything you can use.

"But if the governor does not act as I have described, and the stem seems to be perfectly free and easy in the box, and the governor still acts queerly, starting off

and running fast for a few seconds and then suddenly
concluding to take it easy and away goes the engine
again, see if the governor belt is all right, and if it is
it would be well for you to stop and see if a wheel is
not loose. It might be either the little belt wheel or one
of the little cog wheels. If you find these are all right,
examine the spool on the crank shaft from which the
governor is run, and you will probably find it loose. If
the engine has been run for any length of time, you will
always find the trouble in one of these places; but if it
is a new one, the governor valve might work a little
tight in the valve chamber, and you may have to take it
out and use a little emery paper to take off the rough
projections on the valve. Never use a file on this valve
if you can get emery paper, and I should advise you
always to have some of it with you. It will often come
handy."

This is good advice in regard to any trouble you may
have with an engine. Watch the affected part closely;
think the matter over carefully, and see if you cannot
locate the difficulty before you even stop your engine.
If you find the trouble and know that you have found
it, you will soon be able to correct the defect, and no
time will be lost. At the same time you will not ruin your
engine by trying all sorts of remedies at random in the
thought that you may ultimately hit the right thing. The
chances are that before you do hit the right point, you
will have put half a dozen other matters wrong, and it
will take half a day to get the matter right again.

As there are many different types of governors in
use, it would be impossible to give exact directions for
regulating that would apply to them all; but the following suggestions applying to the Waters governor (one

widely used on threshing engines) will give a general idea of the method for all:

There are two little brass nuts on the top of the stem of the governor, one a thumb nut and the other a loose jam nut. To increase the speed, loosen the jam nut and then turn the thumb nut back slowly, watching the motion of the engine all the time. When the required speed has been obtained, then tighten up as snug as you can with your fingers (not using a wrench). To decrease the speed, loosen the jam nut as before, running it up a few turns, and then turn down the thumb nut till the speed meets your requirements, when the thumb nut is made fast as before. In any case, be very careful not to press down on the stem when turning the thumb nut, as this will make the engine run a little slower than will be the case when your hand has been removed.

If your engine does not start with an open throttle, look to see if the governor stem has not been screwed down tight. This is usually the case with a new engine, which has been screwed down for safety in transportation.

WATER FOR THE BOILER

There is nothing that needs such constant watching and is likely to cause so much trouble if it is not cared for, as the supply of water. Hard well water will coat the inside of the boiler with lime and soon reduce its steaming power in a serious degree, to say nothing of stopping up pipes, cocks, etc. At the same time, rain water that is perfectly pure (theoretically) will be found to have a little acid or alkali in it that will eat through the iron or steel and do equal damage.

However, an engineer must use what water he can. He cannot have it made to order for him, but he must take it from well, from brook, or cistern, or roadside ditch, as circumstances may require. The problem for the engineer is not to get the best water, but to make the best use of whatever water he can get, always, of course, choosing the best and purest when there is such a thing as choosing.

In the first place, all supply pipes in water that is muddy or likely to have sticks, leaves, or the like in it, should be furnished with strainers. If sticks or leaves get into the valve, the expense in time and worry to get them out will be ten times the cost of a strainer.

If the water is rain water, and the boiler is a new one, it would be well to put in a little lime to give the iron a slight coating that will protect it from any acid or alkali corrosion.

If the water is hard, some compound or sal ammonia should be used. No specific directions can be given, since water is made hard by having different substances dissolved in it, and the right compound or chemical is that which is adapted to the particular substance you are to counteract. An old engineer says his advice is to use no compound at all, but to put a hatful of potatoes in the boiler every morning.

Occasionally using rain water for a day or two previous to cleaning is one of the best things in the world to remove and throw down all scale. It beats compounds at every point. It is nature's remedy for the bad effects of hard water.

The important thing, however, is to clean the boiler thoroughly and often. In no case should the lime be allowed to bake on the iron. If it gets thick, the iron or steel is sure to burn, and the lime to bake so hard it will be almost impossible to get it off. But if the boiler is cleaned often, such a thing will not happen.

Mud or sediment can be blown off by opening the valve from the mud drum or the firebox at the bottom of the boiler when the pressure is not over 15 or 20 pounds; and at this pressure much of the lime distributed about the boiler may be blown off. But this is not enough. The inside of the boiler should be scraped and thoroughly washed out with a hose and force-pump just as often as the condition of the water requires it.

In cleaning the boiler, always be careful to scrape all the lime off the top of the fusible plug.

THE PUMP

In order to manage the pump successfully, the young engineer must understand thoroughly its construction as already described. It is also necessary to understand something of the theory of atmospheric pressure, lifting power, and forcing power.

First see that the cocks or globe valves (whichever are used) are open both between the boiler and the pump and between the pump and the water supply. The globe valve next the boiler should *never* be closed, except when examining the boiler check valve. Then open the little pet cock between the two upper horizontal check valves. Be sure that the check valves are in good order, so that water can pass only in one direction. A clear, sharp click of the check valves is certain evidence that the pump is working well. If you cannot hear the click, take a stick or pencil between your teeth at one end, put the other end on the valve, stuff your fingers in your ears, and you will hear the movement of the valve as plainly as if it were a sledge-hammer.

The small drain cock between the horizontal check valves is used to drain hot water out of the pump in starting, for a pump will never work well with hot water in it; and to drain off all water in closing down in cold weather, to prevent damage from freezing. It also assists in testing the working of the pump. In starting up it may be left open. If water flows from the drain cock, we know the pump is working all right, and then close the drain cock. If you are at any time in doubt as to whether water is going into the boiler properly, you may open this drain cock and see if cold water flows freely. If it does, everything is working as it should. If hot water appears, you may know something is wrong. Also, to test the pump, place your hand on the two check valves, and if they are cold, the pump is all right; if they are hot, something is wrong, since the heat must come from the boiler, and no hot water or steam should ever be allowed to pass from the boiler back to the pump.

A stop cock next the boiler is decidedly preferable to a globe valve, since you can tell if it is open by simply looking at it; whereas you must put your hand on a globe valve and turn it. Trouble often arises through inadvertently closing the valve or cock next the boiler, in which case, of course, no water can pass into the boiler, and the pump is likely to be ruined, since the water must get out somewhere. Some part of the pump would be sure to burst if worked against a closed boiler cock or valve.

Should the pump suddenly cease to work or stop, first see if you have any water in the tank. If there is water, stoppage may be due to air in the pump chamber, which can get in only through the stuffing-box. If this is true, tighten up the pump plunger stuffing-box nut a little. If now the pump starts off well, you have found the difficulty; but at the first opportunity you ought to repack the stuffing-box.

If the stuffing-box is all right, examine the supply suction hose. See that nothing is clogging the strainer, and ascertain whether the water is sucked in or not. If it is sucked in and then is forced out again (which you can ascertain by holding your hand lightly over the suction pipe), you may know something is the matter with the first check valve. Probably a stick or stone has gotten into it and prevents it from shutting down.

If there is no suction, examine the second check valve. If there is something under it that prevents its closing, the water will flow back into the pump chamber again as soon as the plunger is drawn back.

You can always tell whether the trouble is in the second check or in the hot water check valve by opening the little drain cock. If hot water flows from it, you may know that the hot water check valve is out of order; if only cold water flows, you may be pretty sure the hot water check is all right. If there is any reason to suspect the hot water check valve, close the stop cock or valve next the boiler before you touch the check in any way. To tamper with the hot water check while the steam pressure is upon it would be highly dangerous, for you are liable to get badly burned with escaping steam or hot water. At the same time, be very sure the stop cock or valve next the boiler is open again before you start the pump.

Another reason for check valves refusing to work besides having something under them, is that the valve may stick in the valve chamber because of a rough place in the chamber, or a little projection on the valve. Light tapping with a wrench may remedy the matter. If that does not work, try the following plan suggested by an old engineer*: "Take the valve out, bore a hole in a board about one-half inch deep, and large enough to permit the valve to be turned. Drop a little emery dust in this hole If you haven't any emery dust, scrape some grit from a whetstone. If you have no whetstone, put some fine sand or gritty soil in the hole, put the valve on top of it, put your brace on the valve and turn it vigorously for a few minutes, and you will remove all roughness."

Sometimes the burr on the valve comes from long use; but the above treatment will make it as good as new.

INJECTORS

All injectors are greatly affected by conditions, such as the lift, the steam pressure, the temperature of the water, etc. An injector will not use hot water well, if at all. As the lift is greater, the steam pressure required to start is greater, and at the same time the highest steam pressure under which the injector will work at all is greatly decreased. The same applies to the lifting of warm water: the higher the temperature, the greater the steam pressure required to start, and the less the steam pressure which can be used as a maximum.

It is important for the sake of economy to use the right sized injector. Before buying a new injector, find out first how much water you need for your boiler, and then buy an injector of about the capacity required, though of course an injector must always have a maximum capacity in excess of what will be required.

---

*J. H. Haggard.

If the feed water is cold, a good injector ought to start with 25 pounds steam pressure and work up to 150 pounds for a 2-foot lift. If the lift is eight feet, it will start at 30 pounds and work up to 130. If the water is heated to 100 degrees Fahrenheit it will start for a 2-foot lift with 26 pounds and work up to 120 pounds, or for an 8-foot lift, it will start with 33 pounds and work up to 100. These figures apply to the single tube injector. The double tube injector should work from 14 pounds to 250, and from 15 to 210 under same conditions as above. The double tube injector is not commonly used on farm engines, however.

Care should be taken that the injector is not so near the boiler as to become heated, else it will not work. If it gets too hot, it must be cooled by pouring cold water on the outside, first having covered it with a cloth to hold the water. If the injector is cool, and the steam pressure and lift are all right, and still the injector does not work, you may be sure there is some obstruction somewhere. Shut off the steam from the boiler, and run a fine wire down through the cone valve or cylinder valve, after having removed the cap or plug nut.

Starting an injector always requires some skill, and injectors differ. Some start by manipulating the steam valve; some require that the steam be turned on first, and then the water turned on in just the right amount, usually with a quick short twist of the supply valve. Often some patience is required to get just the right turn on it so that it will start.

Of course you must be sure that all joints are air-tight, else the injector will not work under any conditions.

Never use an injector where a pump can be used, as the injector is much more wasteful of steam. It is for an emergency or to throw water in a boiler when engine is not running.

No lubricator is needed on an injector.

THE HEATER

The construction of the heater has already been explained. It has two check valves, one on the side of the pump and one on the side of the boiler, both opening toward the boiler. The exhaust steam is usually at a temperature of 215 to 220 degrees when it enters the heater chamber, and heats the water nearly or quite to boiling point as it passes through. The injector heats the water almost as hot.

The heater requires little attention, and the check valves seldom get out of order.

The pump is to be used when the engine is running, and the injector when the engine is closed down. The pump is the more economical; but when the engine is not working the exhaust steam is not sufficient to heat the water in the heater; and pumping cold water into the boiler will quickly bring down the pressure and injure the boiler.

ECONOMICAL FIRING

The management of the fire is one of the most important things in running a steam engine. On it depend two things of the greatest consequence—success in getting up steam quickly and keeping it at a steady pressure under all conditions; and economy in the use of fuel. An engineer who understands firing in the most economical way will probably save his wages to his employer over the engineer who is indifferent or unscientific about it. Therefore the young engineer should give the subject great attention.

First, let us consider firing with coal. All expert engineers advise a "thin" fire. This means that you should have a thin bed of coals, say about four inches thick, all over the grate. There should be no holes or dead places in this, for if there are any, cold air will short-circuit into the fire flues and cool off the boiler.

The best way of firing is to spread the coal on with a small hand shovel, a very little at a time, scattering it well over the fire. Another way, recommended by some, is to have a small pile of fresh fuel at the front of the grate, pushing it back over the grate when it is well lighted. To manage this well will require some practice and skill, and for a beginner, we recommend scattering small shovels-ful all over the fire. All lump coal should be broken to a uniform size. No piece larger than a man's fist should be put in a firebox.

Seldom use the poker above the fire, for nothing has such a tendency to put out a coal fire as stirring it with a poker above. And when there is a good glow all over the grate below, the poker is not needed below. When the grate becomes covered with dead ashes, they should be cautiously but fully removed, and clinkers must be lifted out with the poker from above, care being exercised to cover up the holes with live coals.

Hard coal if used should be dampened before being put on the fire.

When the fire is burning a little too briskly, close the draft but do not tamper with the fire itself. Should it become important on a sudden emergency to check the fire at any time quickly, never dash water upon it, but rather throw plenty of fresh fuel upon it. Fresh fuel always lowers the heat at first. If all drafts are closed tight, it will lower the heat considerably for quite a time.

In checking a fire, it must be remembered that very sudden cooling will almost surely crack the boiler. If there is danger of an explosion it may be necessary to draw the fire out entirely; but under no circumstances should cold water be thrown on. After drawing the fire close all doors and dampers.

FIRING WITH WOOD

Always keep the fire door shut as much as possible, as cold air thus admitted will check the fire and ruin the boiler.

Firing with wood is in many ways the exact reverse of firing with coal. The firebox should be filled full of wood at all times. The wood should be thrown in in every direction, in pieces of moderate size, and as it burns away, fresh pieces should be put in at the front so that they will get lighted and ready to burn before being pushed back near the boiler. It often helps a wood fire, too, to stir it with a poker. Wood makes much less ash than coal, and what little accumulates in the grate will not do much harm. Sometimes green wood will not burn because it gets too much cold air. In that case the sticks should be packed as close together as possible, still leaving a place for the air to pass. Also a wood fire, especially one with green wood, should be kept up to a high temperature all the time; for if it is allowed to drop down the wood will suddenly cease to burn at all.

FIRING WITH STRAW

In firing with straw it is important to keep the shute full of straw all the time so that no cold air can get in on top of the fire. Don't push the straw in too fast, either, but keep it moving at a uniform rate, with small forkfulls. Now and then it is well to turn the fork over and run it down into the fire to keep the fire level. Ashes may be allowed to fill up in rear of ash box, but fifteen inches should be kept clear in front to provide draft. The brick arch may be watched from the side opening in the firebox. and should show a continuous stream of white flame coming over it. If too much straw is forced in, that will check the flame. The flame should never be checked. If damp straw gets against the ends of the flues, it should be scraped off with the poker from side door. Clean the tubes well once a day. The draft must always be kept strong enough to produce a white heat, and if this cannot be done otherwise, a smaller nozzle may be used on the exhaust pipe; but this should be avoided when possible, since it causes back pressure on the engine. Never let the front end of the boiler stand on low ground. Engine should be level, or front end high, if it has a firebox locomotive boiler; if a return flue boiler, be careful to keep it always level. In burning straw take particular notice that the spark screen in stack does not get filled up.

THE ASH PIT

In burning coal it is exceedingly important that the ashes be kept cleaned out, as the hot cinders falling down on the heap of ashes almost as high as the grate will overheat the grate in a very short time and warp it all out of shape, so ruining it.

With wood and straw, on the contrary, an accumulation of ashes will often help and will seldom do any harm, because no very hot cinders can drop down below the grates, and the hottest part of the fire is some distance above the grates.

STARTING A FIRE

You must make up your mind that it will take half an hour to an hour or so to get up steam in any boiler that is perfectly cold. The metal expands and shrinks a great deal with the heat and cold, and a sudden application of heat would ruin a boiler in a short time. Hence it is necessary for reasons of engine economy to make changes of temperature, either cooling off or heating up, gradually.

First see that there is water in the boiler.

Start a brisk fire with pine kindlings, gradually putting on coal or wood, as the case may be, and spreading the fire over the grate so that all parts will be covered with glowing coals.

When you have 15 or 20 pounds of steam, start the blower. As has already been described, the blower is a pipe with a nozzle leading from the steam space of the boiler to the smoke stack, and fitted with a globe valve. The force of the steam drives the air out of the stack, causing a vacuum, which is immediately filled by the hot gases from the firebox coming through the boiler tubes. Little is to be gained by using the blower with less than 15 pounds of steam, as the blower has so little strength below that, that it draws off about as much steam as is made and nothing is gained.

The blower is seldom needed when the engine is working, as the exhaust steam should be sufficient to keep the fire going briskly. If it is not, you should conclude that something is the matter. There are times, however, when the blower is required even when the engine is going. For example, if you are working with very light load and small use of steam, the exhaust may be insufficient to keep up the fire; and this will be especially true if the fuel is very poor. In such a case, turn on the blower very slightly. But remember that you are wasting steam if you can get along without the blower.

Examine the nozzle of the blower now and then to see that it does not become limed up, or turned so as to direct the steam to one side of the stack, where its force would be wasted.

Beware, also, of creating too much draft; for too much draft will use up fuel and make little steam.

SMOKE

Coal smoke is nothing more or less than unburned carbon. The more smoke you get, the less will be the heat from a given amount of fuel. Great clouds of black smoke from an engine all the time are a very bad sign in an engineer. They show that he does not know how to fire. He has not followed the directions already given, to have a thin, hot fire, with few ashes under his grate. Instead, he throws on great shovelsful of coal at a time, and has the coal up to the firebox door. His fuel is always making smoke, which soon clogs up the smoke flues and lessens the amount of steam he is getting. If he had kept his fire very "thin," but very hot, throwing on a small hand shovel of coal at a time, seldom poking his fire except to lift out clinkers or clean away dead ashes under the grate, and keeping his ashpit free from ashes, there would be only a little puff of black smoke when the fresh coal went on, and then the smoke would quickly disappear, while the fire flues would burn clean and not get clogged up with soot.

It is important, however, to keep the small fire flues especially well cleaned out with a good flue cleaner; for all accumulation of soot prevents the heat from passing through the steel, and so reduces the heating capacity of the boiler. Cleaning the tubes with a steam blower is never advisable, as it forms a paste on the tube that greatly impairs its commodity.

SPARKS

With coal there is little danger of fires caused by sparks from the engine. What sparks there are are heavy and dead, and will even fall on a pile of straw without setting it on fire. On a very windy day, however, when you are running your engine very hard, especially if it is of the direct locomotive boiler type, you want to be careful even with coal.

With wood it is very different; and likewise with straw. Wood and straw sparks are always dangerous, and an engine should never be run for threshing with wood or straw without using a spark-arrester.

It sometimes happens that when coal is used it will give out, and you will be asked to finish your job with wood. In such a case, it is the duty of an engineer to state fully and frankly the danger of firing with wood without a spark arrester, and he should go on only when ordered to do so by the proprietor, after he has been fully warned. In that case all responsibility is shifted from the engineer to the owner.

THE FUSIBLE PLUG

The careful engineer will never have occasion to do anything to the fusible plug except to clean the scale off from the top of it on the inside of the boiler once a week, and put in a fresh plug once a month. It is put in merely as a precaution to provide for carelessness. The engineer who allows the fusible plug to melt out is by that very fact marked as a careless man, and ought to find it so much the harder to get a job.

As has already been explained, the fusible plug is a plug filled in the middle with some metal that will melt at a comparatively low temperature. So long as it is covered with water, no amount of heat will melt it, since the water conducts the heat away from the metal and never allows it to rise above a certain temperature. When the plug is no longer covered with water, however,—in short, when the water has fallen below the danger line in the boiler— the metal in the plug will fuse, or melt, and make an opening through which the steam will blow into the firebox and put out the fire. However, if the top of the fusible plug has been allowed to become thickly coated with scale, this safety precaution may not work and the boiler may explode. In any case the fusible plug is not to be depended on.

At the same time a good engineer will take every precaution, and one of these is to keep the top of the plug well cleaned. Also he will have an extra plug all ready and filled with composition metal, to put in should the plug in the boiler melt out. Then he will refill the old plug as soon as possible. This may be done by putting a little moist clay in one end to prevent the hot metal from running through, and then pouring into the other end of the plug as much melted metal as it will hold. When cold, tamp down solidly.

LEAKY FLUES

One common cause of leaky flues is leaving the fire door open so that currents of cold air will rush in on the heated flues and cause them, or some other parts of the boiler, to contract too suddenly. The best boiler made may be ruined in time by allowing cold currents of air to strike the heated interior. Once or twice will not do it; but continually leaving the fire door open will certainly work mischief in the end.

Of course, if flues in a new boiler leak, it is the fault of the boiler maker. The tubes were not large enough to fill the holes in the tube sheets properly. But if a boiler runs for a season or so and then the flues begin to leak, the chances are that it is due to the carelessness of the engineer. It may be he has been making his fires too hot; it may be leaving the firebox door open; it may be running the boiler at too high pressure; it may be blowing out the boiler when it is too hot; or blowing out the boiler when there is still some fire in the firebox; it may be due to lime encrusted on the inside of the tube sheets, causing them to overheat. Flues may also be made to leak by pumping cold water into the boiler when the water inside is too low; or pouring cold water into a hot boiler will do it. Some engineers blow out their boilers to clean them, and then being in a hurry to get to work, refill them while the metal is hot. The flues cannot stand this, since they are thinner than the shell of the boiler and cool much more quickly; hence they will contract much faster than the rest of the boiler and something has to come loose.

Once a flue starts to leaking, it is not likely to stop till it has been repaired; and one leaky flue will make others leak.

Now what shall you do with a leaky flue?

To repair a leaky flue you should have a flue expander and a calking tool, with a light hammer. If you are small enough you will creep in at the firebox door with a candle in your hand. First, clean off the ends of the flues and flue sheet with some cotton waste. Then force the expander into the leaky flue, bringing the shoulder well up against the end of the flue. Then drive in the tapering pin. Be very careful not to drive it in too far, for if you expand the flue too much, you will strain the flue sheet and cause other flues to leak. You must use your judgment and proceed cautiously. It is better to make two or three trials than to spoil your boiler by bad work. The roller expander is preferable to the Prosser in the hands of a novice. The tube should be expanded only enough to stop the leak. Farther expanding will only do injury.

When you think the flue has been expanded enough, hit the pin a side blow to loosen it. Then turn the expander a quarter round, and drive in the pin again. Loosen up and continue till you have turned the expander entirely around.

Finally remove the expander, and use the calking tool to bead the end. It is best, however, to expand all leaky flues before doing any beading.

The beading is done by placing the guide or gauge inside the flue, and then pounding the ends of the flue down against the flue sheet by light blows. Be very careful not to bruise the flue sheet or flues, and use no heavy blows, nor even a heavy hammer. Go slowly and carefully around the end of each flue; and if you have done your work thoroughly and carefully the flues will be all right. But you should test your boiler before steaming up, to make sure that all the leaks are stopped, especially if there have been bad ones.

There are various ways to testing a boiler. If waterworks are handy, connect the boiler with a hydrant and after filling the boiler, let it receive the hydrant pressure. Then examine the calked flues carefully, and if you see any seeping of water, use your beader lightly till the water stops. In case no waterworks with good pressure are at hand, you can use a hydraulic pump or a good force pump.

The amount of pressure required in testing a boiler should be that at which the safety valve is set to blow off, say 110 to 130 lbs. This will be sufficient.

If you are in the field with no hydrant or force pump handy, you may test your boiler in this way: Take off the safety valve and fill the boiler full of water through the safety valve opening. Then screw the safety back in its place. You should be sure that every bit of space in the boiler is filled entirely full of water, with all openings tightly closed. Then get back in the boiler and have a bundle of straw burned under the firebox, or under the waist of the boiler, so that at some point the water will be slightly heated. This will cause pressure. If your safety valve is in perfect order, you will know as soon as water begins to escape at the safety valve whether your flues are calked tight enough or not.

The water is heated only a few degrees, and the pressure is cold water pressure. In very cold weather this method cannot be used, however, as water has no expansive force within five degrees of freezing.

The above methods are not intended for testing the safety of a boiler, but only for testing for leaky flues. If you wish to have your boiler tested, it is better to get an expert to do it.

CHAPTER V

HOW TO MANAGE A TRACTION ENGINE

A traction engine is usually the simplest kind of an engine made. If it were not, it would require a highly expert engineer to run it, and this would be too costly for a farmer or thresherman contractor. Therefore the builders of traction engines make them of the fewest possible parts, and in the most durable and simple style. Still, even the simplest engine requires a certain amount of brains to manage it properly, especially if you are to get the maximum of work out of it at the lowest cost.

If the engine is in perfect order, about all you have to do is to see that all bearings are properly lubricated, and that the automatic oiler is in good working condition. But as soon as an engine has been used for a certain time, there will be wear, which will appear first in the journals, boxes and valve, and it is the first duty of a good engineer to adjust these. To adjust them accurately requires skill; and it is the possession of that skill that goes to make a real engineer.

Your first attention will probably be required for the cross-head and crank boxes or brasses. The crank box and pin will probably wear first; but both the cross-head and crank boxes are so nearly alike that what is said of one will apply to the other.

You will find the wrist box in two parts. In a new engine these parts do not quite meet. There is perhaps an eighth of an inch waste space between them. They are brought up to the box in most farm engines by a wedge-shaped key. This should be driven down a little at a time as the boxes wear, so as to keep them snug up to the pin, though not too tight.

You continue to drive in the key and tighten up the boxes as they wear until the two halves come tight together. Then you can no longer accomplish anything in this way.

When the brasses have worn so that they can be forced no closer together, they must be taken off and the ends of them filed where they come together. File off a sixteenth of an inch from each end. Do it with care, and be sure you get the ends perfectly even. When you have done this you will have another eighth of an inch to allow for wear.

Now, by reflection you will see that as the wrist box wears, and the wedge-shaped key is driven in, the pitman (or piston arm) is lengthened to the amount that the half of the box farthest from the piston has worn away. When the brasses meet, this will amount to one-sixteenth of an inch.

Now if you file the ends off and the boxes wear so as to come together once more, the pitman will have been shortened one-eighth of an inch; and pretty soon the clearance of the piston in the cylinder will have been offset, and the engine will begin to. pound. In any case, the clearance at one end of the cylinder will be one-sixteenth or one-eighth of an inch less, and in the other end one-sixteenth or one-eighth of an inch more. When this is the case you will find that the engine is not working well.

To correct this, when you file the brasses either of the cross-head box or the crank box you must put in some filling back of the brass farthest from the piston, sufficient to equalize the wear that has taken place, that is, one-sixteenth of an inch each time you have to file off a sixteenth of an inch. This filling may be some flat pieces of tin or sheet copper, commonly called shims, and the process is called shimming. As to the front half of the box, no shims are required, since the tapering key brings that box up to its proper place.

Great care must be exercised when driving in the tapering key or wedge to tighten up the boxes, not to drive it in too hard. Many engineers think this is a sure remedy for "knocking" in an engine, and every time they hear a knock they drive in the crank box key. Often the knock is from some other source, such as from a loose fly wheel, or the like. Your ear is likely to deceive you; for a knock from any part of an engine is likely to sound as if it came from the crank box. If you insist on driving in the key too hard and too often, you will ruin your engine.

In tightening up a key, first loosen the set screw that holds the key; then drive down the key till you think it is tight; then drive it back again, and this time force it down with your fist as far as you can. By using your fist in this way after you have once driven the pin in tight and loosened it again you may be pretty certain you are not going to get it so tight it will cause the box to heat.

WHAT CAUSES AN ENGINE TO KNOCK

The most common sign that something is loose about an engine is "knocking," as it is called. If any box wears a little loose, or any wheel or the like gets a trifle loose, the engine will begin to knock.

When an engine begins to knock or run hard, it is the duty of the engineer to locate the knock definitely. He must not guess at it. When he has studied the problem out carefully, and knows where the knock is, then he may proceed to remedy it. Never adjust more than one part at a time.

As we have said, a knock is usually due to looseness somewhere. The journals of the main shaft may be loose and cause knocking. They are held in place by set bolts and jam nuts, and are tightened by simply screwing up the nuts. But a small turn of a nut may make the box so tight it will begin to heat at once. Great care should be taken in tightening up such a box to be sure not to get it too tight. Once a box begins to cut, it should be taken out and thoroughly cleaned.

Knocking may be due to a loose eccentric yoke. There is packing between the two halves of the yoke, and to tighten up you must take out a thin layer of this packing. But be careful not to take out too much, or the eccentric will stick and begin to slip.

Another cause of knocking is the piston rod loose in the cross-head. If the piston rod is keyed to the cross-head it is less liable to get loose than if it were fastened by a nut; but if the key continues to get loose, it will be best to replace it with a new one.

Unless the piston rod is kept tight in the cross-head, there is liability of a bad crack. A small strain will bring the piston out of the cross-head entirely, when the chances are you will knock out one or both cylinder-heads. If a nut is used, there will be the same danger if it comes off. It should therefore be carefully watched. The best way is to train the ear to catch any usual sound, when loosening of the key or nut will be detected at once.

Another source of knocking is looseness of the cross-head in the guides. Provision is usually made for taking up the wear; but if there is not, you can take off the guides and file them or have them planed off. You should take care to see that they are kept even, so that they will wear smooth with the crosshead shoes.

If the fly-wheel is in the least loose it will also cause knocking, and it will puzzle you not a little to locate it. It may appear to be tight; but if the key is the least bit too narrow for the groove in the shaft, it will cause an engine to bump horribly, very much as too much "lead" will.

LEAD

We have already explained what "lead" is. It is opening of the port at either end of the steam cylinder allowed by the valve when the engine is on a dead centre. To find out what the lead is, the cover of the steam chest must be taken off, and the engine placed at each dead centre in succession. If the lead is greater at one end than it is at the other, the valve must be adjusted to equalize it. As a rule the engine is adjusted with a suitable amount of lead if it is equalized. The correct amount of lead varies with the engine and with the port opening. If the port opening is long and narrow, the lead should obviously be less than if the port is short and wide.

If the lead is insufficient, there will not be enough steam let into the cylinder for cushion, and the engine will knock. If there is too much lead the speed of the engine will be lessened, and it will not do the work it ought. To adjust the lead *de novo* is by no means an easy task.

HOW TO SET A SIMPLE VALVE

In order to set a valve the engine must be brought to a dead centre. This cannot be done accurately by the eye. An old engineer* gives the following directions for

---

*J. H. Maggard.

finding the dead centre accurately. Says he: "First provide yourself with a 'tram.' This is a rod of one-fourth inch iron about eighteen inches long, with two inches at one end bent over to a sharp angle. Sharpen both ends to a point. Fasten a block of hard wood somewhere near the face of the fly-wheel, so that when the straight end of your tram is placed at a definite point in the block, the hooked end will reach the crown of the fly-wheel. The block must be held firmly in its place, and the tram must always touch it at exactly the same point.

"You are now ready to set about finding the dead centre. In doing this, remember to turn the fly-wheel always in the same direction.

"Bring the engine over till it nearly reaches one of the dead centres, but not quite. Make a distinct mark across the cross-head and guides. Also go around to the flywheel, and placing the straight end of the tram: at the selected point on the block of wood, make a mark across the crown or centre of face of the fly-wheel. Now turn your engine past the centre, and on to a point at which the mark on the cross head will once more exactly correspond with the line on the guides, making a single straight line. Once more place the tram as before and make another mark across the crown of the fly-wheel. By use of dividers, find the exact centre between the two marks made on the fly-wheel, and mark this point distinctly with a centre punch. Now bring the fly-wheel to the point where the tram, set with its straight end at the required point on the block of wood, will touch this point with the hooked end, and you will have one of the dead centres.

"Turn the engine over and proceed in the same way to find the other dead centre."

Now, setting the engine on one of the dead centres, remove the cover of the steam chest and proceed to set your valve.

Assuming that the engine maker gave the valve the proper amount of lead in the first place, you can proceed on the theory that it is merely necessary to equalize the lead at both ends. Assume some convenient lead, as one-sixteenth of an inch, and set the valve to that. Then turn the engine over and see if the lead at the other end is the same. If it is the same, you have set the valve correctly. If it is less at the other end, you may conclude that the lead at both ends should be less than one-sixteenth of an inch, and must proceed to equalize it. This you can do by fitting into the open space a little wedge of wood, changing the valve a little until the wedge goes in to just the same distance at each end. Then you may know that the lead at one end is the same as at the other end. You can mark the wedge for forcing it against the metal, or mark it against the seat of the valve with a pencil.

The valve is set by loosening the set screws that hold the eccentric on the shaft. When these are loosened up the valve may be moved freely. When it is correctly set the screws should be tightened, and the relative position of the eccentric on the shaft may be permanently marked by setting a cold chisel so that it will cut into the shaft and the eccentric at the same time and giving it a smart blow with the hammer, so as to make a mark on both the eccentric and the shaft. Should your eccentric slip at any time in the future, you can set your valve by simply bringing the mark on the eccentric so that it will correspond with the mark on the shaft. Many engines have such a mark made when built, to facilitate setting a valve should the eccentric become loose.

These directions apply only to setting the valve of a single eccentric engine.

HOW TO SET A VALVE ON A DOUBLE ECCENTRIC ENGINE

In setting a valve on a reversible or double eccentric engine, the link may cause confusion, and you may be trying to set the valve to run one way when the engine is set to run the other.

The valve on such an engine is exactly the same as on a single eccentric engine. Set the reverse lever for the engine to go forward. Then set the valve exactly as with a single eccentric engine. When you have done so, tighten the eccentric screws so that they will hold temporarily, and set the reverse lever for the engine to go backward. Then put the engine on dead centres and see if the valve is all right at both ends. If it is, you may assume that it is correctly set, and tighten eccentric screws, marking both eccentrics as before.

As we have said, most engines are marked in the factory, so that it is not a difficult matter to set the valves, it being necessary only to bring the eccentric around so that the mark on it will correspond with the mark on the shaft.

You can easily tell whether the lead is the same at both ends by listening to the exhaust. If it is longer at one end than the other, the valve is not properly set.

SLIPPING OF THE ECCENTRIC OR VALVE

If the eccentric slips the least bit it may cause the engine to stop, or to act very queerly. Therefore the marks on the shaft and on the eccentric should be watched closely, and of course all grease and dirt should be kept wiped off, so that they can be seen easily. Then the jam nuts should be tightened up a little from time to time.

If the engine seems to act strangely, and yet the eccentrics are all right, look at the valve in the steam chest. If the valve stem has worked loose from the valve, trouble will be caused. It may be held in place by a nut, and the nut may work off; or the valve may be held by a clamp and pin, and the pin may work loose. Either will cause loss of motion, and perhaps a sudden stopping of the engine.

USE OF THE CYLINDER STEAM COCKS

It is a comparatively simple matter to test a steam cylinder by use of the cylinder cocks. To do this, open both cocks, place the engine on the forward center, and turn on a little steam. If the steam blows out at the forward cock, we may judge that our lead is all right. Now turn the engine to the back center and let on the steam. It should blow out the same at the back cock. A little training of the ear will show whether the escape of steam is the same at both ends. Then reverse the engine, set it on each center successfully, and notice whether the steam blows out from one cock at a time and in the same degree of force.

If the steam blows out of both cocks at the same time, or out of one cock on one center, but not out of the other cock on its corresponding center, we may know something is wrong. The valve does not work properly.

We will first look at the eccentrics and see that they are all right. If they are, we must open the steam chest, first turning off all steam. Probably we shall find that the valve is loose on the valve rod, if our trouble was that the steam blew out of the cock but did not out of the other when the engine was on the opposite center.

If our trouble was that steam blew out of both cocks at the same time, we may conclude either that the cylinder rings leak or else the valve has cut its seat. It will be a little difficult to tell which at first sight. In any case it is a bad thing, for it means loss of power and waste of steam and fuel. To tell just where the trouble is you must take off the cylinder head, after setting the engine on the forward center. Let in a little steam from the throttle. If it blows through around the rings, the trouble is with them; but if it blows through the valve port, the trouble is with the valve and valve seat.

If the rings leak you must get a new set if they are of the self-adjusting type. But if they are of the spring or adjusting type you can set them out yourself; but few engines now use the latter kind of rings, so a new pair will probably be required.

If the trouble is in the valve and valve seat, you should take the valve out and have the seat planed down, and the valve fitted to the seat. This should always be done by a skilled mechanic fully equipped for such work, as a novice is almost sure to make bad work of it. The valve seat and valve must be scraped down by the use of a flat piece of very hard steel, an eighth of an inch thick and about 3 by 4 inches in size. The scraping edge must be absolutely straight. It will be a slow and tedious process, and a little too much scraping on one side or the other will prevent a perfect fit. Both valve and valve seat must be scraped equally. Novices sometimes try to reseat a valve by the use of emery. This is very dangerous and is sure to ruin the valve, as it works into the pores of the iron and causes cutting.

LUBRICATION

A knowledge of the difference between good oil and poor oil, and of how to use oil and grease, is a prime essential for an engineer.

First let us give a little attention to the theory of lubrication. The oil or grease should form a lining between the journal and its pin or shaft. It is in the nature of a slight and frictionless cushion at all points where the two pieces of metal meet.

Now if oil is to keep its place between the bearing and the shaft or pin it must stick tight to both pieces of metal, and the tighter the better. If the oil is light the forces at work on the bearings will force the oil away and bring the metals together. As soon as they come together they begin to wear on each other, and sometimes the wear is very rapid. This is called "cutting." If a little sand or grit gets into the bearing, that will help the cutting wonderfully, and more especially if there is no grease there.

For instance, gasoline and kerosene are oils, but they are so light they will not stick to a journal, and so are valueless for lubricating. Good lubricating oil will cost a little more than cheap oil which has been mixed with worthless oils to increase its bulk without increasing its cost. The higher priced oil will really cost less in the end, because there is a larger percentage of it which will do service. A good engineer will have it in his contract that he is to be furnished with good oil.

Now an engine requires two different kinds of oil, one for the bearings, such as the crank pin, the cross-head and journals, and quite a different kind for lubricating the steam cylinder.

It is extremely important that the steam cylinder should be well lubricated; and this cannot be done direct. The oil must be carried into the valve and cylinder with steam. The heat of the steam, moreover, ranging from about 320 degrees Fahr. for 90 lbs. pressure to 350 degrees for 125 lbs. of pressure, will quickly destroy the efficacy of a poor oil, and a good cylinder oil must be one that will stick to the cylinder and valve seat under this high temperature. It must have staying qualities.

The link reverse is one of the best for its purpose; but it requires a good quality of oil on the valve for it to work well. If the valve gets a little dry, or the poor oil used does not serve its purpose properly, the link will begin to jump and pound. This is a reason why makers are substituting other kinds of reverse gear in many ways not as good, but not open to this objection. If a link reverse begins to pound when you are using good oil, and the oiler is working properly, you may be sure something is the matter with the valve or the gear.

A good engineer will train his ear so that he will detect by simply listening at the cylinder whether everything is working exactly as it ought. For example, the exhaust at each end of the cylinder, which you can hear distinctly, should be the same and equal. If the exhaust at one end is less than it is at the other, you may know that one end of the cylinder is doing more work than the other. And also any little looseness or lack of oil will signify itself by the peculiar sound it will cause.

While the cylinder requires cylinder oil, the crank, cross-head and journals require engine oil, or hard grease. The use of hard grease is rapidly increasing, and it is highly to be recommended. With a good automatic spring grease cup hard grease will be far less likely to let the bearings heat than common oil will. At the same time it will be much easier to keep an engine clean if hard grease is used.

An old engineer* gives the following directions for fitting a grease cup on a box not previously arranged for one: "Remove the journal, take a gouge and cut a clean groove across the box, starting at one corner, about one-eighth of an inch from the point of the box, and cut diagonally across, coming out at the opposite corner on the other end of the box. Then start at the opposite corner and run through as before, crossing the first groove in the center of the box. Groove both halves of the box the same, being careful not to cut out at either end, as this will allow the grease to escape from the box and cause unnecessary waste. The shimming or packing in the box should be cut so as to touch the journal at both ends of the box, but not in the center or between these two points. So when the top box is brought down tight this will form another reservoir for the grease. If the box is not tapped directly in the center for the cup, it will be necessary to cut another groove from where it is tapped into the grooves already made. A box prepared in this way and carefully polished inside, will require little attention if you use good grease."

A HOT BOX

When a box heats in the least degree, it is a sign that for lack of oil or for some other reason the metals are wearing together.

The first thing to do, of course, is to see that the box is supplied with plenty of good oil or grease.

---

*J. H. Maggard.

If this does not cause the box to cool off, take it apart and clean it thoroughly. Then coat the journal with white lead mixed with good oil. Great care should be exercised to keep all dirt or grit out of your can of lead and away from the bearing.

Replace the oil or grease cup, and the box will soon cool down.

THE FRICTION CLUTCH

Nearly all traction engines are now provided with the friction clutch for engaging the engine with the propel-

A. W. Stevens Co. Friction clutch.

ling gear. The clutch is usually provided with wooden shoes, which are adjustable as they wear; and the clutch is thrown on by a lever, conveniently placed.

Before running an engine, you must make sure that the clutch shoes are properly adjusted. Great care must be taken to be sure that both shoes will come in contact with the friction wheel at the same instant; for if one shoe touches the wheel before the other the clutch will probably slip.

The shoes should be so set as to make it a trifle difficult to draw the lever clear back.

To regulate the shoes on the Rumely engine, for example, first throw the friction in. The nut on the top of the toggle connecting the sleeve of the friction with the shoe must then be loosened, and the nut below the shoe tightened up, forcing the shoe toward the wheel. Both shoes should be carefully adjusted so that they will engage the band wheel equally and at exactly the same time.

To use the friction clutch, first start the engine, throwing the throttle gradually wide open. When the engine is running at its usual speed, slowly bring up the clutch until the gearing is fully engaged, letting the engine start slowly and smoothly, without any jar.

Traction engines having the friction clutch are also provided with a pin for securing a rigid connection, to be used in cases of necessity, as when the clutch gets broken or something about it gives out, or you have difficulty in making it hold when climbing hills. This pin is a simple round or square pin that can be placed through a hole in one of the spokes of the band wheel until it comes into a similar opening in the friction wheel. When the pin is taken out, so as to disconnect the wheels, it must be entirely removed, not left sticking in the hole, as it is liable to catch in some other part of the machinery.

MISCELLANEOUS SUGGESTIONS

Be careful not to open the throttle valve too quickly, or you may throw off the driving belt. You may also stir up the water and cause it to pass over with the steam, starting what is called "priming."

Always open your cylinder cocks when you stop, to make sure all water has been drained out of the cylinder; and see that they are open when you start, of course closing them as soon as the steam is let in.

When you pull out the ashes always have a pail of water ready, for you may start a fire that will do no end of damage.

If the water in your boiler gets low and you are waiting for the tank to come up, don't think you "can keep on

Friction Clutch

Aultman & Taylor friction clutch.

a little longer." but stop your engine at once. It is better to lose a little time than run the risk of an explosion that will ruin your reputation as an engineer and cause your employer a heavy expense.

Never start the pump when the water in the boiler is low.

Be sure the exhaust nozzle does not get limed up, and be sure the pipe where the water enters the boiler from the heater is not limed up, or you may split a heater pipe or knock out a check valve.

Never leave your engine in cold weather without draining off all the water; and always cover up your engine when you leave it.

Never disconnect the engine with a leaky throttle.

Keep the steam pressure steady, not varying more than 10 to 15 lbs.

If called on to run an old boiler, have it thoroughly tested before you touch it.

Always close your damper before pulling through a stack yard.

Examine every bridge before you pull on to it.

Do not stop going down a steep grade.

CHAPTER VI

HANDLING A TRACTION ENGINE
ON THE ROAD

It is something of a trick to handle a traction engine on the road. The novice is almost certain to run it into a ditch the first thing, or get stuck on a hill, or in a sand patch or a mudhole. Some attention must therefore be paid to handling a traction engine on the road.

In the first place, never pull the throttle open with a jerk, nor put down the reverse lever with a snap. Handle your engine deliberately and thoughtfully, knowing beforehand just what you wish to do and how you will do it. A traction engine is much like an ox; try to goad it on too fast and it will stop and turn around on you. It does its best work when moving slowly and steadily, and seldom is anything gained by rushing.

The first thing for an engineer to learn is to handle his throttle. When an engine is doing work the throttle should be wide open; but on the road, or in turning, backing, etc., the engineer's hand must be on the throttle all the time and he must exercise a nice judgment as to just how much steam the engine will need to do a certain amount of work. This the novice will find out best by opening the throttle slowly, taking all the time he needs, and never allowing any one to hurry him.

As an engineer learns the throttle, he gradually comes to have confidence in it. As it were, he feels the pulse of the animal and never makes a mistake. Such an engineer always has power to spare, and never wastes any power. He finds that a little is often much better than too much.

The next thing to learn is the steering wheel. It has tricks of its own, which one must learn by practice. Most young engineers turn the wheel altogether too much. If you let your engine run slowly you will have time to turn the wheel slowly, and accomplish just what you want to do. If you hurry you will probably have to do your work all over again, and so lose much more time in the end than if you didn't hurry.

Always keep your eyes on the front wheels of the engine, and do not turn around to see how your load is coming on. Your load will take care of itself if you manage the front wheels all right, for they determine where you are to go.

In making a hard turn, especially, go slow. Then you will run no chance of losing control of your engine, and you can see that neither you nor your load gets into a ditch.

GETTING INTO A HOLE

You are sure sooner or later to get into a hole in the road, for a traction engine is so heavy it is sure to find any soft spot in the road there may be.

As to getting out of a hole, observe in the first place that you must use your best judgment.

First, never let the drive wheels turn round without doing any work. The more they spin round without helping you, the worse it will be for you.

Your first thought must be to give the drive wheels something they can climb on, something they can stick to. A heavy chain is perhaps the very best thing you can put under them. But usually on the road you have no chain handy. In that case, you must do what you can. Old hay or straw will help you; and so will old rails or any old timber.

Spend your time trying to give your wheels something to hold to, rather than trying to pull out. When the wheels are all right, the engine will go on its way without any trouble whatever. And do not half do your work of fixing the wheels before you try to start. See that both wheels are secure before you put on a pound of steam. Make sure of this the first time you try, and you will save time in the end. If you fix one wheel and don't fix the other, you will probably spoil the first wheel by starting before the other is ready.

Should you be where your engine will not turn, then you are stuck indeed. You must lighten your load or dig a way out.

BAD BRIDGES

A traction engine is so heavy that the greatest care must be exercised in crossing bridges. If a bridge floor is worn, if you see rotten planks in it, or liability of holes, don't pull on to that bridge without taking precautions.

The best precaution is to carry with you a couple of planks sixteen feet long, three inches thick in the middle, tapering to two inches at the ends; also a couple of planks eight feet long and two inches thick, the latter for culverts and to help out on long bridges.

Before pulling on to a bad looking bridge, lay down your planks, one for each pair of wheels of the engine to run on. Be exceedingly careful not to let the engine drop off the edge of these planks on the way over, or pass over the ends on to the floor of the bridge. If one pair of planks is too short, use your second pair.

Another precaution which it is wise to take is to carry fifty feet of good, stout hemp rope, and when you come to a shaky bridge, attach your separator to the engine by this rope at full length, so that the engine will have crossed the bridge before the weight of the separator comes upon it.

Cross a bad bridge very slowly. Nothing will be gained by hurrying. There should especially be no sudden jerks or starts.

SAND PATCHES

A sandy road is an exceedingly hard road to pull a load over.

In the first place, don't hurry over sand. If you do you are liable to break the footing of the wheels, and then you are gone.

In the second place, keep your engine as steady and straight as possible, so that both wheels will always have an equal and even bearing. They are less liable to slip if you do. It is useless to try to "wiggle" over a sand patch. Slow, steady, and even is the rule.

If your wheels slip in sand, a bundle of straw or hay, especially old hay, will be about the best thing to give them a footing.

HILLS

In climbing hills take the same advice we have given you all along: Go slow. Nothing is gained by rushing at a hill with a steam engine. Such an engine works best when its force is applied steadily and evenly, a little at a time.

If you have a friction clutch, as you probably will have, you should be sure it is in good working order before you attempt to climb hills. It should be adjusted to a nicety, as we have already explained. When you come to a bad hill it would probably be well to put in the tight gear pin; or use it altogether in a hilly country.

When the friction clutch first came into use, salesmen and others used to make the following recommendation (a recommendation which we will say right here is bad). They said, when you come to an obstacle in the road that you can't very well get your engine over, throw off your, friction clutch from the road wheels, let your engine get under good headway running free, and then suddenly put on the friction clutch and jerk yourself over the obstacle.

Now this is no doubt one way to get over an obstacle; but no good engineer would take his chances of spoiling his engine by doing any such thing with it. Some part of it would be badly strained by such a procedure; and if this were done regularly all through a season, an engine would be worth very little at the end of the season.

CHAPTER VII

POINTS FOR THE YOUNG ENGINEER

QUESTIONS AND ANSWERS

THE BOILER

Q. How should water be fed to a boiler?

A. In a steady stream, by use of a pump or injector working continuously and supplying just the amount of water required. By this means the water in the boiler is maintained at a uniform level, and produces steam most evenly and perfectly.

Q. Why should pure water be used in a boiler?

A. Because impure water, or hard water, forms scales on the boiler flues and plates, and these scales act as nonconductors of heat. Thus the heat of the furnace is not able to pass easily through the boiler flues and plates to the water, and your boiler becomes what is called "a hard steamer."

Q. What must be done to prevent the formation of scale?

A. First, use some compound that will either prevent scale from forming, or will precipitate the scale forming substance as a soft powder that can easily be washed off. Sal soda dissolved in the feed water is recommended, but great care should be exercised in the use of sal soda not to use too much at a time, as it may cause a boiler to foam. Besides using a compound, clean your boiler often and regularly with a hand hose and a force pump, and soak it out as often as possible by using rain water for a day or two, especially before cleaning. Rain water will soften and bring down the hard scale far better than any compound.

Q.   How often should you clean your boiler?

A.   As often as it needs it, which will depend upon the work you do and the condition of the water. Once a week is usually often enough if the boiler is blown down a little every day. If your water is fairly good, once a month will be often enough. A boiler should be blown off about one gauge at a time two or three times a day with the blow-off if the water is muddy.

Q.   How long should the surface blow-off be left open?

A.   Only for a few seconds, and seldom longer than a minute. The surface blow-off carries off the scum that forms on the water, and other impurities that rise with the scum.

Q.   How do you clean a boiler by blowing off?

A.   When the pressure has been allowed to run down open the blow-off valve at the bottom of the boiler and let the water blow out less than a minute, till the water drops out of sight in the water gauges, or about two and one-half inches. Blown off more is only a waste of heat and fuel.

Q.   What harm will be done by blowing off a boiler under a high pressure of steam?

A.   The heat in the boiler while there is such a pressure will be so great that it will bake the scale on the inside of the boiler, and it will be very difficult to remove it afterward. After a boiler has been blown off the scale should be for the most part soft, so that it can be washed out by a hose and force pump.

Q.   Why should a hot boiler never be filled with cold water?

A.   Because the cold water will cause the boiler to contract more in some places than in others, and so suddenly that the whole will be badly strained. Leaky flues are made in this way, and the life of a boiler greatly shortened. As a rule a boiler should be filled only when the metal and the water put into it are about at the same temperature.

Q. After a boiler has been cleaned, how should the manhole and manhole plates be replaced?

A. They are held in position by a bolt passing through a yoke that straddles the hole; but to be steam and water tight they must have packing all around the junction of the plate with the boiler. The best packing is sheet rubber cut in the form of a ring just the right size for the bearing surface. Hemp or cotton packing are also used, but they should be free from all lumps and soaked in oil. Do not use any more than is absolutely needed. Be careful, also, to see that the bearings of the plate and boiler are clean and smooth, with all the old packing scraped off. Candle wick saturated with red lead is next best to rubber as packing.

Q. What are the chief duties of an engineer in care of a boiler?

A. First, to watch all gauges, fittings, and working parts, to see that they are in order; try the gauge cocks to make sure the water is at the right height; try the safety valve from time to time to be sure it is working; see that there are no leaks, that there is no rusting or wearing of parts, or to replace parts when they do begin to show wear; to examine the check valve frequently to make sure no water can escape through it from the boiler; take precautions against scale and stoppage of pipes by scale; and keep the fire going uniformly, cleanly, and in an economical fashion.

Q. What should you do if the glass water gauge breaks?

A.   Turn off the gauge cocks above and below, the lower one first so that the hot water will not burn you. You may put in a new glass and turn on gauge cocks at once. Turn on the lower or water cock first, then the upper or steam cock. You may go on without the glass gauge, however, using the gauge cocks or try cocks every few minutes to make sure the water is at the right height, neither too high nor too low.

Q.   Why is it necessary to use the gauge cocks when the glass gauge is all right?

A.   First, because you cannot otherwise be sure that the glass gauge is all right; and, secondly, because if you do not use them frequently they are likely to become scaled up so that you cannot use them in case of accident to the glass gauge.

Q.   If a gauge cock gets leaky, what should be done?

A.   Nothing until the boiler has cooled down. Then if the leak is in the seat, take it out and grind and refit it; if the leak is where the cock is screwed into the boiler, tighten it up another turn and see if that remedies the difficulty. If it does not you will probably have to get a new gauge cock.

Q.   Why not screw up a gauge cock while there is a pressure of steam on?

A.   The cock might blow out and cause serious injury to yourself or some one else. Make it a rule never to fool with any boiler fittings while there is a pressure of steam on the boiler. It is exceedingly dangerous.

Sometimes a gauge cock gets broken off accidentally while the boiler is in use. If such an accident happens, bank the fire by closing the draft and covering the fire with fresh fuel or ashes. Stop the engine and let the water blow out of the hole till only steam appears; then try to plug the opening with a long whitewood or poplar, or even a pine stick (six or eight feet long), one end of which you have whittled down to about the size of the hole. When the steam has been stopped the stick may be cut off close to the boiler and the plug driven in tight. If necessary you may continue to use the boiler in this condition until a new cock can be put in.

Q.   What should you do when a gauge cock is stopped up?

A.   Let the steam pressure go down, and then take off the front part and run a small wire into the passage, working the wire back and forth until all scale and sediment has been removed.

Q.   What should you do when the steam gauge gets out of order.

A.   If the steam gauge does not work correctly, or you suspect it does not, you may test it by running the steam up until it blows off at the safety valve. If the steam gauge does not indicate the pressure at which the safety valve is set to pop off, and you have reason to suppose the safety valve is all right, you may conclude that there is something the matter with the steam gauge. In that case either put in a new one, or, if you have no extra steam gauge on hand, shut down your boiler and engine till you can get your steam gauge repaired. Sometimes this can be done simply by adjusting the pointer, which may have got loose, and you can test it by attaching it to another boiler which has a steam gauge that is all right and by which you can check up yours. It is VERY DANGEROUS to run your boiler without a steam gauge, depending on the safety valve. Never allow the slightest variation in correctness of the steam gauge without repairing it at once. It will nearly always be cheaper in these days to put in a new gauge rather than try to repair the old one.

Q. What should you do if the pump fails to work?

A. Use the injector.

Q. What should you do if there is no injector?

A. Stop the engine at once and bank the fire with damp ashes, especially noting that the water does not fall below the bottom of the glass gauge. Then examine the pump. First see if the plunger leaks air; if it is all right, examine the check valves, using the little drain cock as previously explained to test the upper ones, for the valves may have become worn and will leak; third, if the check valves are all right, examine the supply pipe, looking at the strainer, observing whether suction takes place when the pump is worked, etc. There may be a leak in the suction hose somewhere during its course where air can get in, or it may become weak and collapse under the force of the atmosphere, or the lining of the suction pipe may have become torn or loose. The slightest leak in the suction pipe will spoil the working of the pump. Old tubing should never be used, as it is sure to *give* trouble. Finally, examine the delivery pipe. Close the cock or valve next the boiler, and examine the boiler check valve; notice whether the pipe is getting limed up. If necessary, disconnect the pipe and clean it out with a stiff wire. If everything is all right up to this point, you must let the boiler cool off, blow out the water, disconnect the pipe between the check and the boiler, and thoroughly clean the delivery pipe into the boiler. Stoppage of the delivery pipe is due to deposits of lime from the heating of the water in the heater. Stoppage from this source will be gradual, and you will find less and less water going into your boiler from your pump until none flows at all. From this you may guess the trouble.

Q. How may the communication with the water gauge always be kept free from lime?

A. By blowing it off through the drain cock at the bottom. First close the upper cock and blow off for a few seconds, the water passing through the lower cock; then close the lower cock and open the upper one, allowing the steam to blow through this and the drain cock for a few seconds. If you do this every day or oftener you will have no trouble.

O. Should the water get low for any reason, what should be done?

A. Close all dampers tight so as to prevent all draft, and bank the fire with fresh fuel or with ashes (damp ashes are the best if danger is great). Then let the boiler cool down before putting in fresh water. Banking the fire is better than drawing or dumping it, as either of these make the heat greater for a moment or two, and that additional heat might cause an explosion. Dashing cold water upon the fire is also very dangerous and in every way unwise. Again, do not open the safety valve, for that also, by relieving some of the pressure on the superheated water, might cause it to burst suddenly into steam and so cause an explosion.

Q. Under such circumstances, would you stop the engine?

A. No; for a sudden checking of the outflow of steam might bring about an explosion. Do nothing but check the heat as quickly and effectively as you can by banking or covering the fires.

Q. Why not turn on the feed water?

A. Because the crown sheet of the boiler has become overheated, and any cold water coming upon it would cause an explosion. If the pump or injector are running, of course you may let them run, and the boiler will gradually refill as the heat decreases. Under such circumstances low water is due to overheating the boiler.

Q.   Would not the fusible plug avert any disaster from low water?

A.   It might, and it might not. The top of it is liable to get coated with lime so that the device is worthless. You should act at all times precisely as if there were no fusible plug. If it ever does avert an explosion you may be thankful, but averting explosions by taking such means as we have suggested will be far better for an engineer's reputation.

Q.   Would not the safety valve be a safeguard against explosion?

A.   No; only under certain conditions. It prevents too high a pressure for accumulating in the boiler when there is plenty of water; but when the water gets low the safety valve may only hasten the explosion by relieving some of the pressure and allowing superheated water to burst suddenly into steam, thus vastly expanding instantly.

Q.   Should water be allowed to stand in the boiler when it is not in use?

A.   It is better to draw it off and clean the boiler, to prevent rusting, formation of scale, hardening of sediment, etc., if boiler is to be left for any great length of time.

Q.   What should you do if a grate bar breaks or falls out?

A.   You should always have a spare grate bar on hand to put in its place; but if you have none you may fill the space by wedging in a stick of hard wood cut the right shape to fill the opening. Cover this wood with ashes before poking the fire over it, and it will last for several hours before it burns out. You will find it exceedingly difficult to keep up the fire with a big hole in the grate that will let cold air into the furnace and allow coal to drop down.

In case the grate is of the rocker type the opening may be filled by shaping a piece of flat iron, which can be set in without interfering with the rocking of the grate; or the opening may be filled with wood as before if the wood is covered well with ashes. Of course the use of wood will prevent the grate from rocking and the poker must be used to clean.

Q. Why should an engineer never start a boiler with a hot fire, and never let his fire get hotter than is needed to keep up steam?

A. Both will cause the sheets to warp and the flues to become leaky, because under high heat some parts of the boiler will expand more rapidly than others. For a similar reason, any sudden application of cold to a boiler, either cold water or cold air through the firebox door, will cause quicker contraction of certain parts than other parts, and this will ruin a boiler.

Q. How should you supply a boiler with water?

A. In a regular stream continually. Only by making the water pass regularly and gradually through the heater will you get the full effect of the heat from the exhaust steam. If a great deal of water is pumped into the boiler at one time, the exhaust steam will not be sufficient to heat it as it ought. Then if you have a full boiler and shut off the water supply, the exhaust steam in the heater is wasted, for it can do no work at all. Besides, it hurts the boiler to allow the temperature to change, as it will inevitably do if water is supplied irregularly.

WHATEVER YOU DO, NEVER ATTEMPT TO TIGHTEN A SCREW OR CALK A BOILER UNDER STEAM PRESSURE. IF ANYTHING IS LOOSE IT IS LIABLE TO BLOW OUT IN YOUR FACE WITH DISASTROUS CONSEQUENCES.

Q.   If boiler flues become leaky, can an ordinary person tighten them?

A.   Yes, if the work is done carefully. See full explanation previously given, p. 17. Great care should be taken not to expand the flues too much, for by so doing you are likely to loosen other flues and cause more leaks than you had in the first place. Small leaks inside a boiler are not particularly dangerous, but they should be remedied at the earliest possible moment, since they reduce the power of the boiler and put out the fire. Besides, they look bad for the engineer.

Q.   How should flues be cleaned?

A.   Some use a steam blower; but a better way is to scrape off the metal with one of the many patent scrapers, which just fill the flue, and when attached to a rod and worked back and forth a few times the whole length of the flue do admirable service.

Q.   What harm will dirty flues do?

A.   Two difficulties arise from dirty flues. If they become reduced in size the fire will not burn well. Then, the same amount of heat will do far less work because it is so much harder for it to get through the layer of soot and ashes, which are non-conductors.

Q.   What would you do if the throttle broke?

A.   Use reverse lever.

CHAPTER VIII

POINTS FOR THE YOUNG ENGINEER.— (CONT.)

QUESTIONS AND ANSWERS

THE ENGINE

Q. What is the first thing to do with a new engine?

A. With some cotton waste or a soft rag saturated with benzine or turpentine clean off all the bright work; then clean every bearing, box and oil hole, using a force pump with air current first, if you have a pump, and then wiping the inside out clean with an oily rag, using a wire if necessary to make the work thorough. If you do not clean the working parts of the engine thus before setting it up, grit will get into the bearings and cause them to cut. Parts that have been put together need not be taken apart; but you should clean everything you can get at, especially the oil holes and other places that may receive dirt during transportation.

After the oil holes have been well cleaned, the oil cups may be wiped off and put in place, screwing them in with a wrench.

Q. What kind of oil should you use?

A. Cylinder oil only for the cylinder; lard oil for the bearings, and hard grease if your engine is provided with hard grease cup for the cross-head and crank. The only good substitute for cylinder oil is pure beef suet tried out. Merchantable tallow should never be used, as it contains acid.

Q. Can fittings be screwed on by hand only?

A.   No; all fittings should be screwed up tight with a wrench.

Q.   When all fittings are in place, what must be done before the engine can be started?

A.   See that the grates in the firebox are in place and all right; then fill the boiler with clean water until it shows an inch to an inch and a half in the water gauge. Start your fire, and let it burn slowly until there is a pressure in the boiler of 10 or 15 lbs. Then you can turn on the blower to get up draft. In the meantime fill all the oil cups with oil; put grease on the gears; open and close all cocks to see that they work all right; turn your engine over a few times to see that it works all right; let a little steam into the cylinder with both cylinder cocks open— just enough to show at the cocks without moving the engine—and slowly turn the engine over, stopping it on the dead centers to see if the steam comes from only one of the cylinder cocks at a time, and that the proper one; reverse the engine and make the same test. Also see that the cylinder oiler is in place and ready for operation. See that the pump is all right and in place, with the valve in the feedpipe open and also the valve in the supply pipe.

By going over the engine in this way you will notice whether everything is tight and in working order, and whether you have failed to notice any part which you do not understand. If there is any part or fitting you do not understand, know all about it before you go ahead.

Having started your fire with dry wood, add fuel gradually, a little at a time, until you have a fire covering every part of the grate. Regulate the fire by the damper alone, never opening the firebox door even if the fire gets too hot.

Q.   In what way should the engine be started?

A. When you have from 25 to 40 lbs. of pressure open the throttle valve a little, allowing the cylinder cocks to be open also. Some steam will condense at first in the cold cylinder, and this water must be allowed to drain off. See that the crank is not on a dead center, and put on just enough steam to start the engine. As soon as it gets warmed up, and only dry steam appears at the cocks, close the cylinder cocks, open the throttle gradually till it is wide open, and wait for the engine to work up to its full speed.

Q. How is the speed of the engine regulated??

A. By the governor, which is operated by a belt running to the main shaft. The governor is a delicate apparatus, and should be watched closely. It should move up and down freely on the stem, which should not leak steam. If it doesn't work steadily, you should stop the engine and adjust it, after watching it for a minute or two to see just where the difficulty lies.

Q. Are you likely to have any hot boxes?

A. There should be none if the bearings are all clean and well supplied with oil. However, in starting a new engine you should stop now and then and examine every bearing by laying your hand upon it. Remember the eccentric, the link pin, the cross-head, the crank pin. If there is any heat, loosen the boxes up a trifle, but only a very little at a time. If you notice any knocking or pounding, you have loosened too much, and should tighten again.

Q. What must you do in regard to water supply?

A. After the engine is started and you know it is all right, fill the tank on the engine and start the injector. It may take some patience to get the injector started, and you should carefully follow the directions previously given and those which apply especially to the type of injector used. Especially be sure that the cocks admitting the water through the feed pipe and into the boiler are open.

Q. Why are both a pump and an injector required on an engine?

A. The pump is most economical, because it permits the heat in the exhaust steam to be used to heat the feed water, while the injector heats the water by live steam. There should also be an injector, however, for use when the engine is not working, in order that the water in the boiler may be kept up with heated water. If a cross-head pump is used, of course, it will not operate when the engine is not running; and in case of an independent pump the heater will not heat the water when the engine is not running because there is little or no exhaust steam available. There is an independent pump (the Marsh pump) which heats the water before it goes into the boiler, and this may be used when the engine is shut down instead of the injector.

Q. What is the next thing to test?

A. The reversing mechanism. Throw the reverse lever back, and see if the engine will run equally well in the opposite direction. Repeat this a few times to make sure that the reverse is in good order.

Q. How is a traction engine set going upon the road?

A. Most traction engines now have the friction clutch. When the engine is going at full speed, take hold of the clutch lever and slowly bring the clutch against the band wheel. It will slip a little at first, gradually engaging the gears and moving the outfit. Hold the clutch lever in one hand, while with the other you operate the steering wheel. By keeping your hand on the clutch lever you may stop forward motion instantly if anything goes wrong. When the engine is once upon the road, the clutch lever may set in the notch provided for it, and the engine will go at full speed. You can then give your entire attention to steering.

Q. What should you do if the engine has no friction clutch?

A. Stop the engine, placing the reversing lever in the center notch. Then slide the spur pinion into the gear and open the throttle valve wide. You are now ready to control the engine by the reversing lever. Throw the lever forward a little, bringing it back, and so continue until you have got the engine started gradually. When well under way throw the reverse lever into the last notch, and give your attention to steering.

Q. How should you steer a traction engine?

A. In all cases the same man should handle the throttle and steer the engine. Skill in steering comes by practice, and about the only rule that can be given is to go slow, and under no circumstances jerk your engine about. Good steering depends a great deal on natural ability to judge distances by the eye and power by the feel. A good engineer must have a good eye, a good ear, and a good touch (if we may so speak). If either is wanting, success will be uncertain.

Q. How should an engine be handled on the road?

A. There will be no special difficulty in handling an engine on a straight, level piece of road, especially if the road is hard and without holes. But when you come to your first hill your troubles will begin.

Before ascending a hill, see that the water in the boiler does not stand more than two inches in the glass gauge. If there is too much water, as it is thrown to one end of the engine by the grade it is liable to get into the steam cylinder. If you have too much water, blow off a little from the bottom blow-off cock.

In descending a hill never stop your engine for a moment, since your crown sheet will be uncovered by reason of the water being thrown forward, and any cessation in the jolting of the engine which keeps the water flowing over the crown sheet will cause the fusible plug to blow out, making delay and expense.

Make it a point never to stop your engine except on the level.

Before descending a hill, shut off the steam at the throttle, and control the engine by the friction brake; or if there is no brake, do not quite close the throttle, but set the reverse lever in the center notch, or back far enough to control the speed. It is seldom necessary to use steam in going down hill, however, and if the throttle is closed even with no friction brake, the reverse may be used in such a way as to form an air brake in the cylinder.

Get down to the bottom of a hill as quickly as you can.

Before descending a hill it would be well to close your dampers and keep the firebox door closed tight all the time. Cover the fire with fresh fuel so as to keep the heat down.

The pump or injector must be kept at work, however, since as you have let the water down low, you must not let it fall any lower or you are likely to have trouble.

In ascending a hill, do just the reverse, namely: Keep your fire brisk and hot, with steam pressure ascending; and throw the reverse lever in the last notch, giving the engine all the steam you can, else you may get stuck. If you stop you are likely to overheat forward end of fire tubes. You are less liable to get stuck if you go slowly than if you go fast. Regulate speed by friction clutch.

CHAPTER IX

POINTS FOR THE YOUNG ENGINEER.—(CONT.)

MISCELLANEOUS

Q. What is Foaming?

A. The word is used to describe the rising of water in large bubbles or foam. You will detect it by noticing that the water in the glass gauge rises and falls, or is foamy. It is due to sediment in the boiler, or grease and other impurities in the feed supply. Shaking up the boiler will start foaming sometimes; at other times it will start without apparent cause. In such cases it is due to the steam trying to get through a thick crust on the surface of the water.

Q. How may you prevent foaming?

A. It may be checked for a moment by turning off the throttle, so giving the water a chance to settle. It is generally prevented by frequently using the surface blow-off to clear away the scum. Of course the water must be kept as pure as possible, and especially should alkali water be avoided.

Q. What is priming?

A. Priming is not the same as foaming, though it is often caused by foaming. Priming is the carrying of water into the steam cylinder with the steam. It is caused by various things beside foaming, for it may be found when the boiler is quite clean. A sudden and very hot fire may start priming. Priming sometimes follows lowering of the steam pressure. Often it is due to lack of capacity in the boiler, especially lack of steam space, or lack of good circulation.

Q. How can you detect priming?

A. By the clicking sound it makes in the steam cylinder. The water in the gauge will also go up and down violently. There will also be **a** shower of water from the exhaust.

Q. What is the proper remedy for priming?

A. If it is due to lack of capacity in the boiler nothing can be done but *get* a new boiler. In other cases it may be remedied by carrying less water in the boiler when that can be done safely, by taking steam from a different point in the steam dome, or if there is no dome by using a long dry pipe with perforation at the end.

A larger steam pipe may help it; or it may be remedied by taking out the top row of flues.

Leaky cylinder rings or a leaky valve may also have something to do with it. In all cases these should be made steam tight. If the exhaust nozzle is choked up with grease or sediment, clean it out.

A traction engine with small steam ports would prime quickly under forced speed.

Q. How would you bank your fires?

A. Push the fire as far to the back of the firebox as possible and cover it over with very fine coal or with dry ashes. As large a portion as possible of the grate should be left open, so that the air may pass over the fire. Close the damper tight. By banking your fires at night you keep the boiler warm and can get up steam more quickly in the morning.

Q. When water is left in the boiler with banked fire in cold weather, what precautions ought to be taken?

A. The cocks in the glass water gauge should be closed and the drain cock at the bottom opened, for fear the water in the exposed gauge should freeze. Likewise all drain cocks in steam cylinder and pump should be opened.

Q. How should a traction engine be prepared for laying up during the winter?

A. First, the outside of the boiler and engine should be thoroughly cleaned, seeing that all gummy oil or grease is removed. Then give the outside of the boiler and smokestack a coat of asphalt paint, or a coat of lampblack and linseed oil, or at any rate a doping of grease.

The outside of the boiler should be cleaned while it is hot, so that grease, etc., may be easily removed while soft.

After the outside has been attended to, blow out the water at low pressure and thoroughly clean the inside in the usual way, taking out the handhole and manhole plates, and scraping off all scale and sediment.

After the boiler has been cleaned on the inside, fill it nearly full of water, and pour upon the top a bucket of black oil. Then let the water out through the blow-off at the bottom. As the water goes down it will have a coating of oil down the sides of the boiler.

All the brass fittings should be removed, including gauge cocks, check valves, safety valve, etc. Disconnect all pipes that may contain water, to be sure none remains in any of them. Open all stuffing boxes and take out packing, for the packing will cause the parts they surround to rust.

Finally, clean out the inside of the firebox and the fire flues, and give the ash-pan a good coat of paint all over, inside as well as out.

The inside of the cylinder should be well greased, which can be done by removing the cylinder head.

See that the top of the smoke stack Is covered to keep out the weather.

All brass fittings should be carefully packed and put away in a dry place.

A little attention to the engine when you put it up will save twice as much time when you take it out next season, and besides save many dollars of value in the life of the engine.

Q.   How should belting be cared for?

A.   First, keep belts free from dust and dirt.

Never overload belts.

Do not let oil or grease drip upon them.

Never put any sticky or pasty grease on a belt.

Never allow any animal oil or grease to touch a rubber belt, since it will destroy the life of the rubber.

The grain or hair side should run next the pulley, as it holds better and is not so likely to slip.

Rubber belts will be greatly improved if they are covered with a mixture of black lead and litharge, equal parts, mixed with boiled oil, and just enough japan to dry them quickly. This mixture will do to put on places that peel.

Q.   What is the proper way to lace a belt?

A.   First, square the ends with a proper square, cutting them off to a nicety. Begin to lace in the middle, and do not cross the laces on the pulley side. On that side the lacings should run straight with the length of the belt.

The holes in the belt should be punched if possible with an oval punch, the long diameter coinciding with the length of the belt. Make two rows of holes in each end of the belt, so that the holes in each row will alternate with those in preceding row, making a zigzag. Four holes will be required for a three-inch belt in each end, two holes in each row; in a six-inch belt, place seven holes in each end, four in the row nearest the end.

To find the length of a belt when the exact length cannot be measured conveniently, measure a straight line from the center of one pulley to the center of the other. Add together half the diameter of each pulley, and multiply that by 31/4 (3.1416). The result added to twice the distance between the centers will give the total length of the belt.

A belt will work best if it is allowed to sag just a trifle.

The seam side of a rubber belt should be placed outward, or away from the pulley.

If such a belt slips, coat the inside with boiled linseed oil or soap.

Cotton belting may be preserved by painting the pulley side while running with common paint, afterward applying soft oil or grease.

If a belt slips apply a little oil or soap to the pulley side.

Q. How does the capacity of belts vary?

A. In proportion to width and also to the speed. Double the width and you double the capacity; also, within a certain limit, double the speed and you double the capacity. A belt should not be run over 5,000 feet per minute. One four-inch belt will have the same capacity as two two-inch belts.

Q. How are piston rods and valve rods packed so that the steam cannot escape around them?

A. By packing placed in stuffing-boxes. The stuffing is of some material that has a certain amount of elasticity, such as lamp wick, hemp, soap stone, etc., and certain patent preparations. The packing is held in place by a gland, as it is called, which acts to tighten the packing as the cap of the stuffing-box is screwed up.

Q.   How would you repack a stuffing-box?

A.   First remove the cap and the gland, and with a proper tool take out all the old packing. Do not use any rough instrument like a file, which is liable to scratch the rod, for any injury to the smooth surface of the rod, will make it leak steam or work hard.

If patent packing is used, cut off a sufficient number of lengths to make the required rings. They should be exactly the right length to go around inside the stuffing-box. If too long, they cannot be screwed up tight, as the ends will press together and cause irregularities. If too short, the ends will not meet and will leak steam. Cut the ends diagonally so that they will make a lap joint instead of a square one. When the stuffing-box has been filled, place the gland in position and screw up tight. Afterwards loosen the nuts a trifle, as the steam will cause the packing to expand, usually. The stuffing-box should be just as loose as it can be and not allow leakage of steam. If steam leaks, screw up the box a little tighter. If it still leaks, do not screw up as tight as you possibly can, but repack the box. If the stuffing-box is too tight, either for the piston rod or valve steam, it will cause the engine to work hard, and may groove the rods and spoil them.

If hemp packing is used, pull the fibres out straight and free, getting rid of all knots and lumps. Twist together a few of the fibres, making three cords, and braid these three cords together and soak them with oil or grease, wind around the rod till stuffing-box is sufficiently full, replace the gland, and screw up as before.

Stuffing-box for water piston of pump may be packed as described above, but little oil or grease will be needed.

Never pack the stuffing-box too tight, or you may flute the rod and spoil it.

Always keep the packing in a clean place, well covered up, never allowing any dust to get into it, for the dust or grit is liable to cut the rod.

CHAPTER X

ECONOMY IN RUNNING A FARM ENGINE

It is something to be able to run a farm engine and keep out of trouble. It is even a great deal if everything runs smoothly day in and day out, if the engine looks clean, and you can always develop the amount of power you need. You must be able to do this before you can give the fine points of engineering much consideration.

When you come to the point where you are always able to keep out of trouble, you are probably ready to learn how you can make your engine do more work on less fuel than it does at present. In that direction the best of us have an infinite amount to learn. It is a fact that in an ordinary farm engine only about 4 per cent of the coal energy is actually saved and used for work; the rest is lost, partly in the boiler, more largely in the engine. So we see what a splendid chance there is to save.

If we are asked where all the lost energy goes to, we might reply in a general sort of way, a good deal goes up the smokestack in smoke or unused fuel; some is radiated from the boiler in the form of heat and is lost without producing any effect on the steam within the boiler; some is lost in the cooling of the steam as it passes to the steam cylinder; some is lost in the cooling of the cylinder itself after each stroke; some is lost through the pressure on the back of the steam valve, causing a friction that requires a good deal of energy in the engine to overcome; some is lost in friction in the bearings, stuffing-boxes, etc. At each of these points economy may be practiced if the engineer knows how to do it. We offer a few suggestions.

THEORY OF STEAM POWER

As economy is a scientific question, we cannot study it intelligently without knowing something of the theory of heat, steam and the transmission of power. There will be nothing technical in the following pages; and as soon as the theory is explained in simple language, any intelligent person will know for himself just what he ought to do in any given case.

First, let us define or describe heat according to the scientific theory. Scientists suppose that all matter is made up of small particles called molecules, so small that they have never been seen. Each molecule is made up of still smaller particles called atoms. There is nothing smaller than an atom, and there are only about sixty-five different kinds of atoms, which are called elements; or rather, any substance made up of only one kind of atom is called an element. Thus iron is an element, and so is zinc, hydrogen, oxygen, etc. But a substance like water is not an element, but a compound, since its molecules are made up of an atom of oxygen united with two atoms of hydrogen. Wood is made up of many different kinds of atoms united in various ways. Air is not a compound, but a mixture of oxygen, nitrogen and a few other substances in small quantities.

The reason why air is a mixture and not a compound is an interesting one, and brings us to our next point. In order to form a compound, two different kinds of atoms must have an attraction for each other. There is no attraction between oxygen and nitrogen; but there is great attraction between oxygen and carbon, and when they get a chance they rush together like long separated overs. Anthracite coal is almost pure carbon.

So is charcoal. Soft coal consists of carbon with which various other things are united, one of them, being hydrogen. This is interesting and important, because it accounts for a curious thing in firing up boilers with soft coal. We have already said that water is oxygen united with hydrogen. When soft coal burns, not only does the carbon unite with oxygen, but the hydrogen unites with oxygen and forms water, or steam. While the boilers are cold they will condense the water or steam in the smoke, just as a cold plate in a steamy room will condense water from the steamy air, so sweating.

Now the scientists suppose that two or three atoms stick together by reason of their attraction for each other and form molecules. These molecules in turn stick together and form liquids and solids. The tighter they stick, the harder the substance. At the same time, these molecules are more or less loose, and are constantly moving back and forth. In a solid like iron they move very little; but a current of electricity through iron makes the molecules move in a peculiar way. In a liquid like water, the molecules cling together very loosely, and may easily be pulled apart. In any gas, like air or steam, the molecules are entirely disconnected, and are constantly trying to get farther apart.

Heat, says the scientist, is nothing more or less than the movement of the molecules back and forth. Heat up a piece of iron in a hot furnace, and the molecules keep getting further and further apart, and the iron gets softer and softer, till it becomes a liquid. If we take some liquid like water and heat it, the molecules get farther and farther apart, till the water boils, as we say, or turns into steam. As steam the molecules have broken apart entirely, and are beating back and forth so rapidly that they have a tendency to push each other farther and farther apart. This pushing tendency is the cause of steam pressure. It also explains why steam has an expansive power.

Heat, then, is the movement of the molecules back and forth. There are three fixed ranges in which they move; the small range makes a solid; the next range makes a liquid; the third range makes a gas, such as steam. These three states of matter as affected by heat are very sharp and definite. The point at which a solid turns to a liquid is called the melting point. The melting point of ice is 32° Fahr. The point at which it turns to a gas is called the boiling point. With water that is 212° Fahr. The general tendency of heat is to push apart, or expand; and when the heat is taken away the substances contract.

Let us consider our steam boiler. We saw that some different kinds of atoms have a strong tendency to rush together; for example, oxygen and carbon. The air is full of oxygen, and coal and wood are full of carbon. When they are raised to a certain temperature, and the molecules get loose enough so that they can tear themselves away from whatever they are attached to, they rush together with terrible force, which sets all surrounding molecules to vibrating faster than ever. This means that heat is given out.

Another important thing is that when a solid changes to a liquid, or a liquid to a gas, it must take up a certain amount of heat to keep the molecules always just so far apart. That heat is said to become latent, for it will not show in a thermometer, it will not cause anything to expand, nor will it do any work. It merely serves to hold the molecules just so far apart.

HOW ENERGY IS LOST

We may now see some of the ways in which energy is lost. First, the air which goes into the firebox consists of nitrogen as well as oxygen. That nitrogen is only in the way, and takes heat from the fire, which it carries out at the smokestack.

Again, if the air cannot get through the bed of coals easily enough, or there is not enough of it so that every atom, of carbon, etc., will find the right number of atoms of oxygen, some of the atoms of carbon will be torn off and united with oxygen, and the other atoms of carbon, left without any oxygen to unite with, will go floating out at the smokestack as black smoke. Also, the carbon and the oxygen cannot unite except at a certain temperature, and when fresh fuel is thrown on the fire it is cold, and a good many atoms of carbon after being loosened up, get cooled off again before they have a chance to find an atom of oxygen, and so they, too, go floating off and are lost.

If the smoke could be heated up, and there were enough oxygen mixed with it, the loose carbon would still burn and produce heat, and there would be an economy of fuel. This has given rise to smoke consumers, and arranging two boilers, so that when one is being fired the heat from the other will catch the loose carbon before it gets away and burn it up.

So we have these points:

1. Enough oxygen or air must get into a furnace so that every atom of carbon will have its atom of oxygen. This means that you must have a good draft and that the air must have a chance to get through the coal or other fuel.

2. The fuel must be kept hot enough all the time so that the carbon and oxygen can unite. Throwing on too

much cold fuel at one time will lower the heat beyond the economical point and cause loss in thick smoke.

3. If the smoke can pass over a hot bed of coals, or through a hot chamber, the carbon in it may still be burned. This suggests putting fuel at the front of the fire-box, a little at a time, so that its smoke will have to pass over a hot bed of coals and the waste carbon will be burned. When the fresh fuel gets heated up, it may be pushed farther back.

From a practical point of view these points mean, No dead plates in a furnace to keep the air from going through coal or wood; a thin fire so the air can get through easily; place the fresh fuel where its smoke will have a chance to be burned; and do not cool off the furnace by putting on much fresh fuel at a time.

(Later we will give more hints on firing.)

HOW HEAT IS DISTRIBUTED

We have described heat as the movement of molecules back and forth at a high rate of speed. If these heated molecules beat against a solid like iron, its molecules are set in motion, one knocks the next, and so on, just as you push one man in a crowd, he pushes the next, and so on till the push comes out on the other side. So heat passes through iron and appears on the other side. This is called "conduction."

All space is supposed to be filled with a substance in which heat, light, etc., may be transmitted, called the ether. When the molecules of a sheet of iron are heated, or set vibrating, they transmit the vibration through the air, or ether. This is called "radiation." Heat is "conducted" through solid and liquid substances, and "radiated" through gases.

Now some substances conduct heat readily, and some do so with the greatest difficulty. Iron is a good conductor; carbon, or soot on the flues of a boiler, and lime or scale on the inside of a boiler, are very poor conductors. So the heat will go through the iron and steel to the water in a boiler quickly and easily, and a large per cent of the heat of the furnace will get to the water in a boiler. When a boiler is old and is clogged with soot and coated with lime, the heat cannot get through easily, and goes off in the smokestack. The air coming out of the smokestack will be much hotter; and that extra heat is lost.

Iron is a good radiator, too. So if the outer shell of a boiler is exposed to the air, a great deal of heat will run off into space and be lost. Here, then, is where you need a non-conductor, as it is called, such as lime, wood, or the like.

Economy says, cover the outside of a boiler shell with a non-conductor. This may be brickwork in a set boiler; in a traction boiler it means a jacket of wood, plaster, hair, or the like. The steam pipe, if it passes through outer air, should be covered with felt; and the steam cylinder ought to have its jacket, too.

At the same time all soot and all scale should be scrupulously cleaned away.

PROPERTIES OF STEAM

As we have already seen, steam is a gas. It is slightly blue in color, just as the water in the ocean is blue, or the air in the sky.

We must distinguish between steam and vapor. Vapor is small particles of water hanging in the air. They seem to stick to the molecules composing the air, or hang there in minute drops. Water hanging in the air is, of course, water still. Its molecules do not have the movement that the molecules of a true gas do, such as steam is. Steam, moreover, has absorbed latent heat, and has expansive force; but vapor has no latent heat, and no expansive force. So vapor is dead and lifeless, while steam is live and full of energy to do work.

When vapor gets mixed with steam it is only in the way; it is a sort of dead weight that must be carried; and the steam power is diminished by having vapor mixed with it.

Now all steam as it bubbles up through water in boiling takes up with it a certain amount of vapor. Such steam is called "wet" steam. When the vapor is no longer in it, the steam is called "dry" steam. It is dry steam that does the best work, and that every engineer wants to get.

While water will be taken up to great heights in the air and form clouds, in steam it will not rise very much, and at a certain height above the level of the water in a boiler the steam will be much drier than near the surface. For this reason steam domes have been devised, so that the steam may be taken out at a point as high as possible above the water in the boiler, and so be as dry as possible. Also "dry tubes" have been devised, which let the steam pass through many small holes that serve to keep back the water to a certain extent.

However, there will be more or less moisture in all steam until it has been superheated, as it is called. This may be done by passing it through the hot part of the furnace, where the added heat will turn all the moisture in the steam into steam, and we shall have perfectly dry steam.

The moment, however, that steam goes through a cold pipe, or one cooled by radiation, or goes into a cold cylinder, or a cylinder cooled by radiation, some of the steam will turn to water, or condense, as it is called. So we have the same trouble again.

Much moisture passing into the cylinder with the steam is called "priming." In that case the dead weight of water has become so great as to kill a great part of the steam power.

HOW TO USE THE EXPANSIVE POWER OF STEAM

We have said that the molecules in steam are always trying to get farther and farther apart. If they are free in the air, they will soon scatter; but if they are confined in a boiler or cylinder they merely push out in every direction, forming "pressure."

When steam is let into the cylinder it has the whole accumulated pressure in the boiler behind it, and of course that exerts a strong push on the piston. Shut off the boiler pressure and the steam in the cylinder will still have its own natural tendency to expand. As the space in the cylinder grows larger with the movement of the piston from end to end, the expansive power of the steam becomes less and less, of course. However, every little helps, and the push this lessened expansive force exerts on the piston is so much energy saved. If the full boiler pressure is kept on the piston the whole length of the stroke, and then the exhaust port is immediately opened, all this expansive energy of the steam is lost. It escapes through the exhaust nozzle into the smokestack and is gone. Possibly it cannot get out quickly enough, and causes back pressure on the cylinder when the piston begins its return stroke, so reducing the power of the engine.

To save this the skilled engineer "notches up" his reverse lever, as they say. The reverse lever controls the valve travel. When the lever is in the last notch the valve has its full travel. When the lever is in the center notch the valve has no travel at all, and no steam can get into the cylinder; on the other side the lever allows the valve to travel gradually more and more in the opposite direction, so reversing the engine.

As the change from one direction to the other direction is, of course, gradual, the valve movement is shortened by degrees, and lets steam into the cylinder for a correspondingly less time. At its full travel it perhaps lets steam into the cylinder for three-quarters of its stroke. For the last quarter the work is done by the expansive power of the steam.

Set the lever in the half notch, and the travel of the valve is so altered that steam can get into the cylinder only during half the stroke of the piston, the work during the rest of the stroke being done by the expansive force of the steam.

Set the lever in the notch next to the middle notch, or the quarter notch, and steam will get into the cylinder only during a quarter of the stroke of the piston, the work being done during three-quarters of the stroke by the expansive force of the steam.

Obviously the more the steam is expanded the less work it can do. But when it escapes at the exhaust there will be very little pressure to be carried away and lost.

Therefore when the load on his engine is light the economical engineer will "notch up" his engine with the reverse lever, and will use up correspondingly less steam and save correspondingly more fuel. When the load is unusually heavy, however, he will have to use the full power of the pressure in the boiler, and the waste cannot be helped.

THE COMPOUND ENGINE

The compound engine is an arrangement of steam cylinders to save the expansive power of steam at all times by letting the steam from one cylinder where it is at high pressure into another after it exhausts from the first, in this second cylinder doing more work purely by the expansive power of the steam.

The illustration shows a sectional view of a compound engine having two cylinders, one high pressure and one low. The low pressure cylinder is much larger than the high pressure. There is a single plate between them called the center head, and the same piston rod is fitted with two pistons, one for each cylinder. The steam chest does not receive steam from the boiler, but from the exhaust of the high pressure cylinder. The steam from the boiler goes into a chamber in the double valve, from which it passes to the ports of the high pressure cylinder. At the return stroke the exhaust steam escapes into the steam chest, and from there it passes into the low pressure cylinder. There may be one valve riding on the back of another; but the simplest form of compound engine is built with a single double valve, which opens and closes the ports for both cylinders at one movement.

Theoretically the compound engine should effect a genuine economy. In practice there are many things to operate against this. Of course if the steam pressure is low to start with, the amount of pressure lost in the exhaust will be small. But if it is very high, the saving in the low pressure cylinder will be relatively large. If the work can be done just as well with a low pressure, it would be a practical waste to keep the pressure abnormally high in order to make the most of the compound engine.

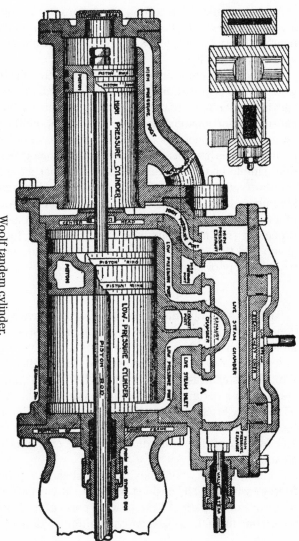

Woolf tandem cylinder.

An engine must be a certain size before the saving of a compound cylinder will be appreciable. In these days nearly all very large engines are compound, while small engines are simple.

Another consideration to be taken into account is that a compound is more complicated and so harder to manage; and when any unfavorable condition causes loss it causes proportionately more loss on a compound than on a simple engine. For these and other reasons compound engines have been used less for traction purposes than simple engines have. It is probable that a skilled and thoroughly competent engineer, who would manage his engine in a scientific manner, would get more out of a compound than out of a simple; and this would be especially true in regions where fuel is high. If fuel is cheap and the engineer unskilled, a compound engine would be a poor economizer.

FRICTION

We have seen that the molecules of water have a tendency to stick in the steam as vapor or moisture. All molecules that are brought into close contact have more or less tendency to stick together, and this is called friction. The steam as it passes along the steam pipe is checked to a certain extent by the friction on the sides of the pipe. Friction causes heat, and it means that the heat caused has been taken from some source of energy. The friction of the steam diminishes the energy of the steam.

So, too, the fly wheel moving against the air suffers friction with the air, besides having to drive particles of air out of its path. All the moving parts of an engine where one metal moves on another suffer friction, since where the metals are pressed very tightly together they have more tendency to stick than when not pressed so tightly. When iron is pressed too tightly, as under the blows of a hammer in a soft state, it actually welds together solidly.

There is a great deal of friction in the steam cylinder, since the packing rings must press hard against the walls of the cylinder to prevent the steam from getting through. There is a great deal of friction between the D valve and its seat, because of the high steam pressure on the back of the valve. There is friction in the stuffing boxes both of the valve and the piston. There is friction at all the bearings.

There are various ways in which friction may be reduced. The most obvious is to adjust all parts so nicely that they will bind as little as possible. The stuffing-boxes will be no tighter than is necessary to prevent leaking of steam; and so with the piston rings. Journal boxes will be tight enough to prevent pounding, but no tighter. To obtain just the right adjustment requires great patience and the keen powers of observation and judgment.

The makers of engines try to reduce friction as much as possible by using anti-friction metals in the boxes. Iron and steel have to be used in shafts, gears, etc., because of the strength that they possess; but there are some metals that stick to each other and to iron and steel much less than iron or steel stick to each other when pressed close together. These metals are more or less soft; but they may be used in boxes and journal bearings. They are called anti-friction metals. The hardest for practical purposes is brass, and brass is used where there is much wear. Where there is less wear various alloys of copper, tin, zinc, etc., may be used in the boxes. One of these is babbit metal, which is often used in the main journal box.

All these anti-friction metals wear out rapidly, and they must be put in so that they can be adjusted or renewed easily.

But the great anti-friction agent is oil.

Oil is peculiar in that while the molecules seem to stick tightly together and to a metal like iron or steel, they roll around upon each other with the utmost ease. An ideal lubricator is one that sticks so tight to the journal that it forms a sort of cushion all around it, and prevents any of its molecules coming into contact with the molecules of the metal box. All the friction then takes place between the different molecules of oil, and this friction is a minimum.

The same principle has been applied to mechanics in the ball bearing. A number of little balls roll around between the journal and its box, preventing the two metals from coming into contact with each other; while the balls, being spheres, touch each other only at a single point, and the total space at which sticking can occur is reduced to a minimum.

As is well known, there is great difference in oils. Some evaporate, like gasoline and kerosene, and so disappear quickly. Others do not stick tightly to the journal, so are easily forced out of place, and the metals are allowed to come together. What is wanted, then, is a heavy, sticky oil that will not get hard, but will always form a good cushion between bearings.

Steam cylinders cannot be oiled directly, but the oil must be carried to the steam chest and cylinder in the steam. A good cylinder oil must be able to stand a high temperature. While it is diffused easily in the steam, it must stick tightly to the walls of the steam cylinder and to the valve seat, and keep them lubricated. Once it is stuck to the metal, the heat of the steam should not evaporate it and carry it away.

Again, a cylinder oil should not have any acid in it which would have a tendency to corrode the metal. Nearly all animal fats do have some such acid. So tallow and the like should not be placed where they can corrode iron or steel. Lard and suet alone are suitable for use on an engine.

When it comes to lubricating traction gears, other problems appear. A heavy grease will stick to the gears and prevent them from cutting; but it will stick equally to all sand and grit that may come along, and that, working between the cogs, may cut them badly. So some engineers recommend the use on gears of an oil that does not gather so much dirt.

The friction of the valve on its seat due to the pressure of the steam on its back has given rise to many inventions for counteracting it. The most obvious of these is what is called "the balanced valve." In the compound engine, where the steam pressure is obtained upon both sides of the valve, it rides much more lightly on its seat—so lightly, indeed, that when steam pressure is low, as in going down hill or operating under a light load, plunger pistons must be used to keep the valve down tight on its seat.

The poppet valves were devised to obviate the undue friction of the D valve; but the same loss of energy is to a certain extent transferred, and the practical saving is not always equal to the theoretical. On large stationary engines rotary valves and other forms, such as are used on the Corliss engine, have come into common use; but they are too complicated for a farm engine, which must be as simple as possible, with least possible liability of getting out of order.

CHAPTER XI

ECONOMY IN RUNNING A FARM ENGINE.— (CONT.)

PRACTICAL POINTS

The first practical point in the direction of farm engine economy is to note that the best work can be done only when every part of the engine and boiler are in due proportion. If the power is in excess of the work to be done there is loss; if the grate surface is too large cold air gets through the fuel and prevents complete combustion, and if the grate surface is too small, not enough air gets in; if the steaming power of the boiler is too large, heat is radiated away that otherwise could be saved, for every foot of exposed area in the boiler is a source of loss; if the steaming power of the boiler is too low for the work to be done, it requires extra fuel to force the boiler to do its work, and any forcing means comparatively large loss or waste. It will be seen that not only must the engine and boiler be built with the proper proportions, but they must be bought with a nice sense of proportion to the work expected of them. This requires excellent judgment and some experience in measuring work in horsepowers.

GRATE SURFACE AND FUEL

The grate surface in a firebox should be not less than two-thirds of a square foot per horsepower, for average size traction engines. If the horsepower of an engine is small, proportionately more grate surface will be needed; if it is large, the grate surface may be proportionately much smaller. An engine boiler 7x8x200 rev., with 100 lbs. pressure, should have a great surface not less than six square feet, and seven would be better. In a traction engine there is always a tendency to make the grate surface as small as possible, so that the engine will not be cumbersome.

Another reason why the grate surface should be sufficiently large is that forced draft is a bad thing, since it has a tendency to carry the products of combustion and hot gases through the smokestack and out into space before they have time to complete combustion and especially before the heat of the gases has time to be absorbed by the boiler surface. A large grate surface, then, with a moderate draft, is the most economical.

The draft depends on other things, however. If a great deal of fine fuel is thrown on a fire, the air must be forced through, because it cannot get through in the natural way. This results in waste. So a fire should be as open as possible. Coal should be "thin" on the grates; wood should be thrown in so that there will be plenty of air spaces; straw should be fed in just so that it will burn up completely as it goes in. Moderate size coal is better than small or fine. Dust in coal checks the draft. A good engineer will choose his fuel and handle his fire so that he can get along with as little forced draft as possible.

In a straw burning engine a good circulation of air can be obtained, if the draft door is just below the straw funnel, by extending the funnel into the furnace six inches or so. This keeps the straw from clogging up the place where the air enters and enables it to get at the fuel so much more freely that the combustion is much more complete.

We have already suggested that in firing with coal, the fresh fuel be deposited in front, so that the smoke will have to pass over live coals and so the combustion will be more complete. Then when the coal is well lighted it can be poked back over the other portions of the grate. This method has another advantage, in that the first heating is usually sufficient to separate the pure coal from the mineral substances which form, clinkers, and most of the clinkers will be deposited at that one point in the grate. Here they can easily be lifted out, and will not seriously interfere with the burning of the coal as they would if scattered all over the grate. Clinkers in front can easily be taken out by hooking the poker over them toward the back of the firebox and pulling them up and to the front. They often come out as one big mass which can be easily lifted out.

The best time to clean the grate is when there is a good brisk fire. Then it will not cause steam to go down. Stirring a fire does little good. For one thing, it breaks up the clinkers and allows them to run down on the grate bars when they stick and finally warp the bars. If the fire is not stirred the clinkers can be lifted out in large masses. Stirring a fire also creates a tendency to choke up or coke, and interferes with the even and regular combustion of the coal at all points.

The highest heat that can be produced is a yellow heat. When there is a good yellow heat, forced draft will only carry off the heat and cause waste. It will not cause still more rapid combustion. When the heat is merely red, increased draft will raise the temperature. Combustion is not complete until the flame shows yellow. However, if the draft is slight and time is given, red heat will be nearly as effective, but it will not carry the heated gases over so large a part of the heating surface of the boiler. With a very large grate surface, red heat will do very well. Certainly it will be better than a forced draft, or an effort at heating beyond the yellow point.

BOILER HEATING SURFACE

The heat of the furnace does its work only as the heated gases touch the boiler surface. The iron conducts the heat through to the water, which is raised to the boiling point and turned into steam.

Now the amount of heat that the boiler will take up is directly in proportion to the amount of exposed surface and to the time of exposure. If the boiler heating surface is small, and the draft is forced so that the gases pass through rapidly, they do not have a chance to communicate much heat.

Also if the heating surface is too large, so that it cannot all be utilized, the part not used becomes a radiating surface, and the efficiency of the boiler is impaired.

Practice has shown that the amount of heating surface practically required by a boiler is 12 to 15 square feet per horsepower. In reckoning heating surface, all area which the heated gases touch is calculated.

Another point in regard to heating surface in the production of steam is this, that only such surface as is exposed to a heat equal to turning the water into steam is effective. If there is a pressure of 150 lbs. the temperature at which the water would turn to steam would be 357 degrees, and any gases whose temperature was below 357 degrees would have no effect on the heating surface except to prevent radiation. Thus in a return flue boiler the heated gases become cooled often to such an extent before they pass out at the smokestack that they do not help the generation of steam. Yet a heat just below 357 degrees would turn water into steam under 149 lbs. pressure. Though it has work in it, the heat is lost.

Another practical point as to economy in large heating surface is that it costs money to make, and is cumbersome to move about. It may cost more to move a traction engine with large boiler from place to place than the saving in fuel would amount to. So the kind of roads and the cost of fuel must be taken into account and nicely balanced.

However, it may be said that a boiler with certain outside dimensions that will generate 20 horsepower will be more economical than one of the same size that will generate only 10 horsepower. In selecting an engine, the higher the horsepower for the given dimensions, the more economical of both fuel and water.

The value of heating surface also depends on the material through which the heat must penetrate, and the rapidity with which the heat will pass. We have already pointed out that soot and lime scale permit heat to pass but slowly and if they are allowed to accumulate will greatly reduce the steaming power of a boiler for a given consumption of fuel. Another point is that the thinner the iron or steel, the better will the heat get through even that. So it follows that flues, being thinner, are better conductors than the sides of the firebox. Long flues are better than short ones in that the long ones allow less soot, etc., to accumulate than the short ones do, and afford more time for the boiler to absorb the heat of the gases.

Again, we have stated that heating surface is valuable only as it is exposed to the gases at a sufficiently high temperature. Some boilers have a tendency to draw the hot gases most rapidly through the upper flues, while the lower flues do not get their proportion of the heat. This results in a loss, for the heat to give its full benefit should be equally distributed.

To prevent the heat being drawn too rapidly through upper flues, a baffle plate may be placed in the smoke box just above the upper flues, thus preventing them from getting so much of the draft.

Again, if the exhaust nozzle is too low down, the draft through the lower flues may be greater than through the upper. This is remedied by putting a piece of pipe on the exhaust to raise it higher in the smokestack.

EXPANSION AND CONDENSATION

We have already pointed out that economy results if we hook up the reverse lever so that the expansive force of the steam has an opportunity to work during half or three-quarters of the stroke.

One difficulty arising from this method is that the walls of the cylinder cool more rapidly when not under the full boiler pressure. Condensation in the cylinder is a practical difficulty which should be met and overcome as far as possible.

High speed gives some advantage. A judicious use of cushion helps condensation somewhat also, because when any gas like steam or air is compressed, it gives off heat, and this heat in the cushion will keep up the temperature of the cylinder. This cannot be carried very far, however, for the back pressure of cushion will reduce the energy of the engine movement.

LEAD AND CLEARANCE

Too much clearance will detract from the power of an engine, as there is just so much more waste space to be filled with hot steam. Too little clearance will cause pounding.

Likewise there will be loss of power in an engine if the lead is too great or too little. The proper amount of lead differs with conditions. A high speed engine requires more than a low speed, and if an engine is adjusted for a certain speed, it should be kept uniformly at that speed, as variation causes loss. The more clearance an engine has the more lead it needs. Also the quicker the valve motion, the less lead required. Sometimes when a large engine is pulling only a light load and there is no chance to shorten the cut-off, a turn of the eccentric disk for a trifle more lead will effect some economy.

Cut-off should be as sharp as possible. A slow cut-off in reducing pressure before cut-off is complete, causes a loss of power in the engine.

THE EXHAUST

If the exhaust from the cylinder does not begin before the piston begins its return stroke, there will be back pressure due to the slowness with which the valve opens. The exhaust should be earlier in proportion to the slowness of the valve motion, and also in proportion to the speed of the engine, since the higher the speed the less time there is for the steam to get out. It follows that an engine whose exhaust is arranged for a low speed cannot be run at a high speed without causing loss from back pressure.

In using steam expansively the relative proportion between the back pressure and the force of the steam is of course greater. So in using steam expansively the back pressure must be at a minimum, and this is especially true in the compound engine. So many things affect this, that it becomes one of the reasons why it is hard to use a compound engine with as great economy as theory would indicate.

Another thing, the smallness of the exhaust nozzle in the smokestack affects the back pressure. The smaller the nozzle, the greater the draft a given amount of steam will create; but the more back pressure there will be, due to the inability of the exhaust steam to get out easily. So the exhaust nozzle should be as large as circumstances will permit. It is a favorite trick with engineers testing the pulling power of their engines to remove the exhaust nozzle entirely for a few minutes when the fire is up. The back pressure saved will at once show in the pulling power of the engine, and every one will be surprised. Of course the fire couldn't be kept going long without the nozzle on. We have already pointed out that a natural draft is better than a forced one. Here is another reason for it.

LEAKS

Leaks always cause a waste of power. They may usually be seen when about the boiler; but leaks in the piston and valve will often go unnoticed.

It is to be observed that if a valve does not travel a short distance beyond the end of its seat, it will wear the part it does travel on, while the remaining part will not wear and will become a shoulder. Such a shoulder will nearly always cause a leak in the valve, and besides will add the friction, and otherwise destroy the economy of the engine.

Likewise the piston will wear part of the cylinder and leave a shoulder at either end if it does not pass entirely beyond the steam-tight portion of the inside of the cylinder. That it may always do this and yet leave sufficient clearance, the counterbore has been devised. All good engines are bored larger at each end so that the piston will pass beyond the steam-tight portion a trifle at the end of each stroke. Of course it must not pass far enough to allow any steam to get through.

Self-setting piston rings are now generally used. They are kept in place by their own tension. There will always be a little leakage at the lap. The best lap is probably a broken joint rather than a diagonal one. Moreover, as the rings wear they will have a tendency to get loose unless they are thickest at a point just opposite to the lap, since this is the point at which it is necessary to make up for the tension lost by the lapping.

CHAPTER XII

DIFFERENT TYPES OF ENGINES

STATIONARY

So far we have described and referred exclusively to the usual form of the farm traction engine, which is nearly always the simplest kind of an engine, except in one particular, namely, the reverse which gives a variable cut-off. Stationary engines, however, are worked under such conditions that various changes in the arrangement may

D. June & Co.'s stationary four-valve engine.

be made which gives economy in operating, or other desirable qualities. We will now briefly describe some of the different kinds of stationary engines.

THROTTLING AND AUTOMATIC CUT-OFF TYPES

Engines may be divided into two classes, namely, throttling and automatic cut-off engines. The throttling engine regulates the speed of the engine by cutting off the supply of steam from the boiler, either by the hand of the engineer on the throttle or by a governor working a special throttling governor valve. Railroad locomotives are throttling engines, and moreover they have no governor, the speed being regulated by the engineer at the throttle valve. Traction engines are usually throttling engines provided with a governor.

An automatic cut-off engine regulates its speed by a governor connected with the valve, and does it by shortening the time during which steam can enter the cylinder. This is a great advantage, in that the expansive power of steam is given a chance to work, while in the throttling engine steam is merely cut off. The subject has been fully discussed under "Economy in Running a Farm Engine." An automatic cut-off engine is much the most economical.

While on traction engines the governor is usually of the ball variety, on stationary engines improved forms of governors are also placed in the fly wheel, and work in various ways, according to the requirements of the valve gear.

THE CORLISS ENGINE

The Corliss engine is a type now well known and made by many different manufacturers. It is considered one of the most economical stationary engines made, but cannot be used for traction purposes. It may be compound, and may be used with a condenser. It cannot be used as a high speed engine, since the valves will not work rapidly enough.

The peculiarity of a Corliss engine is the arrangement of the valves. It has four valves instead of one, and they are of the semi-rotary type. They consist of a small, long cylinder which rocks back and forth, so as to close and open the port, which is rather wide and short compared to other types. There is a valve at each end of the cylinder opening usually into the clearance space, to admit steam; and two more valves below the cylinder for the exhaust. These exhaust valves allow any water of condensation *to* run out of the cylinder. Moreover, as the steam when it leaves the cylinder is much colder than when it enters, the exhaust always cools the steam ports, and when the same ports are used for exhaust and admission the fresh steam has to pass through ports that have been cooled and cause condensation. In the Corliss engine the exhaust does not have an opportunity to cool the live steam ports and the condensation is reduced. This works considerable economy.

Also the Corliss valves have little friction from steam pressure on their own backs, since the moment they are lifted from their seats they work freely. The valves are controlled by a governor so as to make the automatic cut-off engine.

The Corliss type of frame for engine is often used on traction engines and means the use of convex shoes on cross-head and concave ways or guides. In locomotive type, cross-head slides in four square angle guides.

THE HIGH SPEED ENGINE

A high speed engine means one in which the speed of the piston back and forth is high, rather than the speed of rotation, there being sometimes a difference. High speed engines came into use because of the need of such to run dynamos for electric lighting. Without a high speed engine an intermediate gear would have to be used, so as to increase the speed of the operating shaft. In the high speed engine this is done away with.

As an engine's power varies directly as its speed as well as its cylinder capacity or size, an engine commonly used for ten horsepower would become a twenty horsepower engine if the speed could be doubled. So high speed engines are very small and compact, and require less metal to build them. Therefore they should be much cheaper per horsepower.

A high speed engine differs from a low speed in no essential particular, except the adjustment of parts. A high steam pressure must be used; a long, narrow valve port is used, so that the full steam pressure may be let on quickly at the beginning of the stroke when the piston is reversing its motion and needs power to get started quickly on its return; the slide valve must be used, since the semi-rotary Corliss would be too wide and short for a quick opening. Some high speed engines are built which use four valves, as does the Corliss. The friction of the slide valve is usually "balanced" in some way, either by "pressure plates" above the valve, which prevent the steam from getting at the top and pressing the valve down, or by letting the steam under the valve, making it slide on narrow strips, since the pressure above would then be reduced in proportion with the smallness of the bearing surface below, and if the bearing surface were very small the pressure above would be correspondingly small, perhaps only enough to keep the valve in place. Some automatic cut-off gear is almost always used. A high speed engine may attain 900 revolutions per minute, 600 being common. In many ways it is economical.

CONDENSING AND NON-CONDENSING

In the traction engine the exhaust is used in the smokestack to help the draft, since the smokestack must necessarily be short. A stationary engine is usually provided with a boiler set in brickwork, and a furnace with a high chimney, which creates all the draft needed.

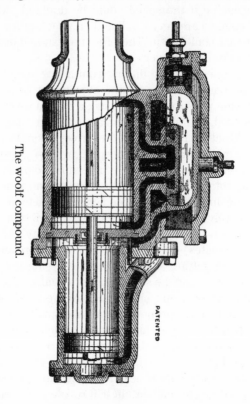

The woolf compound.

In other words, the heated gases wasted in a traction engine are utilized to make the draft.

It then becomes desirable to save the power in the exhaust steam in some way. Some of this can be used to heat the feed water, but only a fraction of it.

Now when the exhaust steam issues into the air it must overcome the pressure of the atmosphere, nearly 15 lbs. to the square inch, which is a large item to begin with. This can be saved by letting the steam exhaust into a condenser, where a spray of cold water or the like suddenly condenses the steam so that a vacuum is created. There is then no back pressure on the exhaust steam, theoretically. Practically a perfect vacuum cannot be created, and there is a back pressure of 2 or 3 lbs. per square inch. By the use of a condenser a back pressure of about 12 lbs. is taken off the head of the piston on its return stroke, a matter of considerable economy. But an immense amount of water is required to run a condenser, namely, 20 times as much for a given saving of power as is required in a boiler to make that power. So condensers are used only where water is cheap.

COMPOUND AND CROSS-COMPOUND

We have already explained the economy effected by the compound engine, in which a large low pressure cylinder is operated by the exhaust from a small high pressure cylinder. In the cut used for illustration the low pressure cylinder is in direct line with the high pressure cylinder, and one piston rod connects both pistons. This arrangement is called the "tandem." Sometimes the low pressure cylinder is placed by the side of the high pressure, or at a distance from it, and operates another piston and connecting rod. By using a steam chest to store the exhaust steam and varying the cut-off of the two cylinders, the crank of the low pressure may be at an angle of 90 degrees with the crank of the high pressure, and there can be no dead center.

When a very high pressure of steam is used the exhaust from the low pressure cylinder may be used to operate a third cylinder; and the exhaust from that to operate a fourth. Engines so arranged are termed triple and quadruple expansion engines, or multiple expansion.

The practical saving of a compound engine when its value can be utilized to the full is 10 per cent to 20 per cent. Small engines are seldom compounded, large engines nearly always.

CHAPTER XIII

GAS AND GASOLINE ENGINES

The gas and gasoline engines (they are exactly the same except that one generates the gas it needs from gasoline, while the other takes common illuminating gas, the use of gas or gasoline being interchangeable on the same engine by readjustment of some of the parts) are operated on a principle entirely different from steam. While they are arranged very much as a steam engine, the power is given by an explosion of gas mixed with air in the cylinder. Instead of being a steady pressure like that furnished by steam, it is a sudden pressure given to one end of the piston usually once in four strokes or two revolutions, one stroke being required to draw the gasoline in, the second to compress it, the third to receive the effect of the explosion (this is the only power stroke), the fourth to push out the burned gases preparatory to admitting a new charge. The fact that force is given the cylinder at such wide intervals makes it necessary to have an extra heavy flywheel to keep the engine steady, and the double cylinder engine which can give a stroke at least every revolution is still better and is indispensable when the flywheel cannot be above a certain weight.

For small horsepowers, such as are required for pumping, feed grinding, churning, etc., the gas engine is so much more convenient and so very much cheaper in operation than the small steam engine that it is safe to say that within a very few years the gas engine will have completely displaced the small steam engine. In fact, the discovery of the gas engine permits the same economies for the small engine that the progress in steam engineering has made possible for the large steam engine. As yet the gas engine has made little or no progress against the large steam plant, with its Corliss engine, its triple expansion, its condenser, and all the other appliances which are not practicable with the small engine.

COMPARISON OF STEAM AND GAS ENGINES

The following points prepared by an experienced farm engine manufacturer will show clearly the advantages of the gas engine over the steam engine for general use about a farm:

In the first place, the farmer uses power, as a rule, at short intervals, and also uses small power. Should he install a steam engine and wish power for an hour or two, it would be necessary for him to start a fire under the boiler and get up steam before he could start the engine. This would take at least an hour. At the end of the run he would have a good fire and good steam pressure, but no use for it, and would have to let the fire die out and the pressure run down. This involves a great waste of water, time and fuel. With a gasoline engine he is always ready and can start to work within a few minutes after he makes up his mind to do so, and he does not have to anticipate his wants in the power line for half a day. Aside from this, in some states, notably Ohio, the law compels any person operating an engine above ten horsepower to carry a steam engineer's license. This does not apply to a gasoline engine.

Again, the gasoline engine is as portable as a traction engine, and can be applied to all the uses of a traction engine and to general farm use all the rest of the year. At little expense it can be fitted up to hoist hay, to pump water, to husk and shell corn, to saw wood, and even by recent inventions to plowing. It is as good about a farm as an extra man and a team of horses.

A gasoline engine can be run on a pint of gasoline per hour for each horsepower, and as soon as the work is done there is no more consumption of fuel and the engine can be left without fear, except for draining off the water in the water jacket in cold weather. A steam engine for farm use would require at least four pounds of coal per horsepower per hour, and in the majority of cases it would be twice that, taking into consideration the amount of fuel necessary to start the fire and that left unburned after the farmer is through with his power. If you know the cost of crude gasoline at your point and the cost of coal, you can easily figure the exact economy of a gasoline engine for your use. To the economy of fuel question may be added the labor or cost of pumping or hauling water.

The only point wherein a farmer might find it advantageous to have a steam plant would be where he is running a dairy and wished steam and hot water for cleansing his creamery machinery. This can be largely overcome by using the water from the jackets which can be kept at a temperature of about 175 degrees, and if a higher temperature is needed he can heat it with the exhaust from the engine. The time will certainly come soon when no farmer will consider himself up to date until he has a gasoline engine.

Some persons unaccustomed to gasoline may wonder if a gasoline engine is as safe as a steam engine. The fact is, they are very much safer, and do not require a skilled engineer to run them. The gasoline tank is usually placed outside the building, where the danger from an explosion is reduced to a minimum. The only danger that may be encountered is in starting the engine, filling the supply tank when a burner near at hand is in flame, etc. Once a gasoline engine is started and is supplied with gasoline, it may be left entirely alone without care for hours at a time without danger and without adjustment.

With a steam engine there is always danger, unless a highly skilled man is watching the engine every moment. If the water gets a little low he is liable to have an explosion; if it gets a little too high he may knock out a cylinder head in his engine; the fire must be fed every few minutes; the grates cleaned. There is always something to be done about a steam engine.

So here is another point of great saving in a gasoline engine, namely, the saving of one man's time. The man who runs the gasoline engine may give nearly all his time to other work, such as feeding a corn-sheller, a fodder chopper, or the like.

Kerosene may also be used in the same way with a special type of gas engine.

The amounts of fuel required of the different kinds possible in a gas engine are compared as follows by Roper:

Illuminating gas, 17 to 20 cubic feet per horsepower per hour.

Pittsburg natural gas, as low as 11 cubic feet.

74° gasoline, known as stove gasoline, one-tenth of a gallon.

Refined petroleum, one-tenth of a gallon.

If a gas producing plant using coal supplies the gas, one pound of coal per horsepower per hour is sufficient on a large engine.

DESCRIPTION OF THE GAS OR GASOLINE ENGINE

The gas engine consists of a cylinder and piston, piston rod, cross-head, connecting rod, crank and flywheel, very similar to those used in the steam engine.

There is a gas valve, an exhaust valve, and in connection with the gas valve a self-acting air valve. The gas valve and the exhaust valve are operated by lever arm or cam worked from the main shaft, arranged by spiral gear or the like so that it gets one movement for each two revolutions of the main shaft. Such an engine is called "four cycle" (meaning one power stroke to each four strokes of the piston), and works as follows:

As the piston moves forward the air and fuel valves are simultaneously opened and closed, starting to open just as the piston starts forward and closing just as the piston completes its forward stroke. Gas and air are simultaneously sucked into the cylinder by this movement. As the cylinder returns it compresses the charge taken in during the forward stroke until it again reaches back center. The mixture in the Otto engine is compressed to about 70 pounds per square inch. Ignition then takes place, causing the mixture to explode and giving the force from which the power is derived. As the crank again reaches its forward center the piston uncovers a port which allows the greater part of the burnt gases to escape. As the piston comes back, the exhaust valve is opened, enabling the piston to sweep out the remainder of the burnt gases. By the time the crank is on the back center the exhaust valve is closed and the engine is ready to take another charge, having completed two revolutions or four strokes. The side shaft which performs the functions of opening and closing the valves, getting its motion in the Columbus engine by a pair of spiral gears, makes but one revolution to two of the crank shaft.

FAIRBANKS, MORSE & CO.'S GASOLINE ENGINE.

A is engine cylinder. H is gasoline supply tank located outside of building and under ground. I is air-suction pipe. E is gasoline pump. O is suction pipe from gasoline tank. N is pipe from pump E, leading to reservoir P. Q is igniter tube. R is chimney surrounding tube. T is tank supplying Bunsen burner for heating tube.

Gas engines are governed in various ways. One method is to attach a ball governor similar to the Waters on the steam engine. When the speed is too high, the balls go out, and a valve is closed or partly closed, cutting off the fuel supply. Since the engine takes in fuel only once in four strokes, the governing cannot be so close as on the steam engine, since longer time must elapse before the governor can act.

Another type of governor operates by opening the exhaust port and holding it open. The piston then merely draws in air through the exhaust port, but no gas. This is called the "hit or miss" governing type. One power stroke is missed completely.

The heat caused by the explosion within the cylinder is very great, some say as high as 3,000 degrees. Such a heat would soon destroy the oil used to lubricate the cylinder and make the piston cut, as well as destroying the piston packing. To keep this heat down the cylinder is provided with a water jacket, and a current of water is kept circulating around it to cool it off.

When gas is used, the gas is passed through a rubber bag, which helps to make the supply even. It is admitted to the engine by a valve similar to the throttle valve on an engine.

Gasoline is turned on by a similar valve or throttle. It does not have to be gasefied, but is sucked into the cylinder in the form of a spray. As soon as the engine is started, the high heat of the cylinder caused by the constant explosions readily turns the gasoline to gas as it enters. The supply tank of gasoline is placed outside the building, or at a distance, and stands at a point below the feed. A small pump pumps it up to a small box or feed tank, which has an overflow pipe to conduct any superfluous gasoline back to the supply tank. In the gasoline box or feed tank a conical-shaped basin is filled with gasoline to a certain height, which can be regulated. Whatever this conical basin contains is sucked into the cylinder with the air. By regulating the amount in the basin the supply of gasoline in the cylinder can be regulated to the amount required for any given amount of work. In the Columbus engine this regulation is accomplished by screwing the overflow regulator up or down.

There are two methods of igniting the charge in the cylinder in order to explode it. One is by what is called a gasoline or gas torch. A hollow pin or pipe is fixed in the top of the cylinder. The upper part of this pin or pipe runs up into a gasoline or gas lamp of the Bunsen type where it is heated red hot. When the gas and air in the cylinder are compressed by the back stroke of the piston, some of the mixture is forced up into this pipe or tube until it comes in contact with the heated portion and is exploded, together with the rest of the charge in the cylinder. Of course this tube becomes filled with burnt gases which must be compressed before the explosive mixture can reach the heated portion, and no explosion is theoretically possible until the piston causes compression to the full capacity of the cylinder. The length of the tube must therefore be nicely regulated to the requirements of the particular engine used.

The other method is by an electric spark from a battery. Two electrodes of platinum or some similar substance are placed *in* the compression end of the cylinder. The spark might be caused by bringing the electrodes sufficiently near together at just the right moment, but the more practical and usual way is to break the current, closing it sharply by means of a lever worked by the gearing at just the moment the piston is ready to return after compressing the charge. The electric spark is by long odds the most desirable method of ignition, being safer and easier to take care of, but it requires some knowledge of electricity and electric connection to keep it always in working order.

OPERATION OF GAS AND GASOLINE ENGINES

To all intents and purposes the operation of a gas or gasoline engine is the same as that of a steam engine with the care of the boiler eliminated. The care of the engine itself is practically the same, though the bearings are relatively larger in a gasoline or gas engine and do not require adjustment so often. Some manufacturers will tell you that a gas engine requires no attention at all. Any one who went on that theory would soon ruin his engine. To keep a gasoline engine in working order so as to get the best service from it and make it last as long as possible, you should give it the best of care.

An engine of this kind needs just as much oiling and cleaning as a steam engine. All bearings must be lubricated and kept free from dirt, great care must be taken that the piston and cylinder are well lubricated. In addition, the engineer must see that the valves all work perfectly tight, and when they leak in any way they must be taken out and cleaned. Usually the valve seats are cast separate from the cylinder, so that they can be removed and ground when they have worn.

Also the water jacket must be kept in order so that the cylinder cannot become too hot.

STARTING A GASOLINE ENGINE

It is something of a trick to get a gasoline or gas engine started—especially a gasoline engine—and some skill must be developed in this or there will be trouble. This arises from the fact that when an engine has not been running the cylinder is cold and does not readily gasefy the gasoline. At best only a part of a charge of gasoline can be gasefied, and if the cylinder is very cold indeed the charge will not explode at all till the cylinder is warmed up.

When preparing to start an engine, first see that the nuts or studs holding cylinder head to cylinder are tight, as the heating and cooling of the cylinder are liable to loosen them. Then oil all bearings with a hand oil can, and carefully wipe off all outside grease.

When all is ready, work the gasoline pump to *get* the air out of the feed pipes and fill the reservoir.

First, the engine must be turned so that the piston is as far back as it will go, and to prevent air being pressed back the exhaust must be held open, or a cock in priming cup on top of cylinder opened.

If gasoline priming is needed, the gasoline must be poured into the priming cup after closing the cock into the cylinder, for it would do no good to merely let the gasoline run down into the cylinder in a cold stream: it must be sprayed in. If the exhaust has been held open, and the priming charge of gasoline is to be drawn in through the regular supply pipe and valve, the exhaust should be closed and the throttle turned on to a point indicated by the manufacturer of the engine.

We suppose that the igniter is ready to work. If the hot tube is used, the tube should be hot; if the electric igniter is used, the igniter bar should be in position to be snapped so as to close the circuit and cause a spark when the charge has been compressed.

If all is ready, open the cock from which the supply of gasoline is to be obtained, and at the same time turn the engine over so as to draw the charge into the cylinder. If a priming cock has been opened, that must be closed by hand as soon as the cylinder is filled and the piston ready to return for compression. If the regular feed is used, the automatic valve will close of itself.

Bring the flywheel over to back center so that piston will compress the charge. With the flywheel in the hand, bring the piston back sharply two or three times, compressing the charge. This repeated compression causes a little heat to be liberated, which warms up the cylinder inside. If the cylinder is very cold this compression may be repeated until the cylinder is sufficiently warm to ignite. When performing this preparatory compression the piston may be brought nearly up to the dead center but not quite. At last bring it over the dead center, and just as it passes over, snap the electric ignition bar. If an explosion follows the engine will be started.

If the hot tube is used, the flywheel may be brought around sharply each time so that the piston will pass the dead center, as an explosion will follow complete compression. If the explosion does not follow, the flywheel may be turned back again and brought up sharply past the dead center. Each successive compression will warm up the cylinder a little till at last an explosion will take place and the engine will be started.

More gasoline will be needed to start in cold weather than in warm, and the starting supply should be regulated accordingly. Moreover, when the engine gets to going, the cylinder will warm up, more of the gasoline will vaporize, and a smaller supply will be needed. Then the throttle can be turned so as to reduce the supply.

After the engine is started, the water jacket should be set in operation, and you should see that the cylinder lubrication is taking place as it ought.

As the above method of starting the engine will not always work well, especially in cold weather, what are called "self-starters" are used. They are variously arranged on different engines, but are constructed on the same general principle. This is, first, to pump air and gasoline into the cylinder instead of drawing it in by suction. Sometimes the gasoline is forced in by an air compression tank. The engine is turned just past the back center, care having been taken to make sure that the stroke is the regular explosion stroke. This may be told by looking at the valve cam or shaft. If an electric igniter is used, it is set ready to snap by hand. If the tube igniter is used, a detonator is arranged in the cylinder, to be charged by the head of a snapping parlor match which can be exploded by hand. Holding the flywheel with one hand with piston just past back center, fill the compressed end of the cylinder by working the pump or turning on the air in compression tank till you feel a strong pressure on the piston through the flywheel. Then snap igniter or detonator and the engine is off. If throttle valve has not been opened, it may now be immediately opened.

The skill comes in managing the flywheel with one hand, or one hand and a foot, and the igniter, etc., with the other hand. Care must be exercised not to get caught when the flywheel starts off. The foot must never be put through the arm of the wheel, the wheel merely being held when necessary by the ball of the big toe, so that if the flywheel should start suddenly it would merely slip off the toe without carrying the foot around or unbalancing the engineer. Until one gets used to it, it is better to have some one else manage the flywheel, while you look after the gasoline supply, igniter, etc. When used to it, one man can easily start any gasoline engine up to 15 horsepower.

WHAT TO DO WITH A GASOLINE ENGINE WHEN IT DOESN'T WORK.

Questions and Answers

Q. If the engine suddenly stops, what would you do?

A. First, see that the gasoline feed is all right, plenty of gasoline in the tank, feed pipe filled, gasoline pump working, and then if valves are all in working order. Perhaps there may be dirt in the feed reservoir, or the pipe leading from it may be stopped up. If everything is right so far, examine the valves to see that they work freely and do not get stuck from lack of good oil, or from use of poor oil. Raise them a few times to see if they work freely. Carefully observe if the air valve is not tight in sleeve of gas valve.

Q. What would be the cause of the piston's sticking in the cylinder?

A. Either it was not properly lubricated, or it got too hot, the heat causing it to expand.

Q. Are boxes on a gasoline engine likely to get hot?

A. Yes, though not so likely as on a steam engine. They must be watched with the same care as they would be on a steam engine. If the engine stops, turn it by hand a few times to see that it works freely without sticking anywhere.

Q. Is the electric sparking device likely to get out of order?

A. Yes. You can always test it by loosening one wire at the cylinder and touching it to the other to see that a spark passes between them. If there is no spark, there is trouble with the battery.

Q. How should the batteries be connected up?

A. A wire should pass from carbon of No. 1 to copper of No. 2; from carbon of No. 2 to copper of No. 3, etc., always from copper to carbon, never from carbon to carbon or copper to copper. Wire from last carbon to spark coil and from coil to switch, and from switch to one of the connections on the engine. Wire from copper of No. 1 to the other connection on the engine. In wiring, always scrape the ends of the wire clean and bright where the connection is to be made with any other metal.

Q. What precautions can be taken to keep batteries in order?

A. The connections between the cells can be changed every few days, No. 1 being connected with No. 3, No. 3 with No. 5, etc., alternating them, but always making a single line of connection from one connection on cylinder to first copper, from the carbon of that cell to copper of next cell, and so on till the circuit to the cylinder is completed. When the engine is not in operation, always throw out the switch, to prevent possible short circuiting. If battery is feeble at first, fasten wires together for half an hour at engine till current gets well started.

Q. Is there likely to be trouble with the igniter inside cylinder?

A. There may be. You will probably find a plug that can be taken out so as to provide a peep hole. Never put your eye near this hole, for some gasoline may escape and when spark is made it will explode and put out your eye. Always keep the eye a foot away from the hole. Practice looking at the spark when you know it is all right and no gasoline is near, in order that you may get the right position at which to see the spark in case of trouble. In any case, always take pains to force out any possible gas before snapping igniter to see if the spark works all right.

Q. If there is no spark, what should be done?

A. Clean the platinum points. This may be done by throwing out switch and cutting a piece of pine one-eighth of an inch thick and one-half inch wide, and rubbing it between the points. It may be necessary to push cam out a trifle to compensate for wear.

Q. How can you look into peep hole without endangering eyesight?

A. By use of a mirror.

Q. If the hot tube fails to work, what may be done?

A. Conditions of atmosphere, pressure, etc., vary so much that the length of the tube cannot always be determined. If a tube of the usual length fails to work, try one a little longer or shorter, but not varying over 11/2 inches.

Q. When gas is used, what may interfere with gas supply?

A. Water in the gas pipes. This is always true of gas pipes not properly drained, especially in cold weather when condensation may take place. If water accumulates, tubes must be taken apart and blown out, and if necessary a drain cock can be put in at the lowest point.

Q. What trouble is likely to be had with the valves?

A. In time the seats will wear, and must be taken out and ground with flour or emery.

Q. Should the cylinder of a gasoline engine be kept as cool as it can be kept with running water?

A. No. It should be as hot as the hand can be borne upon it, or about 100 degrees. If it is kept cooler than this the gasoline will not gasefy well. If a tank is used, the circulation in the tank will justify the temperature properly. The water may be kept at 175 degrees of temperature, and used for hot water heating. The exhaust gases are also hot and may be used for heating by carrying in pipes coiled in a hot water heater.

Q. Are water joints likely to leak?

A. Yes. The great heating given the cylinder is liable to loosen the water joints. They are best packed with asbestos soaked in oil, sheets 1-16 inch thick. Old packing should always be thoroughly cleaned off when new packing is put in.

Q. How may the bearings be readjusted when worn?

A. Usually there are liners to adjust bearing. In crank box adjust as in steam engine by tightening the key.

Q. If you hear a loud explosion in the exhaust pipe after the regular explosion, should you be alarmed?

A. No. All gas or gasoline engines give them at times and they are harmless. If the gas or gasoline fed to the engine is not sufficient to make an explosive mixture, the engine will perhaps miss the explosion, and live gas will go into the exhaust pipe. After two or three of these have accumulated an explosion may take place and the burned gases coming out of the port as hot flames will explode the live gas previously exhausted. Any missing of the regular explosion by the engine, through trouble with battery, or the like, will cause the same condition.

Q. When you get exhaust pipe explosions, what should you do?

A. Turn on the fuel till the exhaust is smoky. Then you know you have fuel enough and more than enough. If the explosions still continue, conclude that the igniter spark is too weak, or does not take place.

Q. What precaution must be taken in cold weather?

A. The water must be carefully drained out of jacket.

Q. Will common steam engine cylinder oil do for a gasoline engine?

A. No. The heat is so great that only a special high grade mineral oil will do. Any oil containing animal fat will be worse than useless.

Q. How can you tell if right amount of gas or gasoline is being fed to engine to give highest power?

A. Turn on as much as possible without producing smoke. A smokeless mixture is better than one which causes smoke.

Q. If you have reason to suppose gas may be in the cylinder, should you try to start cylinder?

A. No. Empty the gas all out by turning the engine over a few times by hand, holding exhaust open if necessary.

Q. How long will a battery run without recharging?

A. The time varies. Usually not over three or four months.

Q. Is it objectionable to connect an electric bell with an engine battery?

A. Certainly. Never do it.

Q. If your engine doesn't run, how many things are likely to be the trouble?

A. Not more than four—compression, spark, gas supply, valves.

GAS AND GASOLINE ENGINES—
CONTINUED

Explanation of Principles

Reference has already been made to the gas engine, and a general description is given of this interesting and useful machine, but as no detailed explanation is there given of the principles controlling its action, it was deemed wise by the author to place before his readers, additional matter pertaining to the subject, and in doing so an effort has been made to present it in as simple and plain a manner as possible, in order that it may be easily within the comprehension of all. As the gas, and gasoline engine are practically identical in principle, the same explanation and illustrations will, with the exception of a few minor details, apply to both.

All gas engines in practical use at the present time are two or four cycle, as herein described, or are modifications of these forms. Of these two types the four cycle machine has by far been the more generally adopted. We will now explain and illustrate the principles involved and the difference of action in the two types, taking up the four cycle first.

GAS ENGINES

Before proceeding farther, it will be well to explain the meaning of the word cycle as used in this connection.

A CYCLE means a round of time or a round of events necessary to produce a certain result.

As applied to the gas engine it means the round of movements or events to get one "explosion" as it is commonly called, or on impulse. In other words one working stroke.

In the four cycle engine we have four distinct movements or events to get one "explosion."

Beginning with Fig. I we show the piston C, starting on the first or downward stroke, drawing in, by suction, the charge or mixture of air and fuel through valve A.

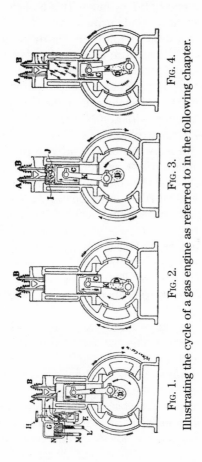

Fig. 4. Fig. 3. Fig. 2. Fig. 1.

Illustrating the cycle of a gas engine as referred to in the following chapter.

This is the SUCTION or INTAKE stroke and is the first movement or event.

The valve A is now closed by its spring, and in Fig. 2 we show the piston C returning, and compressing the charge into a small space, called the compression space, in the upper end of the cylinder. This is the COMPRESSION stroke and is the second movement.

In Fig. 3 we show piston C almost to the top or end of the stroke. The mixture of air and gas is now under compression and at this point the electric spark is made which fires the charge. By the time the mixture is ignited, and the crank D reaches the center, the heat of the burning charge expands with great force and drives piston C down on its POWER stroke. This is the third movement.

In Fig. 4 we show piston C nearly to the end of the stroke. As much of the power of the heat expansion has been delivered as can be obtained, and at this point the exhaust valve B is forced open by a cam, and the remaining heat rushes out. Valve B is held open by the cam while piston C travels back to the top driving out the foul, spent gases. This is the SCAVENGING stroke and is the fourth and last movement in the cycle of operation. When the piston C, reaches the top, valve B closes and the engine is ready to begin the cycle or round of events over again.

We thus see that in a four cycle engine we have four movements—one cycle—giving it the short and well fitted name "four cycle."

Two revolutions of the flywheel are used to get the four movements of the piston. Many four cycle engines are made horizontal, that is, with the cylinder lying down instead of standing up as shown in the illustrations. The movements or actions, however, remain the same.

The two cycle engine draws in a charge, compresses it, fires it, and exhausts the burned gases, but it is all done in two strokes or movements, hence we call it a two cycle engine. In Figs. 5 and 6 we illustrate the action of the ordinary two cycle machine.

In Fig. 5 the air is drawn in at A, and fuel from B, as regulated by needle valve C. This mixture is drawn into the crankcase as the piston goes up, and a charge that is in the cylinder above the piston is compressed at the same time, thus drawing in a charge and compressing

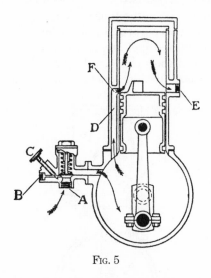

FIG. 5

one, in one stroke or movement. When the piston reaches the top the compressed charge is fired, and the piston is driven down, delivering the power of the burning charge and also compressing the fresh charge just drawn into the crankcase. As the piston nears the end of the stroke, it uncovers the port E and the engine exhausts. An instant later the inlet port F is uncovered as the piston moves on, and the fresh charge now under compression in the crankcase rushes up through the inlet port F, and is turned upward by the projection on the end of the piston.

As the new charge rushes in it is expected to drive out the burned gases at the exhaust port E. So we see the down stroke combines both the power and scavenging events, while the up stroke combines the intake and compression events. Two movements do here what four are required to do in the four cycle engine. This type is known as the two port two cycle engine.

In Fig. 6 air and fuel are drawn in at port A by the vacuum the piston produces, in the crankcase, on its upward stroke. By this construction the check valve,

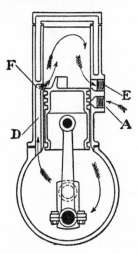

Fig. 6

shown in Fig. 5, is not needed. Otherwise the action illustrated in Fig. 6 is the same as in Fig. 5. The type shown in Fig. 6 is known as the three port two cycle engine.

There are many other forms for the mechanical construction of the two-cycle engine, but they all follow the same general principles we have here illustrated and described.

Owing to its simple construction and an impulse every two strokes or every revolution, the two cycle engine has proven very attractive to hundreds of inventors, and a great variety of designs have been built. Simplicity of construction is much to be desired in any machine if it produces the results we want. The results demanded of a gas engine are economy in the use of fuel and reliability of action.

The four cycle engine, in spite of its valves and gears, has given so much better practical results as a rule, that it has been adopted by most of the builders. The two cycle engine finds its best applications in service where the load and speed are comparatively uniform.

Comparing the machines we have illustrated, we see a charge drawn in every revolution by the two cycle construction. Part of the charge may be lost, by leakage through the bearings, when the piston comes down compressing it in the crankcase.

The charge next passes to the cylinder and as the piston returns to the top it compresses the charge a second time, a loss of net power to the engine.

As the two cycle engine, when working properly, takes a charge and burns one every revolution, it would seem, at first thought, that it should give twice as much power as a four cycle machine of the same cylinder dimensions.

Owing to the losses we have mentioned, and the fact that the scavenging (driving out of the burned gases) is uncertain, the two cycle develops only 20 per cent to 50 per cent more power than the same size four cycle cylinder. As the bearings wear, the loss from the crankcase is apt to become greater, and the port action also changes slightly.

One construction of the two cycle engine avoids crankcase compression, and the losses thereby, by a design similar to steam engine practice. The engine is made with a crosshead and piston rod. The end of the cylinder, usually left open by gas engine builders, is closed and fitted with a stuffing box through which the piston rod works exactly like the steam engine. This closed end of the cylinder is used, instead of the crankcase, for handling the mixture or charge.

The principles involved in the two cycle engine depend largely on a certain velocity for the moving or transferring of the charge from the receiving and first compression chamber to the cylinder; hence as wide a variation in the speed cannot be permitted as with the four cycle machine.

FUELS FOR THE GAS ENGINE

Gas in its natural form, as found in some places, is the most convenient fuel known for the gas engine, especially for stationary work.

Only a few sections, however, are so favored, and in other places, and for portable or traction work the gas for the engine must be made or produced artificially from the most available substance.

At the present time gasoline is used more than anything else owing to the ease with which it is carburetted, or converted into gas just as it is needed by the engine.

Natural gas and gasoline have been used so much more than other fuels that the expression "gas and gasoline engines" as frequently used is taken by some to mean two distinct types of engines.

The expression as heretofore explained is misleading, for the general principles are precisely the same. The only difference in construction is in the compression used, and in the mixer or device for feeding the fuel; it is a simple matter to change a gas engine to gasoline, or a gasoline engine to gas.

When we say "gas engine" we have covered the ground, and we understand that in using gasoline, oil, coal, etc., proper means must be provided for converting the fuel used into gas.

Many fuels will produce gas which may be used in the gas engine. Among these are coal, crude oil, coal oil or kerosene, gasoline, wood alcohol, spirits of various kinds, etc.

A great many of the possible fuels are out of the question because of price, and others involve difficulties in the way of generating or producing the gas as needed and of proper quality. Some gases also involve objectionable features in the burning or combustion, as for example, the gas from crude oil (a very cheap fuel) carries with it a carbon element that is deposited on the head of the piston and on the walls of the compression space, making it necessary to clean these parts frequently.

The difficulties in generating or producing gases, and in burning them to produce power, are being rapidly overcome by new improvements and methods, and the advantages of the gas engine are increasing thereby.

Gas from coal is proving to be a very cheap fuel for gas engines for heavy and stationary work.

One ton of coal used in this way, as proved by actual practical results, will do two to three times the work that it will do by burning it under a steam boiler.

As this book has reference, more especially, to gas engines for light portable and traction work, we will pay particular attention to gasoline as the available fuel possessing the most advantages.

Carbureters or Mixers.—Several different types of mixers or carbureters for gasoline are in common use, the principle we illustrate in Fig. I, probably being most generally used.

G is an overflow chamber holding the gasoline at a certain level in standpipe F, as indicated by the dotted line N. Gasoline flows into the chamber, G, from a pump, through pipe, L, and the overflow goes back to the tank through pipe M.

As the air is drawn into the cylinder, through the air regulator E, it pulls gasoline with it from standpipe F, the amount being regulated by the needle valve H.

This may be called the constant level overflow system, and is generally built in as part of the engine proper, in the plain, heavy engines now common for stationary work.

If a float was placed in chamber G, operating a gasoline inlet valve, and the gasoline tank was placed higher than the chamber, we would then have the float feed carbureter system. The float would hold the gasoline at a certain level in the standpipe just as the overflow in Fig. 1.

The float feed carbureter is generally a separate part or adjunct to the engine, and it is common for this part to be made by the manufacturer of parts or specialties.

Figure 7 is a cross-section of a float feed carbureter. The float M controls the valve O, and holds the gasoline at a certain level in the spray nozzle L. The air supply in starting, or at slow speed of the motor comes through the narrow passage I, and in passing the spray nozzle L, it draws a small quantity of gasoline as regulated positively by needle valve A. J is the connection to the engine and K is a throttle to enable the operator to

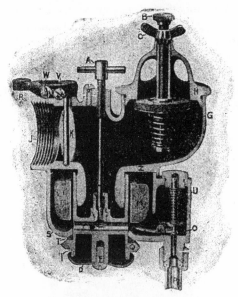

FIG. 7

control the quantity of mixture admitted to the cylinder. The air valve F, is held to its seat by a light spring G, with tension adjusted by screw B and locking device C.

As the throttle K is opened, admitting more mixture to the engine, the air valve A opens wider, admitting more air. As the suction on the gasoline spray nozzle

L is greater, more gasoline is drawn, thus keeping the proportionate mixture approximately right under the throttle, and at the various engine speeds. Both air and gasoline are thus automatically measured, under the varying conditions.

Figure 8 shows a float carbureter with a different principle.

A is the connection; B, gasoline needle valve; C, constant air inlet; D, compensating, or automatic air valve, with spring tension regulated by E; F is the gasoline pipe connection; G is a throttle in the air passage C; H, float chamber; I, needle valve control lever; J, cam, operating mixture throttle lever; L, nut for adjusting lever to position desired.

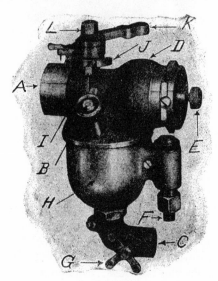

FIG. 8

In starting, the constant air passage C, is partly closed, to secure more suction on the gasoline. As the throttle, operated by lever K is opened, the cam J moves lever I, gradually opening gasoline needle valve B, admitting more gasoline in proportion to the increased air supply coming through the air valve D.

In this carbureter the air supply is automatically controlled, but the gasoline is positively regulated.

It is evident that the gasoline tank must be higher than the carbureter to supply gasoline to the flat valve by gravity.

The gasoline may be supplied from a lower level by air pressure in the gasoline tank, but as this is complicating the mechanism of the outfit, it is rarely used.

It will be observed that the general principles of the overflow and feed float systems are the same.

Another very common method of feeding gasoline is by means of a mixing valve as illustrated in Fig. 5.

The gasoline, regulated by the needle valve C, feeds to the seat of the mixing valve. When the piston draws in air from A, the mixing valve is lifted, gasoline flows in and is mixed with the air.

At the end of the suction stroke the valve closes, shutting off the gasoline. The gasoline may be supplied to the valve by gravity or by air pressure, the same as with the float feed system.

A fault with the ordinary mixing valve is found in the fact that, as the valve closes, the fuel remaining on the seat is projected backward by the angle of the seat, causing the valve to "spit" or "slobber" the fuel.

In all these systems of feeding, the gasoline is so volatile that, by mixing with the air as it is drawn in and in passing into the heated cylinder, it is carburetted or vaporized, so that by the time the spark is made, the mixture is formed and ready to be ignited.

Fuel gas for gas engines is made from the heavy crude oils by subjecting the oil to heat in a special producer apparatus, that makes a gas vapor from the volatile parts of the oil, while separating and retaining the heavy, solid matter.

Too much gas in the cylinder will not burn for want of sufficient air, just the same as a furnace fire will not burn if the dampers are closed.

Too much fuel turned on in starting is a frequent cause of a gas engine refusing to start. In this case close the needle valve, and turn the engine until the surplus fuel has been driven out.

No matter what kind of fuel gas is used, the principle of feeding to the engine remains the same—A CERTAIN QUANTITY OF GAS WITH THE RIGHT AMOUNT OF AIR, must be taken for each "explosion" or impulse.

It must also be remembered that solid and liquid fuels *must be converted into gas* before the engine will run.

In using gasoline the natural heat of the air is generally depended upon for vaporizing, or making enough gas to start the engine. In the winter season the air frequently does not possess the required warmth, or heat for starting, often causing trouble to the inexperienced operator. In this case the necessary heat to supply the first charges of gas must be provided.

After the first few "explosions" there will be enough heat in the engine cylinder to vaporize the gasoline in the coldest weather.

Gases vary a great deal in the heat power or heat units possessed, and for this reason different gases will increase, or decrease the power of an engine of given size.

Gas from gasoline is very powerful, furnishing another excellent reason for its common use with gas engines.

Compression.—Compressing the charge or mixture of air and gas, before igniting it increases the force of the "explosion."

The higher the compression can be successfully carried, the greater will be the power derived from a given amount of fuel.

The compression heats the mixture rapidly, and, if the compression is carried too high, this heat will fire the charge before time for the spark, and before the piston reaches the end of the stroke. This would cause some of the force to be applied in the backward direction, and cause the engine to "pound" or perhaps stop.

The amount of compression that can be successfully used, depends on, first, the kind of fuel gas that is to be used; second, the speed for which the engine is designed, and third, the uniform heat of the cylinder walls and head at all times.

Different gases require higher or lower compression to obtain the best results, as for example, natural gas will admit of much higher compression than the gas from gasoline.

An engine built for high speed will carry a higher compression than could be used at low speed. The piston coming up to the end of the stroke so much faster enables the crank to pass the center before the impulse begins, even though the charge should be self-ignited from the heat of the high compression. Reliable, even temperature of the cylinder walls and head is of great importance for a high compression, for if the walls become overheated at times, the compression heat will be greater; if the compression is already up to the limit, this extra heat will cause pre-ignition, or firing of the charge too soon.

While it is desirable, from the standpoint of economy in fuel, and maximum or greatest power for a given cylinder dimension, to use the highest compression possible, yet no rule can be given that will fit different makes of engines for the reasons given above.

The manufacturer must be depended upon for the highest compression practical in his particular engine, to suit the design, speed, and fuel to be used.

As we are referring to gasoline as the most convenient and practical fuel for light portable, and traction work, we might say that a fair average compression for this fuel would be 60 lbs. gauge pressure, but the reader will understand that it may be more or less depending on the conditions as stated. This would be equal to about five atmospheres—that is, the volume of the cylinder and compression space would be squeezed up into a space one-fifth the total volume.

This amount of compression will, under proper conditions, give about 300 lbs. per square inch, heat expansive force, or "explosive" pressure at the moment of complete ignition.

A compression of 40 lbs. gauge pressure will give an initial "explosive" force of about 225 lbs. per square inch, so we see, as stated, that the net working force increases as we increase the compression.

The gauge pressure of the compression may be determined by the method described under the heading "How to Test the Condition of an Engine."

If a different gas fuel, from that for which the engine was sold, is to be used, it would be advisable to write the manufacturer of the engine as to the proper compression, as shown by factory tests, and how to change the compression space to best advantage. This will save much time and trouble in experimenting to obtain the best possible results.

The efficiency and economy of a gas engine depends in large measure on perfect compression, and any leakages in rings, valves, packings or porous cylinder walls, directly affect the working of the machine.

The building of an engine for the highest possible compression is a matter of close and careful study for the designer only, hence we will not go into details of construction.

In the operator's hands any make of engine must be carefully guarded against leakages of compression, if the highest possible efficiency and fuel economy are desired.

Ignition Apparatus.—In the early stages of the development of the gas engine, the charge of air and gas was ignited by a hot tube. This tube, with the outer end closed, was screwed into position on the engine and connected with the compression space. The tube was enclosed by a casing lined with heavy asbestos, and was kept at an intense heat by a gas fire within the casing. A part of the charge or mixture was forced into the tube by the compression stroke when *it* would be ignited by the fierce heat of the outer closed end.

This system, clumsy and crude in the light of late improvements, is known as hot tube ignition. It required time in starting to properly heat the tube; it was wasteful in the use of fuel, and the fire to heat the tube was a source of danger. Waste of fuel was due to maintaining the fire to heat the tube, and to the fact that the time of ignition was not under perfect control. Tubes burning out frequently added to the troubles.

The ignition or firing of the charge by an electric spark, under control at all times, is one of the great improvements in the gas engine, and has had much to do with bringing the machine into favor with power users.

The spark is made on the inside of the cylinder, thus eliminating the danger of fire with the hot tube. By this improvement the gas engine became a safe power generating machine in the strictest sense of the word.

Electric ignition has come into such general use that the hot tube is now seldom made, unless for emergency use and most manufacturers do not furnish it at all.

There are two systems of electric ignition in general use, viz: the primary, or make and break, and the secondary or jump spark. Both of these systems must have a source of electric current; a coil for storing, and discharging the current, a device for making and breaking the circuit, and an igniter or spark plug as the case may be.

The source of electric current for either system may be a battery, or a generator driven by the engine. Where a generator is used it is generally considered necessary to have batteries for starting, and switch over to the generator after the engine gets up speed. Most generators require more speed, to furnish the necessary current, than the operator would be able, or willing to give it in starting the engine.

The spark coil acts as a sort of reservoir to store up current when the circuit is made, and to discharge it when the circuit is broken, and this discharge between two points, inside the compression space, makes the spark that fires the charge.

With the make and break system, the circuit is made and broken inside the compression space, giving this system its name.

The contact, or make and break points are set in a block or carrier, the whole forming a device known as the igniter.

The contact points are called electrodes, one of which is made stationary and insulated by a non-conducting material from the other parts of the engine. The other electrode is movable, and the mechanism of the engine causes it to form a contact, inside the compression space, with the insulated stationary electrode, just an instant before time for the spark. During this very short time of contact the current from the battery or generator flows through and charges the coil. At the right moment the movable electrode is snapped back, breaking the contact with the insulated electrode, and the current, stored in the coil, is discharged across the gap between the contact points or electrodes, causing the spark. The quicker the break is made the better and stronger will be the spark produced.

The spark coil for primary or make and break ignition consists of a bundle or core of soft iron wire around which is wound a quantity of insulated copper wire called a primary winding. The current from this coil is a primary current, which explains why make and break ignition is also called primary ignition.

For the secondary or jump spark system the spark coil receives another winding, of several thousand feet of fine insulated wire, called the secondary winding.

This makes a jump spark, or high tension coil as the secondary winding carries a current of high voltage. The electric circuit for this system of ignition is interrupted at any suitable, convenient point on the engine, and causes a spark to jump between two stationary points inside the combustion chamber.

These two points are carried by a spark plug that is screwed into an opening to the combustion chamber, and one of the points must be insulated so the current will pass around and jump the gap provided. The device for interrupting the circuit in the jump spark system is a timer, sometimes called the "commutator" and is shown in Fig. 9 at D. The break of the contact points of the timer must be very quick, and produces a spark at the gap between the points of the spark plug.

It has become quite common to provide the spark coil with an automatic vibrator; the instant the timer makes the circuit, the automatic vibrator sets up a vibrating motion producing a string of sparks at the plug instead of one. With the automatic vibrator, the very quick parting of the timer contact is not essential.

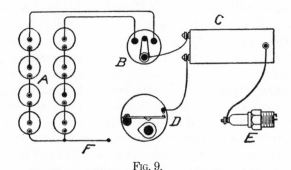

FIG. 9.

Wiring diagram—jump spark ignition.

A—Batteries.

B—Switch.

C—Jump Spark Coil.

D—Timer or "Commutator."

E—Spark Plug.

F—Ground Wire to Engine.

The merits of the vibrator as against the plain jump spark coil have been much discussed. Some authorities claim the plain coil is less liable to get out of order, not having a delicate vibrator adjustment; that making one good spark does the work, which is all that is required.

It is claimed for the vibrator coil that a more sure and rapid ignition is obtained; that the delicate adjustment of the vibrator is a simple matter, and not a disadvantage in the hands of the intelligent operator, and that the necessary quick make and break of the circuit, being made automatically, insures perfect ignition at any speed of the engine. As the vibrator coil has come into general use it must be conceded that the majority of users think it has advantages which overbalance its disadvantages.

The illustration, Fig. 9, shows how to connect two sets of batteries to a switch so one set may be used while the other is out of circuit—thus holding an extra battery in readiness for immediate service should the set in use fail.

When the switch B, is in the central position as shown, both batteries are out of circuit.

In connecting up a jump spark ignition outfit it will generally be found that the manufacturer of the coil has marked the terminals or binding posts. "Bat." stamped on the coil means to attach the battery to that binding post. "Com." means the connection to the timer or "commutator," while "Sec." denotes the terminal of the secondary winding, to be connected to the spark plug.

Should there be two secondary terminals on the coil, one may be connected to the terminal marked "com." This is usually done within the coil, leaving but three outside connections as shown in Fig. 9.

The switch is placed between the battery and the coil and it is understood that the "Bat." connection on the coil is carried to the switch and then on to the battery.

A generator, made for jump spark ignition may be connected to the switch instead of one of the batteries, similar to the connections for primary ignition, illustrated in Fig. 10, which shows the wiring for a make and break ignition outfit, using a battery for starting and connections to switch the generator into circuit as soon as the engine gets up the speed necessary to make the generator deliver the required current.

Some generators are advertised as furnishing current at a very low speed and thereby dispensing with the battery. Most generators, however, will require more speed than the operator would be able or willing to give it in starting the engine.

It is important, in connecting up an ignition outfit, to see that the wire terminals, and binding posts are clean and that the connections are made secure. The ground wires may be attached at any convenient point on the engine, but paint and grease must be removed to secure a good circuit.

Flexible wiring is less liable to break at the point of connection and cause trouble. A solid copper wire will frequently break close to the binding post and remain in position apparently sound.

Spark coils are manufactured by specialists and their manufacture, on a large scale, has been so well developed that the selling price is too low for the engine manufacturer to think of making his own coils.

The wire, from the secondary winding of the jump spark coil to the spark plug, should be heavily insulated, or care must be taken to keep it clear of other parts that would complete a circuit, owing to the high voltage current that would leak and short circuit through a light insulation. This is a frequent cause of trouble with jump spark ignition. A light or faulty insulation on the secondary wire will often deceive the inexperienced operator.

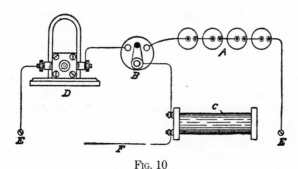

FIG. 10

Wiring diagram—battery and generator. Make and break ignition.

A—Battery.

B—Switch.

C—Simple Primary Coil.

D—Magneto or Generator.

EE—Ground Wires to Engine.

F—Wires to Stationary Electrode of Igniter.

A bare wire, from the secondary binding post on the coil to the plug, would be better than a defective insulation, for with the bare wire every one would know it must not touch other objects that would conduct the current.

There is a great deal of discussion among gas engine builders and users concerning the relative merits of the two systems of electric ignition, but as both give good service and satisfaction under proper conditions, it is a matter that can be decided only in individual cases. It is argued for the make and break system that the low voltage current is less liable to short circuit; that the coil is much cheaper, and less liable to go wrong, and that a bigger and better spark is made.

For the jump spark ignition a great advantage is claimed by doing away with the movable electrode, its wear and consequent leakage from the combustion chamber; that the time of the spark can be easily retarded and advanced at will to suit all speeds and conditions, and that if the coil and high tension current are handled intelligently they will not fail, but will go on indefinitely doing their work faithfully.

It is generally considered that jump spark ignition is better suited for high speed engines, while the low speed, heavily constructed engines commonly used for stationary work are usually equipped with the make and break system. So far as practical application is concerned either system can be applied to suit any condition.

As both are good, the reader will be left to decide for himself as his own experience or preference may direct.

Timing the Valves and Spark—The valves of the gas engine are almost universally of the poppet variety, and are operated by cams and springs which produce a very quick opening and closing action. In order to obtain a high efficiency in the working of the engine it is necessary that the valves open, and that the spark occur at the proper moment, to produce the best results. The inlet valve A, shown in Fig. I, is of the automatic type, being opened by the suction stroke of the piston. While many gas engines are built this way, it is quite common to open the inlet, as well as the exhaust valve mechanically, or by means of a cam operated by the engine.

The automatic inlet valve, as its name implies, is self timed, opening at the beginning, and closing at the end of the suction stroke.

The cams to open all mechanically operated valves must be set or timed with reference to the position of the crank and piston. The exhaust valve should be opened about 40° before the crank on its power stroke reaches center.

In an engine with 6″ stroke, the piston would be about 11/16″ from the end of the stroke.

This last part of the stroke is not effective in delivering power to the crank shaft, and the exhaust valve is opened thus early to get rid of the remaining heat as soon as it becomes useless and thus have the cylinder in better shape to receive the next charge. The exhaust valve should not close before the end of the scavenging stroke, and not later than 20° past dead center.

If the inlet valve is operated mechanically, the cam should be set to open and close the valve when the crank has passed the dead centers 10°, to 20° according to the speed of the engine.

The late closing of the inlet valve on high speed engines is to allow the inertia or moving force of the incoming charge to increase the power of the cylinder by increasing the amount or volume of mixture taken in. Some claim the crank may pass the center 30° to 40° before the mixture stops coming in, although the piston has traveled back on the compression stroke one-half inch or more. The possible advantage is a slight increase of power from a cylinder of given size.

The timing of the ignition is of much greater importance than was realized for many years after the gas engine came into use. Although a proper mixture under compression fires easily and burns rapidly, yet it requires a little space of time, and the spark must occur far enough ahead of the center so the charge will be aflame, and the expansion taking place when the piston and crank start on the power stroke. If the spark comes too late, a part of the effective impulse stroke is lost, while if the spark is made too early, the heat expansion begins before the crank reaches the center and some of the power is thus delivered in a backward direction. This will cause the engine to "pound" or perhaps stop, if the ignition is very much too early.

The correct time for the spark depends on the fuel used, and speed of the engine. At high speeds the spark must be advanced or made further ahead of the center to give the necessary time for ignition, while at low speeds the spark must be retarded or made later.

It is necessary to provide high speed engines with a device for retarding the spark in starting, and changing to the advanced position after the engine gets up speed.

For very high speeds the spark must be produced somewhere from 60° to 90° ahead of center and this position, with the slow speed in starting, would deliver all the power in a backward direction, causing the engine to "kick."

Owing to the greatly varying speeds used it is impossible to give a set rule for the correct point of ignition, but the proper timing of the spark can be readily determined by a little experimenting. The operator will soon learn the correct position by observing the results of early or late ignition.

It is needless to say, that if the spark is too far advanced in starting the operator will soon find it out, for the engine is sure to make a "kick" about it.

A gas engine will run with the valves and spark considerably out of time, but its full power and efficiency will not be developed unless the timing is right.

As the inlet and exhaust valves, in proper turn, only open every second revolution of the crank shaft (with the four-cycle engine), the reader will understand that the cams are located on the back geared shaft, which runs at just one-half the speed of the crankshaft.

In timing the valves the question naturally arises when is the crank exactly at the end of the stroke or on "dead center?"

The crank travels a considerable distance at each end of the stroke, with but little perceptible movement of the piston and this fact gives considerable range in setting the valves while not greatly affecting the results.

Some users, especially of small engines, guess at the center by noting the piston's movement, but for the benefit of readers who insist on *knowing* when the crank is at center, we illustrate in Fig. 11 the following method:

With the crank turned to one side of center, as shown, insert a rod A, through an opening in the head of the engine allowing the rod to rest against the piston. Mark on the rod at B to show the distance to the piston and also mark the balance wheel at a fixed, stationary pointer C provided on the engine. Now turn the engine until the crank is on the other side of center as shown by the dotted lines. This position is determined by bringing the piston to the same distance from point B as shown by the mark we placed on the rod A.

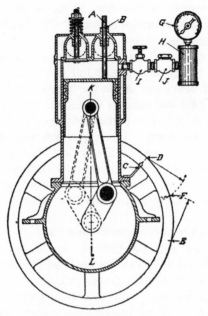

Fig. 11

Now make another mark on the balance wheel at the stationary pointer C. The two marks D and E on the balance wheel are at equal distances from the central position for the crank, and it follows that in bisecting or equally dividing the distance between the marks D and E and turning the engine so the central mark F, comes to the stationary pointer C, we have thus brought the crank to the "dead" center.

The opposite center is determined in a like manner. The crank is thus brought at each end of the piston stroke exactly to the center line K-L. Having established the center we can readily calculate the degrees from this for the opening and closing of the valves. The circumference of the wheel is always equal to 360 degrees. If we divide 360 by the circumference in inches we will know how many degrees in each inch. To find a point 40 degrees from "dead" center divide 40 degrees by the number of degrees in an inch of the circumference. The result will be the number of inches from center to the point desired.

In the illustration, Fig. 11, it will be noticed that one of the valves has been removed to insert the rod A through the valve stem guide. By using a rod that fits the guide the two positions of the piston, at equal distance from the center can be accurately determined. For engines that do not have the valves in the head any other openings, such as for the spark plug or an igniter, may be used, but it would be advisable to use a special plug, or plate to fit the opening through which a hole, to fit rod A, may be drilled.

The above method for locating dead center is the same that is generally used for the steam engine except that the mark B on the rod A is made on the crosshead and guides.

As the gas engine ordinarily has no crosshead the process we have described and illustrated will be found equally effective and simple, while, as with the steam engine, it is mechanically correct.

TESTING THE CONDITION OF A GAS ENGINE

First see that the valves are correctly timed.

The next thing to know is that the fuel reaches the mixing chamber or carbureter. Now look after the compression to see if there is any serious leakage through the rings, valves or packed joints.

Oil the engine thoroughly using care to know that the cylinder walls are well lubricated with good gas engine oil, then as a quick, ordinary compression test, sufficient for practical purposes, the engine is revolved bringing the piston up quickly on the compression stroke and holding it at the highest point of compression to see how soon the pressure will disappear. This may properly be called "feeling of the compression" and after a little experience the operator will be able to judge pretty accurately as to what results may expected of the engine.

The only recourse when serious leakage through the rings occurs, is new rings, or perhaps re-boring of the cylinder, new piston and new rings.

This is a job for the machinist.

After knowing that the fuel gets to the engine in proper time, and that the compression is all right, next look after the ignition apparatus, a very important part of every gas power machine.

The make and break system may be tested out as follows: Throw in the switch, then detach the wire from the stationary, insulated electrode of the igniter, and scrape it on the binding post from which it was removed. If a spark is produced with the igniter contact points open it will prove the insulation of the stationary electrode to be faulty. Should no spark appear, next close the contact points and scrape the wire again on the binding post. A good spark should now be produced. If not, go over the wiring very carefully to see if all connections are clean and secure, and to look for possible leakage of all the current, or short circuiting as it is commonly called.

Next remove the igniter to see if the contact points are corroded thus preventing the passage of the current.

While having the igniter detached it is a good plan to hold it to the engine and snap or break the contact points apart as when the engine is running. If, with the contact points clean, connections all properly made and no short circuits, a spark is not yet obtained, next look after the source of current (battery or generator as the case may be), and the trouble will soon be located in an exhausted battery, or in case of a generator it may be bad brushes or possible loss of speed if the generator is driven by belt or friction.

Once in a great while the spark coil may fail, but this is a rare occurrence, if the ignition apparatus is kept in a dry place as it should be.

Briefly stated, see that the engine gets the fuel in the proper time; see that there is no serious leakage and see that a good spark is produced at the right time.

These things in proper order and assuming, of course, in case liquid fuel is used, that the proper condition is present for carburetion, or vaporization, the engine is ready to run and may be depended upon.

The routine for testing a jump spark ignition outfit is similar to that just described for the primary or make and break system. By detaching the spark plug and allowing it to rest on the engine, so the circuit will be the same as when the plug is in position, work the circuit interrupting device (if a plain jump spark coil is used), or in case of a vibrator coil, turn the engine until the circuit is made by the timer, when the vibrator, if properly adjusted, will set up the buzzing sound familiar to users of vibrator coils.

A good spark should now appear between the points of the spark plug. If not, detach the wire from the plug, and holding the end of the wire within one-sixteenth to one-eighth inch of some part of the engine again work the trembler or make contact with the timer.

If a spark can now be produced it proves the insulation of the plug faulty, while should no spark appear next look for bad connections, short circuit or further back to the source of current as with the make and break system.

Now, knowing that the engine takes the charge and fires it properly, next see that the cooling system is in working order. If the cooling jacket, or passages formed in the castings of the cylinder and head for allowing the cooling element, oil, water or other liquid to circulate, should become clogged or choked the heat of the cylinder will rise too high, so in testing the condition of a machine we must examine the cooling facilities, and know that sufficient radiation of excess heat is maintained. This means, of course, that the proper circulation of the heat carrying agent (whether it be the water, oil, or air) must be provided.

The compression test described in the fore-going is a quick, offhand way of sizing up the running condition of small and medium sized engines, but it can only give an approximate idea of the amount of the compression. A very good method of obtaining the gauge pressure of the compression is illustrated in Fig. 11.

A pressure gauge G, is attached to a receiving chamber H, which is connected to the compression chamber of the engine by a globe valve I, and check valve J.

Run the engine up to full speed, then throw out the switch and immediately open the valve I. The highest compression pressure will be accumulated in the chamber H, and the gauge will register the pounds per square inch.

The valve I must not be opened while the engine is yet firing the charges, but it should be opened very quickly after the firing has stopped so the compression pressure may be registered at practically the normal running speed of the engine.

This test of the compression is not necessary to the successful care and operation of a gas engine for the manufacturer of the machine has, of course, figured out the best compression for the kind of fuel to be used and the work to be done.

We describe and illustrate the gauge test for the benefit of readers who may wish to make a deeper study of the gas engine, and gas engine principles than is necessary for the ordinary user.

THE SCIENCE OF THRESHING

CHAPTER XIV

HOW TO RUN A THRESHING MACHINE

A threshing machine, though large, is a comparatively simple machine, consisting of a cylinder with teeth working into other teeth which are usually concaved (this primary part really separates the grain from the husk), and rotary fan and sieves to separate grain from chaff, and some sort of stacker to carry off the straw. The common stacker merely carries off the straw by some endless arrangement of slats working in a long box; while the so-called "wind stacker" is a pneumatic device for blowing the straw through a large pipe. It has the advantage of keeping the straw under more perfect control than the common stacker. The separation of the grain from the straw is variously effected by different manufacturers, there being three general types, called apron, vibrating, and agitating.

The following list of parts packed inside the J. I. Case separator (of the agitative type) when it is shipped will be useful for reference in connection with any type of separator:

2 Hopper arms, Right and Left,
1 Hopper bottom,
I Hopper rod with thumb nut,
2 Feed tables,
2 Feed table legs,
2 Band cutter stands and bolts,
1 Large crank shaft,

1 Tailings auger, 1 Elevator spout,
1 Elevator shake arm, complete,
1 Set fish-backs, for straw-rack,
1 Elevator pulley, 529 T.,
1 Beater pulley, 6-inch 1254 T., or 4-inch 1255 T.,
1 Elevator drive pulley

1 Grain auger with 1223 T. pulley and 1154 T., Box,

1 Cylinder pulley to drive crank 4-inch 973 T., or 6-inch 1085 T.,

1 Cylinder pulley to drive fan 1347 T., 1348 T., or 1633 T.,

1 Fan pulley, 1244 T., or 1231 T.,

I Belt tightener, complete, with pulley,

1673 T.,

1 Crank pulley to drive grain auger 1605 T.,

1 Belt reel, 5016 T., or 1642 T., with crank and bolt.

4 Shoe sieves,

4 Shoe rods, with nuts and washers,

1 Conveyor extension,

1 Sheet iron tail board,

2 Tail board castings 1654 T., and 1655 T.

In addition to these are the parts of the stacker.

As each manufacturer furnishes all needed directions for putting the parts together, we will suppose the separator is in working condition.

A new machine should be set up and run for a couple of hours before attempting to thresh any grain. The oil boxes should be carefully cleaned, and all dirt, cinders, and paint removed from the oil holes. The grease cups on cylinder, beater and crank boxes should be screwed down after being filled with hard oil, moderately thin oil being used for other parts of the machine. Before putting on the belts, turn the machine by hand a few times to see that no parts are loose. Look into the machine on straw rack and conveyor.

First connect up belt with engine and run the cylinder only for a time. Screw down the grease cup lugs when necessary, and see that no boxes heat. Take off the tightener pulley, clean out oil chambers and thoroughly oil the spindle. Then oil each separate bearing in turn,

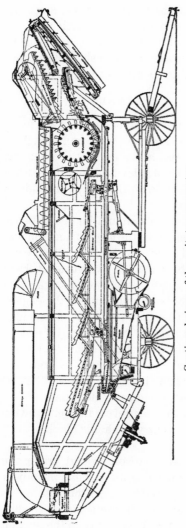

Sectional view of the agitator separator.

seeing that oil hole is clean, and that pulley or journal works freely. The successive belts may then be put on one at a time, until the stacker belt is put on after its pulleys have been oiled. Especially note which belts are to run crossed—usually the main belt and the stacker belt. You can tell by noting which way the machinery must run to keep the straw moving in the proper direction.

Oiling on the first run of a machine is especially important, as the bearings are a trifle rough and more liable to heat than after machine has been used for some time. It is well to oil a shaft while it runs, since the motion helps the oil to work in over the whole surface.

The sieves, concaves, check board and blinds must be adjusted to the kind of grain to be threshed. When they have been so adjusted the machine is ready to thresh.

SETTING SEPARATOR

It is important that the machine be kept perfectly steady, and that it be level from side to side, though its being a little higher or lower at one end or the other may not matter much. If the level sidewise is not perfect the grain will have a tendency to work over to one side. A spirit level should be used.

One or more of the wheels should be set in holes, according to the unevenneas of the ground, and the rear wheels should be well blocked. Get the holes ready, judging as well as possible what will give a true level and a convenient position. Haul the machine into position and see that it is all right before uncoupling the engine. If holes need redigging to secure proper level, machine may be pulled out and backed in again by the engine. When machine is high in front it can easily be leveled when engine or team have been removed, by cramping the front wheels and digging in front of one and behind the other, then pulling the tongue around square.

Block the right hind wheel to prevent the belt drawing machine forward. Always carry a suitable block to have one handy.

In starting out of holes or on soft ground, cramp the front axle around, and it will require only half the power to start that would be required by a straight pull.

In setting the machine, if the position can be chosen, choose one in which the straw will move in the general direction of the wind, but a little quartering, so that dust and smoke from engine will be carried away from the men and the straw stack. In this position there is less danger from fire when wood is used.

THE CYLINDER

The cylinder is arranged with several rows of teeth working into stationary teeth in what is called the concave. It is important that all these teeth be kept tight, and that the cylinder should not work from side to side. The teeth are liable to get loose in a new machine, and should be tightened up frequently. A little brine on each nut will cause it to rust slightly and help to hold it in place. If the cylinder slips endwise even a sixteenth of an inch, the teeth will be so much nearer the concaves on one side and so much farther away from them on the other side. Where they are close, they will crack the grain; where they are wide apart they will let the straw go through without threshing or taking out the grain. So it is important that the cylinder and its teeth run true and steady. If the teeth get bent in any way, they must be straightened.

The speed of the cylinder is important, since its pulley gives motion to the other parts of the machine, and this movement must be up to a certain point to do the work well. A usual speed for the cylinder pulley is 1,075 revolutions per minute, up to 1,150.

There is always an arrangement for adjusting the cylinder endwise, so that teeth will come in the middle. This should be adjusted carefully when necessary. The end play to avoid heating may be about 1-64 of an inch. It may be remembered that the cylinder teeth carry the straw to the concaves, and the concaves do the threshing.

THE CONCAVES

The concaves are to be adjusted to suit the kind of grain threshed. When desiring to adjust concaves, lift them up a few times and drop so as to jar out dust. Wedging a block of wood between cylinder teeth and concaves will in some types of separator serve to bring up concaves when cylinder is slowly turned by hand.

There are from two to six rows of teeth in the concave, and usually the number of rows is adjustable or variable. Two rows will thresh oats, where six are required for flax and timothy. Four rows are commonly used for wheat and barley. The arrangement of rows of teeth and blanks is important. When four rows are used, one is commonly placed well back, one front, blank in the middle. When straw is dry and brittle, cylinder can be given "draw" by placing blank in front. Always use as few teeth and leave them as low as possible to thresh clean, since with more teeth than necessary set higher than required the straw will be cut up and a great deal of chopped straw will get into the sieves, all of which also requires additional power. Sometimes the teeth can be taken out of one row, so that one, three, or five rows may be used. For especially difficult grain like Turkey wheat, a concave with corrugated teeth may be used, in sets of three rows each up to nine rows. The corrugated teeth are used for alfalfa in localities where much is raised.

THE BEATER AND CHECK BOARD

After the cylinder has loosened the grain from the husk and straw, it must still be separated. Some threshers have a grate under the cylinder and behind it. In any case the beater causes the heavy grain to work toward the bottom, and the check board keeps the grain from being carried to rear on top of the straw, where it would not have a chance to become separated. If the grain is very heavy or damp, there may be a tendency for the straw to stick to the cylinder and be carried around too far. In such a case the beater should be adjusted to give more space, and the check board raised to allow the straw to pass to the rear freely.

STRAW RACK

The straw rack and conveyor carry the straw and grain to the rear with a vibratory movement, causing the grain to be shaken out. To do good work the straw rack must move with a sufficient number of vibrations per minute, say 230. A speed indicator on the crank shaft will show the number of vibrations best. Great care must be taken with this part of the thresher, or a great deal of grain will be carried into the straw. The. less the straw is cut up, the better this portion of the machine works; so the smallest practicable number of teeth in the concave should be used.

The crank boxes and pitmans should be adjusted so that there is no pounding. If the rear vibrating arms drop too low they get below the dead center and are liable to break, at any rate causing severe pounding and hard running. To prevent this, the crank boxes can be moved forward by putting leather between them and the posts, or should be otherwise adjusted. The trouble being due to the pitmans having worn short, the pitmans may be lengthened in some way by putting pieces of leather over the end or the like, or new pitmans may be introduced.

THE FAN

The chief difficulty likely to arise with the fan is blowing over grain. To prevent this blinds are usually arranged, which may be adjusted while the machine is running so as to prevent the grain from being blown over. At the same time it is important to clean the grain, so the adjustment should not go to one extreme or the other.

In windy weather the blinds should be closed more on one side than on the other. The speed of the fan must be adjusted to the requirements of the locality.

As much blast should be used as the grain will stand, and heavy feeding requires more wind than light feeding, since the chaff checks the blast to a certain extent.

Care should be taken that the wind board over the grain auger does not get bent, and it should be adjusted so that the strongest part of the blast will come about the middle of the sieve.

SIEVES

There is usually one conveyor sieve, which causes the grain to move along, and shoe sieves, which are required to clean the grain thoroughly. Different kinds of sieves are provided for different kinds of grain, and the proper selection and adjustment of these sieves as to mesh, etc., is of the utmost importance.

Much depends on the way the sieves are set, and on the rate at which the thresher is fed, or the amount of work it is really doing. The best guide is close observation and experience, both your own and that of other threshermen.

CONVEYOR EXTENSION

This carries the coarse chaff from the conveyor sieve to the stacker. The conveyor sieve should be coarse enough to let all the good grain through, as whatever is carried on to the extension must be returned with the tailings to the cylinder. This means so much waste work. The conveyor extension is removable, and should always be tight before machine is started. See that it is.

When necessary, the grain may be run over a screen, which differs from a sieve in that the mesh is small and intended to let dust and small chaff through while the grain does not pass. The refuse from the screen is dropped onto the ground. All screens have a tendency to become clogged, and in this condition obstruct the grain and wind. It is desirable not to use them except when necessary, and if used they should be frequently cleaned.

TAILINGS ELEVATOR

The tailings are carried back to the cylinder by an elevator usually worked with a chain. This chain should be kept tight enough not to unhook, yet not so tight as to bind.

To put the chain into the elevator, tie a weight on a rope and drop it down the lower part of the elevator. The chain may be fastened to the rope and a man at the top can then pull the chain up, while another feeds it in at the bottom. When chain has been drawn up to the top, the rope should be dropped down upper portion of elevator and used at bottom to pull chain down after it has been adjusted over the sprocket. Some one at the bottom should continue to feed the chain in as it is pulled down, so that it will go into the elevator straight. When the chain has been pulled through it may be hooked and adjusted to lower sprocket, and tightened up by screws at top. Turn the chain around once by hand to make sure there are no kinks in it.

The tailings should be small, containing no light chaff and little full-size grain. They are a good indication of how the sieves are working. If much good grain is coming through, see if it gets over the conveyor sieve by way of the extension to the tailings auger, or over the shoe sieve. If the sieves are not right, they may be adjusted in various ways, according to the directions of the manufacturer.

Grain returned in the tailings is liable to get cracked in the cylinder, and much chaff in the tailings chokes the cylinder. For every reason, the tailings should be kept as low as possible.

SELF-FEEDER

The self-feeder is arranged to cut the bands of the sheaves and feed the grain to the cylinder automatically. It has a governor to prevent crowding in too much grain, and usually a change of pulleys for slow or fast feeding, as circumstances may require. In starting a new governor the friction pulley and inside of the band should have paint scraped off, and a little oil should be put on face of friction wheel. The carrier should not start till the machine attains full threshing motion, and to prevent this a few sheaves should be laid upon it. The knife arms should be raised or lowered to adjust them to the size of the sheaves and condition of the grain for cutting bands.

The cranks and carrier shaft boxes should be oiled regularly, but the friction bands should not be oiled after it once becomes smooth.

THE WIND STACKER

The wind stacker is arranged to swing by a hand-wheel or the like, and also automatically.

Great care should be taken not to use the hand moving apparatus when the stacker is set for automatic moving, as a break is liable to follow. There is a clutch to stop the stacker, however. At times it will be more convenient to leave off the belt that causes the automatic movement.

By the use of various pulleys the speed of the stacker may be altered, and it should be run no faster than is necessary to do the work required, which will depend on the character of the straw. Any extra speed used will add to the cost of running the engine and is a loss in economy.

In moving machine with wind stacker in place, care should be taken to see that it rests in its support before machine moves.

The canvas curtain under the decking, used to turn the straw into the hopper, may need a piece of wood fastened to its lower edge to keep it more stiff when stiff rye straw is passing. The bearings of the fan and jack shafts should be kept well lubricated with hard oil, and the bevel gears should be kept well greased with axle grease applied with a stick. Other bearings and worm gear of automatic device should be oiled with soft oil.

The attached stacker is simple in operation, and if it is desired not to use the automatic swinging device but swing by hand, the automatic gear may be thrown out. An independent stacker is managed in much the same way.

ATTACHMENTS

A weigher, bagger, and a high loader are usually used with a separator. Their operation is simple, and depends upon the particular type or make.

BELTING

The care of the belting is one of the most important things about the management of a threshing machine, and success or failure will depend largely on the condition in which the belts are kept. Of course the hair side should be run next the band wheel. Once there was disagreement among engineers on this point, but it has been conclusively proven that belts wear longer this way and get better friction, for the simple reason that the flesh side is more flexible than the hair side, and when on the outside better accommodates itself to the shape of the pulley. If the hair side is outermost, it will be stretched more or less in going around the pulley and in time will crack. Rubber belts must be run with the seam on the outside.

When leather belts become hard they should be softened with neatsfoot oil. A flexible belt is said to transmit considerably more power than a hard one.

Pulleys must be kept in line or the belt will slip off. When pulleys are in line the belt has a tendency to work to the tightest point. Hence pulleys are usually made larger in the middle, which is called "crowning."

Belts on a separator should be looked over every day, and when any lacing is worn, it should be renewed at once. This will prevent breaks during working, with loss of time. Some threshermen carry an extra set of belts to be ready in case anything does break, and they assert that they save money by so doing.

Lacing is not stronger in proportion as it is heavy. If it is heavy and clumsy it gets strained in going round the pulley, and soon gives out. The ideal way to lace a belt is to make it as nearly like the rest of the belt as possible, so that it will go over the pulleys without a jar. The ends of the belt should be cut off square with a try square, and a small punch used for making holes. Holes should be equally spaced, and outside ones not so near the edge as to tear out. The rule is a hole to every inch of the belt, and in a leather belt they may be as close as a quarter of an inch to the ends without tearing out. Other things being equal, the nearer the ends the holes are the better, as belt will then pass over pulley more easily. The chief danger of tearing is between the holes.

A stacker web belt may be laced by turning the ends up and lacing them together flat at right angles to rest of belt. Rubber or cotton belting that does not run over idler or tightener pulleys so that both sides must be smooth may be laced in this way. This lacing lasts two or three times as long with such belts as any other, for the reason that the string is not exposed to wear and there is no straining in passing round pulleys.

The ordinary method of lacing a leather belt is to make the laces straight on the pulley side, all running in the same direction as the movement of the belt, and crossing them on the outside diagonally in both directions. When belts run on pulleys on both sides, as they do on the belt driving beater and crank, and also on wind stacker, a hinge lacing may be made by crossing the lacing around the end of the belt to the next adjacent hole opposite, the lacing showing the same *on* both sides. This allows the belt to bend equally well either way.

The best way to fasten a lacing is to punch a hole where the next row of lace holes would come when the belt is cut off, and after passing the lace through this hole, bring the end around and force it through again, cutting the end off short after it has passed through. This hole must be small enough *to* hold the lace securely, and care should be taken that it is in position to be used as a lace-hole the next time a series of holes is required.

New belts stretch a good deal, and the ends of the lacing should not be cut off short till the stretch is taken out of the belts.

Belting that has got wet will shrink and lacing must be let out before belt is put on again. Tight belts have been known to break the end of a shaft off, and always cause unnecessary friction.

Cotton or Gandy belting should not be punched for lacing, but holes made with a pointed awl, since punching cuts some of the threads and weakens belt.

HOW TO BECOME A GOOD FEEDER

The art of becoming a good feeder will not be learned in a day. The bundles should be tipped well up against the cylinder cap, and flat bundles turned on edge, so that cylinder will take them from the top. It is not hard to spread a bundle, and in fast threshing a bundle may be fed on each side, each bundle being kept pretty well to its own side, while the cylinder is kept full the entire width. A good feeder will keep the straw carrier evenly covered with straw, and will watch the stacker, tailings and grain elevator and know the moment anything goes wrong.

WASTE

No threshing machine will save every kernel of the grain, but the best results can be attained only by care and judgment in operating.

It is easy to exaggerate the loss of grain, for if a very small stream of grain is seen going into the straw it will seem enormous, though it will not amount to a bushel a day. There are practically a million kernels of wheat in a bushel, or 600 handfuls, and even if a handful is wasted every minute, it would not be enough to counterbalance the saving in finishing a job quickly.

Of course, waste must be watched, however, and checked if too great. First determine whether the grain is carried over in the straw or the waste is at the shoe sieve.

If the waste is in the conveyor sieve, catch a handful of the chaff, and if grain is found, see whether the sieve is the proper mesh. Too high a speed will cause the grain to be carried over. If too many teeth are used in the concave, the conveyor sieve will be forced to carry more chaff than it can handle. The blast may be too strong and carry over grain, so adjust the blinds that the blast will be no stronger than is necessary to clean the wheat well and keep sieves free. If grain is still carried over, the conveyor sieve may be adjusted for more open work, but care should be taken not to overwork the shoe sieve. Be careful that the wind board is not bent so that some grain will go into the fan and be thrown out of the machine altogether.

If the grain is not separated from the straw thoroughly, it may be due to "slugging" the cylinder (result of poor feeding), causing a variable motion. It may also be because speed of crank is not high enough. Check board should be adjusted as low as possible to prevent grain being carried on top of straw. See that cylinder and concave teeth are properly adjusted so as not to cut up straw, while at the same time threshing out all the grain. Sometimes heads not threshed out by the cylinder will be threshed out by the fan of the wind stacker, and the fault will be placed on the separating portions instead of on the imperfect cylinder.

Grain passes through the cylinder at the rate of about a mile a minute. The beater reduces this to 1,500 feet per minute. After passing the check board the straw moves about 36 feet per minute. At these three different speeds the straw passes the 17 feet length of the machine in about 25 seconds. The problem is to stop the grain while the straw is allowed to pass out. Evidently there must be a small percentage of loss, and there is always a limit as to what it will pay to try to save. Each man must judge for himself.

BALANCING A CYLINDER

A cylinder should be so balanced that it will come to rest at any point. In a rough way a cylinder may be balanced by placing the journals on two carpenter's squares laid on saw-horses. Gently roll the cylinder back and forth and every time it stops, make a chalk mark on the uppermost bar. If the same bar comes up three times in succession it probably is light, and a wedge should be driven under center band at chalk mark. Continue experimenting until cylinder will come to rest at any point.

COVERING PULLEYS

This is easily done, but care must be taken that the leathers are tight or they will soon come off.

To cover a cylinder pulley, take off what remains of the old cover, pull out the nails, and renew the wedges if necessary. Select a good piece of leather a little wider than face of pulley and about four inches longer than enough to go around. Soak it in water for about an hour. Cut one end square and nail it to the wedges, using nails just long enough to clinch. Put a clamp made of two pieces of wood and two bolts on the leather, block the cylinder to keep it from turning, and by means of two short levers pry over the clamp to stretch the leather. Nail to the next wedges, move the clamp and nail to each in turn, finally nailing to the first one again before cutting off. Trim the edges even with the rim of the pulley.

The same method may be used with riveted covers.

CARE OF A SEPARATOR

A good separator ought to last ten years, and many have been in use twice that time. After the season is over the machine ought to be thoroughly cleaned and stored in a dry place. Dirt on a machine holds moisture and will ruin a separator during a winter if it is left on. It also causes the wood to rot and sieves and iron work to rust.

Once in two years at least a separator ought to have a good coat of first-class coach varnish. Before varnishing, clean off all grease and oil with benzine and see that paint is bright.

At the beginning of the season give the machine a thorough overhauling, putting new teeth in cylinder if any are imperfect, and new slats in stacker web or straw rack if they are needed. Worn boxes should be taken up or rebabbitted, and conveyor and shoe eccentrics replaced if worn out. Tighten nuts, replace lost bolts, leaving the nut always turned square with the piece it rests on. Every separator ought to be covered with a canvas during the season. It will pay.

The right and left sides of a threshing machine are reckoned from the position of the feeder as he stands facing the machine.

In case of fire, the quickest way is to let the engine pull the machine out by the belt. Take blocks away from wheels, place a man at end of tongue to steer, and back engine slowly. If necessary, men should help the wheels to start out of holes or soft places.

Watch the forks of the pitchers to see that none are loose on the handles, especially if a self-feeder is used. A pitchfork in a separator is a bad thing.

CHAPTER XV

QUESTIONS ASKED ENGINEERS WHEN APPLYING FOR A LICENSE*

Q. If you were called on to take charge of a plant what would be your first duty?

A. To ascertain the exact condition of the boiler and all its attachments (safety valve, steam gauge, pump, injector), and engine.

Q. How often would you blow off and clean your boilers if you had ordinary water to use?

A. Once a month.

Q. What steam pressure will be allowed on a boiler 50 inches diameter 3/8 inch thick, 60,000 T. S. 1-6 of tensile strength factor of safety?

A. One-sixth of tensile strength of plate multiplied by thickness of plate, divided by one-half of the diameter of boiler, gives safe working pressure.

Q. How much heating surface is allowed per horse power by builders of boilers?

A. Twelve to fifteen feet for tubular and flue boilers.

Q. How do you estimate the strength of a boiler?

A. By its diameter and thickness of metal.

Q. Which is the better, single or double riveting?

A. Double riveting is from sixteen to twenty per cent stronger than single.

Q. How much grate surface do boiler makers allow per horse power?

A. About two-thirds of a square foot.

---

*Furnished by courtesy of a friend of Aultman & Taylor Co.

Q.   Of what use is a mud drum on a boiler, if any?

A.   For collecting all the sediment of the boiler.

Q.   How often should it be blown out?

A.   Three or four times a day.

Q.   Of what use is **a** steam dome on a boiler?

A.   For storage of dry steam.

Q.   What is the object of a safety valve on a boiler?

A.   To relieve pressure.

Q.   What is your duty with reference to it?

A.   To raise it twice a day and see that it is in good order.

Q.   What is the use of check valve on a boiler?

A.   To prevent water from returning back into pump or injector which feeds the boiler.

Q.   Do you think a man-hole in the shell on top of a boiler weakens it any?

A.   Yes, to a certain extent.

Q.   What effect has cold water on hot boiler plates?

A.   It will fracture them.

Q.   Where should the gauge cock be located?

A.   The lowest gauge cock ought to be placed about an inch and a half above the top row of flues.

Q.   How would you have your blow-off located?

A.   In the bottom of mud-drum or boiler.

Q.   How would you have your check valve arranged?

A.   With a stop cock between check and boiler.

Q.   How many valves are there in a common plunger force pump?

A.   Two or more—a receiving and a discharge valve.

Q.   How are they located?

A.   One on the suction side, the other on the discharge.

Q.   How do you find the proper size of safety valves for boilers?

A. Three square feet of grate surface is allowed for one inch area of spring loaded valves; or two square feet of grate surface to one inch area of common lever valves.

Q. Give the reasons why pumps do not work sometimes?

A. Leak in suction, leak around the plunger, leaky check valve, or valves out of order, or lift too long.

Q. How often ought boilers to be thoroughly examined and tested?

A. Twice a year.

Q. How would you test them?

A. With hammer and with hydrostatic test, using warm water.

Q. Describe the single acting plunger pump; how it gets and discharges its water?

A. The plunger displaces the air in the water pipe, causing a vacuum which is filled by the atmosphere forcing the water therein; the receiving valve closes and the plunger forces the water out through the discharge valve.

Q. What is the most economical boiler-feeder?

A. The (Trix) Exhaust Injector.[*]

Q. What economy is there in the Exhaust Injector?

A. From 15 to 25 per cent saving in fuel.

Q. Where is the best place to enter the boiler with the feed water?

A. Below the water level, but so that the cold water can not strike hot plates. If injector is used this is not so material as feed water is always hot.

Q. What are the principal causes of priming in boilers?

---

[*]   So says one expert. Others may think otherwise.

A.   To high water, not steam room enough, miscon-struction, engine too large for boiler.

Q.   How do you keep boilers clean or remove scale therefrom?

A.   The best "scale solvent" and "feed water puri-fier" is an honest, intelligent engineer who will regularly open up his boilers and clean them thoroughly, soaking boilers in rain water now and then.

Q.   If you found a thin plate, what would you do?

A.   Put a patch on it.

Q.   Would you put it on the inside or outside?

A.   Inside.

Q.   Why so?

A.   Because the action that has weakened the plate will then set on the patch, and when this is worn it can be repeated.

Q.   If you found several thin places, what would you do?

A.   Patch each and reduce the pressure.

Q.   If you found a blistered plate?

A.   Put a patch on the fire side.

Q.   If you found a plate on the bottom buckled?

A.   Put a stay through the center of buckle.

Q.   If you found several of the plates buckled?

A.   Stay each and reduce the pressure.

Q.   What is to be done with a cracked plate?

A.   Drill a hole at each end of crack, caulk the crack and put a patch over it.

Q.   How do you change the water in the boiler when the steam is up?

A.   By putting on more feed and opening the surface blow cock.

Q. If the safety valve was stuck how would you relieve the pressure on the boiler if the steam was up and could not make its escape?

A. Work the steam off with engine after covering fires heavy with coal or ashes, and when the boiler is sufficiently cool put safety valve in working order.

Q. If water in boiler is suffered to get too low, what may be the result?

A. Burn top of combustion chamber and tubes, perhaps cause an explosion.

Q. If water is allowed to get too high, what result?

A. Cause priming, perhaps cause breaking of cylinder covers or heads.

Q. What are the principal causes of foaming in boilers?

A. Dirty and impure water.

Q. How can foaming be stopped?

A. Close throttle and keep closed long enough to show true level of water. If that level is sufficiently high, feeding and blowing off will usually suffice to correct the evil.

Q. What would you do if you should find your water gone from sight very suddenly?

A. Draw the fires and cool off as quickly as possible. Never open or close any outlets of steam when your water is out of sight.

Q. What precautions should you take to blow down a part of the water in your boiler while running with a good fire?

A. Never leave the blow-off valve, and watch the water level.

Q. How much water would you blow off at once while running?

A. Never blow off more than one gauge of water at a time while running.

Q. What general views have you in regard to boiler explosions—what is the greatest cause?

A. Ignorance and neglect are the greatest causes of boiler explosions.

Q. What precaution should the engineer take when necessary to stop with heavy fires?

A. Close dampers, put on injector or pump and if a bleeder is attached, use it.

Q. Where is the proper water level in boilers?

A. A safe water level is about two and a half inches over top row of flues.

Q. What is an engineer's first duty on entering the boiler room?

A. To ascertain the true water level.

Q. When should a boiler be blown out?

A. After it is cooled off, never while hot.

Q. When laying up a boiler what should be done?

A. Clean thoroughly inside and out; remove all oxidation and paint places with released; examine all stays and braces to see if any are loose or badly worn.

Q. What is the last thing to do at night before leaving plant?

A. Look around for greasy waste, hot coals, matches, or anything which could fire the building.

Q. What would you do if you had a plant in good working order?

A. Keep it so, and let well enough alone.

Q. Of what use is the indicator?

A. The indicator is used to determine the indicated power developed by an engine, to serve as a guide in setting valves and showing the action of the steam in the cylinder.

Q. How would you increase the power of an engine?

A. To increase the power of an engine, increase the speed; or get higher pressure of steam, use less expansion.

Q. How do you find the horsepower of an engine?

A. Multiply the speed of piston in feet per minute by the total effective pressure upon the piston in pounds and divide the product by 33,000.

Q. Which has the most friction, a perfectly fitted, or an imperfectly fitted valve or bearing?

A. An imperfect one.

Q. How hot can you get water under atmospheric pressure with exhaust steam?

A. 12 degrees.

Q. Does pressure have any influence on the boiling point?

A. Yes.

Q. Which do you think is the best economy, to run with your throttle wide open or partly shut?

A. Always have the throttle wide open on a governor engine.

Q. At what temperature has iron the greatest tensile strength?

A. About 600 degrees.

Q. In what position on the shaft does the eccentric stand in relation to the crank?

A. The throw of the eccentric should always be in advance of the crank pin.

Q. About how many pounds of water are required to yield one horsepower with our best engines?

A. From 25 to 30.

Q. What is meant by atmospheric pressure?

A. The weight of the atmosphere.

Q.   What is the weight of atmosphere at sea level?

A.   14.7 pounds.

Q.   What is the coal consumption per hour per indicated horsepower?

A.   Varies from one and a half to seven pounds.

Q.   What is the consumption of coal per hour on a square foot of grate surface?

A.   From 10 to 12 pounds.

Q.   What is the water consumption in pounds per hour per indicated horsepower?

A.   From 25 to 60 pounds.

Q.   How many pounds of water can be evaporated with one pound of best soft coal?

A.   From 7 to 10 pounds.

Q.   How much steam will one cubic inch of water evaporate under atmospheric pressure?

A.   One cubic foot of steam (approximately).

Q.   What is the weight of a cubic foot of fresh water?

A.   Sixty-two and a half pounds.

Q.   What is the weight of a cubic foot of iron?

A.   486.6 pounds.

Q.   What is the weight of a square foot of one-half inch boiler plate?

A.   20 pounds.

Q.   How much wood equals one ton of soft coal for steam purposes?

A.   About 4,000 pounds of wood.

Q.   What is the source of all power in the steam engine?

A.   The heat stored up in the coal.

Q.   How is the heat liberated from the coal?

A.   By burning it; that is, by combustion.

Q.   Of what does coal consist?

A.   Carbon, hydrogen, nitrogen, sulphur, oxygen and ash.

Q. What are the relative proportions of these that enter into coal?

A. There are different proportions in different specimens of coal, but the following shows the average per cent: Carbon, 80; hydrogen. 5; nitrogen, 1; sulphur, 2; oxygen 7; ash.

Q. What must be mixed with coal before it will burn?

A. Atmospheric air.

Q. What is air composed of?

A. It is composed of nitrogen and oxygen in the proportion of 77 of nitrogen to 23 of oxygen.

Q. What parts of the air mix with what parts of the coal?

A. The oxygen of the air mixes with the carbon and hydrogen of the coal.

Q. How much air must mix with the coal?

A. 150 cubic feet of air for every pound of coal.

Q. How many pounds of air are required to burn one pound of carbon?

A. Twelve.

Q. How many pounds of air are required to burn one pound of hydrogen?

A. Thirty-six.

Q. Is hydrogen hotter than carbon?

A. Yes, four and one-half times hotter.

Q. What part of the coal gives out the most heat?

A. The hydrogen does part for part, but as there is so much more of carbon than hydrogen in the coal we get the greatest amount of heat from carbon.

Q. In how many different ways is heat transmitted?

A. Three; by radiation, by conduction and by convection.

Q. If the fire consisted of glowing fuel, show how the heat enters the water and forms steam?

A.   The heat from the glowing fuel passes by radiation through the air space above the fuel to the furnace crown. There it passes through the iron of the crown by conduction. There it warms the water resting on the crown, which then rises and parts with its heat to the colder water by conduction till the whole mass of water is heated. Then the heated water rises to the surface and parts with its steam, so a constant circulation of water is maintained by convection.

Q.   What does water consist of?

A.   Oxygen and hydrogen.

Q.   In what proportion?

A.   Eight of oxygen to one of hydrogen by weight.

Q.   What are the different kinds of heat?

A.   Latent heat, sensible heat and sometimes total heat.

Q.   What is meant by latent heat?

A.   Heat that does not affect the thermometer and which expands itself in changing the nature of a body, such as turning ice into water or water into steam.

Q.   Under what circumstances do bodies get latent heat?

A.   When they are passing from a solid state to a liquid or from a liquid to a gaseous state.

Q.   How can latent heat be recovered?

A.   By bringing the body back from a state of gas to a liquid or from that of a liquid to that of a solid.

Q.   What is meant by a thermal unit?

A.   The heat necessary to raise one pound of water by 1 degree Fahrenheit, which is 39 degrees Fahrenheit.

Q.   If the power is in coal, why should we use steam?

A. Because steam has some properties which make it an invaluable agent for applying the energy of the heat to the engine.

Q. What is steam?

A. It is an invisible elastic gas generated from water by the application of heat.

Q. What are its properties which make it so valuable to us?

A. 1.—The ease with which we can condense it. 2.—Its great expansive power. 3.—The small space it occupies when condensed.

Q. Why do you condense the steam?

A. To form a vacuum, and so destroy the back pressure that would otherwise be on the piston and thus get more useful work out of the steam.

Q. What is vacuum?

A. A space void of all pressure.

Q. How do you maintain a vacuum?

A. By the steam used being constantly condensed by the cold water or cold tubes, and the air pump as constantly clearing the condenser out.

Q. Why does condensing the used steam form a vacuum?

A. Because a cubic foot of steam, at atmospheric pressure, shrinks into about a cubic inch of water.

Q. What do you understand by the term horse power?

A. A horse power is equivalent to raising 33,000 pounds one foot per minute, or 550 pounds raised one foot per second.

Q. How do you calculate the horse power of tubular or flue boilers?

A. For tubular boilers, multiply the square of the diameter by length, and divide by four. For flue boilers, multiply the diameter by the length and divide by four; or, multiply area of grate surface in square feet by 11/2.

Q. What do you understand by lead on an engine's valve?

A. Lead on a valve is the admission of steam into the cylinder before the piston completes its stroke.

Q. What is the clearance of an engine as the term is applied at the present time?

A. Clearance is the space between the cylinder head and the piston head with the ports included.

Q. What are considered the greatest improvements on the stationary engine in the last forty years?

A. The governor, the Corliss valve gear and the triple compound expansion.

Q. What is meant by triple expansion engine?

A. A triple expansion engine has three cylinders using the steam expansively in each one.

Q. What is a condenser as applied to an engine?

A. The condenser is a part of the low pressure engine and is a receptacle into which the exhaust enters and is there condensed.

Q. What are the principles which distinguish a high pressure from a low pressure engine?

A. Where no condenser is used and the exhaust steam is open to the atmosphere.

Q. About how much gain is there by using the condenser?

A. 17 to 25 per cent where cost of water is not figured.

Q. What do you understand by the use of steam expansively?

A. Where steam admitted at a certain pressure is cut off and allowed to expand to a lower pressure.

Q. How many inches of vacuum *give* the best results in a condensing engine?

A. Usually considered 25.

Q. What is meant by a horizontal tandem engine?

A. One cylinder being behind the other with two pistons on same rod.

Q. What is a Corliss valve gear?

A. (*Describe the half moon or crab claw gear, or oval arm gear with dash pots.*)

Q. From what cause do belts have the power to drive shafting?

A. By friction or cohesion.

Q. What do you understand by lap?

A. Outside lap is that portion of valve which extends beyond the ports when valve is placed on the center of travel, and inside lap is that portion of valve which projects over the ports on the inside or towards the middle of valve.

Q. What is the use of lap?

A. To give the engine compression.

Q. Where is the dead center of an engine?

A. The point where the crank and the piston rod are in the same right line.

Q. What is the tensile strength of American boiler iron?

A. 40,000 to 60,000 pounds per square inch.

Q. What is very high tensile strength in boiler iron apt to go with?

A. Lack of homogeneousness and lack of toughness.

Q. What is the advantage of toughness in boiler plate?

A. It stands irregular strains and sudden shocks better.

Q.   What are the principal defects found in boiler iron?

A.   Imperfect welding, brittleness, low ductility.

Q.   What are the advantages of steel as a material for boiler plates?

A.   Homogeneity, tensile strength, malleability, ductility and freedom from laminations and blisters.

Q.   What are the disadvantages of steel as a material for boiler plates?

A.   It requires greater skill in working than iron, and has, as bad qualities, brittleness, low ductility and flaws induced by the pressure of gas bubbles in the ingot.

Q.   When would you oil an engine?

A.   Before starting it and as often while running as necessary.

Q.   How do you find proper size of any stay bolts for a well made boiler?

A.   First, multiply the given steam pressure per square inch by the square of the distance between centers of stay bolts, and divide the product by 6,000, and call the answer "the quotient." Second, divide "the quotient" by 7854, and extract the square root of the last quotient; the answer will give the required diameter of stay bolts at the bottom of thread.

Q.   In what position would you place an engine, to take up any slack motion of the reciprocating parts?

A.   Place engine in the position where the least wear takes place on the journals. That is, in taking up the wear of the crank-pin brasses, place the engine on either dead center, as, when running, there is but little wear upon the crank-pin at these points. If taking up the cross-head pin brasses—without disconnecting and swinging the rod—place the engine at half stroke, which is the extreme point of swing of the rod, there being the least wear on the brasses and cross-head pin in this position.

Q. What benefits are derived by using flywheels on steam engines?

A. The energy developed in the cylinder while the steam is doing its work is stored up in the flywheel, and given out by it while there is no work being done in the cylinder—that is, when the engine is passing the dead centers. This tends to keep the speed of the engine shaft steady.

Q. Name several kinds of reducing motions, as used in indicator practice?

A. The pantograph, the pendulum, the brumbo pulley, the reducing wheel.

Q. How can an engineer tell from an indicator diagram whether the piston or valves are leaking?

A. Leaky steam valves will cause the expansion curve to become convex; that is, it will not follow hyperbolic expansion, and will also show increased back pressure. But if the exhaust valves leak also, one may offset the other, and the indicator diagram would show no leak.

A leaky piston can be detected by a rapid falling in the pressure on the expansion curve immediately after the point of cut-off. It will also show increased back pressure.

A falling in pressure in the upper portion of the compression curve shows a leak in the exhaust valve.

Q. What would be the best method of treating a badly scaled boiler, that was to be cleaned by a liberal use of compound?

A. First open the boiler up and note where the loose scale, if any, has lodged. Wash out thoroughly and put in the required amount of compound. While the boiler is in service, open the blow-off valve for a few seconds, two or three times a day, to be assured that it does not become stopped up with scale.

After running the boiler for a week, shut it down, and, when the pressure is down and the boiler cooled off, run the water out and take off the hand-hole plates. Note what effect the compound has had on the scale, and where the disengaged scale has lodged. Wash out thoroughly and use judgment as to whether it is advisable to use a less or greater quantity of compound, or to add a small quantity daily.

Continue the washing out at short intervals, as many boilers have been burned by large quantities of scale dropping on the crown sheets and not being removed.

Q.   If a condenser was attached to a side-valve engine, that had been set to run non-condensing, what changes, if any, would be necessary?

A.   More lap would have to be added to the valve to cut off the steam at an earlier point of the stroke; if not, the initial pressure into the cylinder would be throttled down and the economy, to be gained from running condensing, lessened.

Q.   If you are carrying a vacuum equal to 271/2 inches of mercury, what should the temperature of the water in the hot well be?

A.   108 degrees Fahrenheit.

Q.   Define specific gravity.

A.   The specific gravity of a substance is the number which expresses the relation between the weights of equal volume of that substance, and distilled water of 60 degrees Fahrenheit.

Q.   Find the specific gravity of a body whose volume is 12 cubic inches, and which floats in water with 7 cubic inches immersed.

A.   When a body floats in water, it displaces a quantity of water equal to the weight of the floating body. Thus, if a body of 12 cubic inches in volume floats with 7 cubic inches immersed, 7 cubic inches of water must be equal in weight to 12 cubic inches of the substance and one cubic inch of water to twelve-sevenths cubic inches of the substance.

As specific gravity equals weight of one volume of substance divided by weight of equal volume of water, then specific gravity of the substance in this case equals I divided by twelve-sevenths.

USEFUL INFORMATION

To find circumference of a circle, multiply diameter by 3.1416.

To find diameter of a circle, multiply circumference by 31831.

To find area of a circle multiply square of diameter by 7854.

To find area of a triangle, multiply base by one-half the perpendicular height.

To find surface of a ball, multiply square of diameter by 3.1416.

To find solidity of a sphere, multiply cube of diameter by 5236.

To find side of an equal square, multiply diameter by 8862.

To find cubic inches in a ball multiply cube of diameter by 5236.

Doubling the diameter of a pipe increases its capacity four times.

A gallon of water (U. S. standard) weighs 81-3 pounds and contains 231 cubic inches.

A cubic foot of water contains 71/2 gallons, 1728 cubic inches, and weighs 621/2 pounds.

To find the pressure in pounds per square inch of a column of water multiply the height of the column in feet by 434.

Steam rising from water at its boiling point (212 degrees) has a pressure equal to the atmosphere (14.7 pounds to the square inch).

A standard horse power: The evaporation of 30 lbs. of water per hour from a feed water temperature of 100 degrees F. into steam at 70 lbs. gauge pressure.

To find capacity of tanks any size; given dimensions of a cylinder in inches, to find its capacity in U. S. gallons: Square the diameter, multiply by the length and by 0034.

To ascertain heating surface in tubular boilers, multiply two-thirds of the circumference of boiler by length of boiler in inches and add to it the area of all the tubes.

One-sixth of tensile strength of plate multiplied by thickness of plate and divided by one-half the diameter of boiler gives safe working pressure for tubular boilers. For marine boilers add 20 per cent for drilled holes.

To find the horsepower of an engine, the following four factors must be considered: Mean effective or average pressure on the cylinder, length of stroke, diameter of cylinder, and number of revolutions per minute. Find the area of the piston in square inches by multiplying the diameter by 3.1416 and multiply the result by the steam pressure in pounds per square inch; multiply this product by twice the product of the length of the stroke in feet and the number of revolutions per minute; divide the result by 33,000, and the result will be the horse-power of the engine.

(Theoretically a horsepower is a power that will raise 33,000 pounds one foot in one minute.)

The power of fuel is measured theoretically from the following basis: If a pound weight fall 780 feet in a vacuum, it wall generate heat enough to raise the temperature of one pound of water one degree. Conversely, power that will raise one pound of water one degree in temperature will raise a one pound weight 780 feet. The heat force required to turn a pound of water at 32 degrees into steam would lift a ton weight 400 feet high, or develop two-fifths of one horsepower for an hour. The best farm engine practically uses 35 pounds of water per horsepower per hour, showing that one pound of water would develop only one-thirty-fifth of a horsepower in an hour, or 71-7 per cent of the heat force liberated. The rest of the heat force is lost in various ways, as explained in the body of this book.

The following* will assist in determining the amount of power supplied to an engine:

"For instance, a 1 inch belt of the standard grade with the proper tension, neither too tight or too loose, running at a maximum speed of 800 feet a minute will transmit one horsepower, running 1,600 feet two horsepower and 2,400 feet three horsepower. A 2-inch belt at the same speed, twice the power.

"Now if you know the circumference of your flywheel, the number of revolutions your engine is making and the width of belt, you can figure very nearly the amount of

---

*J. H. Maggard in "Rough and Tumble Engineering.

power you can supply without slipping your belt. For instance, we will say your flywheel is 40 inches in diameter or 10.5 feet nearly in circumference and your engine was running 225 revolutions a minute, your belt would be traveling 225 x 10.5 feet = 2362.5 feet, or very nearly 2,400 feet, and if one inch of belt would transmit three horsepower running this speed, a 6-inch belt would transmit eighteen horsepower, a 7-inch belt twenty-one horsepower, an 8-inch belt twenty-four horsepower, and so on. With the above as a basis for figuring you can satisfy yourself as to the power you are furnishing. To get the best results a belt wants to sag slightly, as it hugs the pulley closer, and will last much longer."

KEYING PULLEYS*

A key must be of equal width its whole length and accurately fit the seats on shaft and in pulley. The thickness should vary enough to make the taper correspond with that of the seat in the pulley. The keys should be driven in tight enough to be safe against working loose. The hubs of most of the pulleys on the machine run against the boxes, and in keying these on, about 1-32 of an inch end play to the shaft should be allowed, because there is danger of the pulley rubbing so hard against the end of the box as to cause it to heat.

A key that is too thin but otherwise fits all right can be made tight by putting a strip of tin between the key and the bottom of the seat in the pulley.

Drawing Keys. If a part of the key stands outside of the hub, catch it with a pair of horseshoe pinchers and

---

*Courtesy J. I. Case Threshing Machine Co., from "Science of Successful Threshing."

pry with them against the hub, at the same time hitting the hub with a hammer so as to drive pulley on. A key can sometimes be drawn by catching the end of it with a claw hammer and driving on the hub of pulley. If pulley is against box and key cut off flush with hub, take the shaft out and use a drift from the inside, or if seat is not long enough to make this possible, drive the pulley on until the key loosens.

BABBITTING BOXES*

To babbitt any kind of a box, first chip out all of the old babbitt and clean the shaft and box thoroughly with benzine. This is necessary or gas will be formed from the grease when the hot metal is poured in and leave "blow holes." In babbitting a *solid box* cover the shaft with paper, draw it smooth and tight, and fasten the lapped ends with mucilage. If this is not done the shrinkage of the metal in cooling will make it fast on the shaft, so that it can't be moved. If this happened it would be necessary to put the shaft and box together in the fire and melt the babbitt out or else break the box to get it off. Paper around the shaft will prevent this and if taken out when the babbitt has cooled the shaft will be found to be just tight enough to run well.

Before pouring the box, block up the shaft until it is in line and in center of the box and put stiff putty around the shaft and against the ends of the box to keep the babbitt from running out. Be sure to leave airholes at each end at the top, making a little funnel of putty around each. Also make a larger funnel around the pouring hole, or, if there is none, enlarge one of the air-holes at the end and pour in that. The metal should be heated until it is just hot enough to run freely and the fire should not be too far away. When ready to pour the box, don't hesitate or stop, but pour continuously and rapidly until the metal appears at the air holes. The oil hole may be stopped with a wooden plug and if this plug extends through far enough to touch the shaft, it will leave a hole through the babbitt so that it will not be necessary to drill one.

A split box is babbitted in the same manner except that strips of cardboard or sheet-iron are placed between the two halves of the box and against the shaft to divide the babbitt. To let the babbitt run from the upper half to the lower, cut four or six V-shaped notches, a quarter of an inch deep, in the edges of the sheet-iron or cardboard that come against the shaft. Cover the shaft with paper and put cardboard liners between the box to allow for adjustment as it wears. Bolt the cap on securely before pouring. When the babbitt has cooled, break the box apart by driving a cold chisel between the two halves. Trim off the sharp edges of the babbitt and with a round-nose chisel cut oil grooves from the oil hole towards the ends of the box and on the slack side of the box or the one opposite to the direction in which the belt pulls.

The ladle should hold six or eight pounds of metal. If much larger it is awkward to handle and if too small it will not keep the metal hot long enough to pour a good box. The cylinder boxes on the separator take from two to three pounds of metal each. If no putty is at hand, clay mixed to the proper consistency may be used. Use the best babbitt you can get for the cylinder boxes. If not sure of the quality, use ordinary zinc. It is not expensive and is generally satisfactory.

MISCELLANEOUS

Lime may be taken out of an injector by soaking it over night in a mixture of one part of muriatic acid and ten parts soft water. If a larger proportion of acid is used it is likely to spoil the injector.

A good blacking for boilers and smokestacks is asphaltum dissolved in turpentine.

To polish brass, dissolve 5 cents' worth of oxalic acid in a pint of water and use to clean the brass. When tarnish has been removed, dry and polish with chalk or whiting.

It is said that iron or steel will not rust if it is placed for a few minutes in a warm solution of washing soda.

Grease on the bottom of a boiler will stick there and prevent the water from conducting away the heat. When steel is thus covered with grease it will soon melt in a hot fire, causing a boiler to burst if the steel is poor, or warping it out of shape if the steel is good.

Sulphate of lime in water, causing scale, may be counteracted and scale removed by using coal oil and sal soda. When water contains carbonate of lime, molasses will remove the scale.

CODE OF WHISTLE SIGNALS

One short sound means to stop.

Two short sounds means the engine is about to begin work.

Three medium short sounds mean that the machine will soon need grain and grain haulers should hurry.

One rather long sound followed by three short ones means the water is low and water hauler should hurry.

A succession of short, quick whistles means distress or fire.

WEIGHT PER BUSHEL OF GRAIN

The following table gives the number of pounds per bushel required by law or custom in the sale of grain in the several states:

	Barley.	Beans.	Buckwheat.	Clover.	Flax.	Millet.	Oats.	Rye.	Shelled Corn	Timothy.	Wheat.
Arkansas	48	60	52	60	..	..	..	56	56	45	60
California	50	..	40	..	..	..	32	54	52		60
Connecticut	..	..	45	..	..	..	32	56	56		56
District of Columbia	47	62	48	60	..	..	32	56	56	45	60
Georgia	40	..	..	60	..	..	..	35	56	45	60
Illinois	48	60	52	60	56	45	32	56	56		60
Indiana	48	60	50	60	..	..	32	56	56	45	60
Iowa	48	60	52	60	56	48	32	56	56	45	60
Kansas	50	60	50	..	..	..	32	..	56	..	60
Kentucky	48	60	52	60	56	..	32	56	56	45	60
Louisiana	32	..	..	..	..	..	32	..	56	..	60
Maine	48	64	48	..	..	..	30	..	56		60
Manitoba	48	..	48	60	56	34	34	56	56	..	60
Maryland	48	64	48	..	..	..	32	56	56	45	60
Massachusetts	48	48	..	..	..	..	32	56	56	..	60
Michigan	48	..	48	60	56	..	32	56	56	..	60
Minnesota	48	60	42	60	..	48	32	56	56	..	60
Missouri	48	60	52	60	56	50	32	56	56	45	60
Nebraska	48	60	52	60	..	..	34	56	56	45	60

New York	48	62	48	60	..	..	32	56	58	44	60
New Jersey	48	..	50	64	..	..	30	56	56	..	60
New Hampshire	..	60	..	..	..	..	30	56	56	..	60
North Carolina	48	..	50	64	..	..	30	56	54	..	60
North Dakota	48	..	42	60	56	..	32	56	56	..	60
Ohio	48	60	50	60	..	..	32	50	56	45	60
Oklahoma	48	..	42	60	56	..	32	56	56	..	60
Oregon	46	..	42	60	..	..	36	56	56	..	60
Pennsylvania	47	..	48	62	..	..	30	56	56	..	60
South Dakota	48	..	52	60	56	50	32	56	56	..	60
South Carolina	48	60	56	60	..	..	33	56	56	..	60
Vermont	48	64	48	..	60	..	32	56	56	42	60
Virginia	48	60	48	64	..	..	32	56	56	45	60
West Virginia	48	60	52	60	..	..	32	56	56	45	60
Wisconsin	48	..	48	60	..	..	32	56	56	..	60

CHAPTER XVI

DIFFERENT MAKES OF TRACTION ENGINES

J. I. CASE TRACTION ENGINES

These engines are among the simplest and at the same time most substantial and durable traction engines on the market. They are built of the best materials throughout, and are one of the easiest engines for a novice to run.

They are of the side crank type, with spring mounting. The engine is supported by a bracket bolted to the side of the boiler, and a pillow block bearing at the firebox end bolted to the side plate of the boiler.

The valve is the improved Woolf, a single simple valve being used, worked by a single eccentric. The eccentric strap has an extended arm pivoted in a wooden block sliding in a guide. The direction of this guide can be so changed by the reverse lever as to vary the cut-off and easily reverse the engine when desired.

The engine is built either with a simple cylinder or with a tandem compound cylinder.

In the operation of the differential gear, the power is first transmitted to spur gear, containing cushion springs, from thence by the springs to a center ring and four bevel pinions which bear equally upon both bevel gears. The whole differential consequently will move together as but one wheel when engine is moving straight forward or backward; but when turning a corner the four pinions revolve in the bevel gears just in proportion to the sharpness of the curve.

There is a friction clutch working on the inside of the flywheel by means of two friction shoes that can be adjusted as they wear.

There is a feed water heater with three tubes in a watertight cylinder into which the exhaust steam is admitted. The three tubes have smaller pipes inside so that the feed water in passing through forms a thin cylindrical ring.

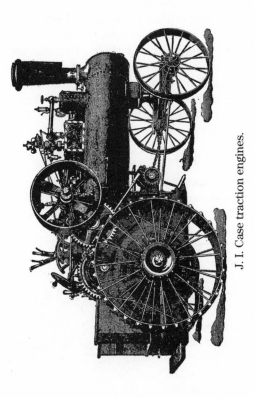

J. I. Case traction engines.

The traction wheels are driven from the rims. The front wheels have a square band on the center of the rim, to prevent slipping sidewise. The smokestack is cast iron in one piece.

The firebox will burn wood, coal or straw, a fire brick arch being used for straw, making this fuel give a uniform heat.

The boiler is of the simple locomotive type, with water leg around the firebox and numerous fire flues connecting the firebox with the smokestack in front. There is safety plug in crown sheet and the usual fittings. The water tank is under the platform. The steering wheel and band wheel are on right side of engine. An independent Marsh pump and injector are used. The Marsh pump is arranged to heat the feed water when exhaust heater cannot be used. The governor is the Waters, the safety valve the Kunkle.

THE FRICK CO.'S TRACTION ENGINE

The most noticeable feature of this engine is that it has a frame mounted on the traction wheels entirely independent of the boiler, thus relieving the boiler of all strain. This is an undeniable advantage, since usually the strain on the boiler is great enough without forcing the boiler to carry the engine and gears.

The gearing to the traction wheels is

The Frick Co.'s traction engine.

simple and direct, and a patent elastic spring or cushion connection is used which avoids sudden strain and possible breakage of gears. Steel traction wheels and riveted spokes. Differential gear in main axle, with locking device when both traction wheels are required to pull out of a hole. The reverse gear is single eccentric, the eccentric turning on the shaft. It is well adapted to using steam expansively. The crown sheet is so arranged as

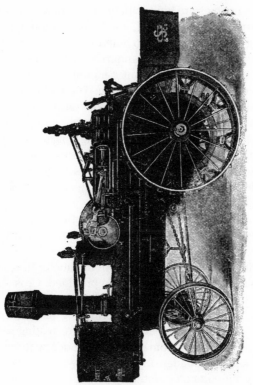

Garr, Scott & Co.'s traction engine.

not to be left bare of water in going up or down hills. Working parts are covered dust proof. Engine has self-oiling features and sight feed lubricator. Friction clutch in flywheel. Safety brake on main axle. Engineer's platform mounted on springs and every part of engine requiring attention can be reached conveniently from platform.

Crank is center type. Cross-head pump is used. Usual fittings.

These engines are built with boiler of locomotive type for burning wood and coal, and of return flue type for burning straw. They are also built of three general types, "Corliss-pattern" frame, "Standard" and "Compound."

The engine is side crank, mounted on brackets attached to the sides of the boiler. The bedplate, cylinder and guides are bored at one operation and cannot get out of alignment. Cylinder has wide ports and free exhaust, and piston has self-setting rings. The genuine link reverse gear is used, as on locomotives, and it undoubtedly has many advantages over any other, including an easily adjustable variable cut-off by correct setting of reverse lever.

The differential gear is heavy and effective. A patent steering attachment, with spiral roll, holds chains taut and gives positive motion. Friction clutch is mounted on engine shaft and connects with the hub of the pinion on this shaft. Rigid pinion is also provided. Cross-head pump and injector are used, and Pickering governor with improved spring speeder, permitting quick and easy change of speed; also Sawyer's lever for testing safety. Steam passes direct from dome to cylinder, without loss from cooling or condensing. The steel water tank can be filled by a jet pump operated by steam.

D. JUNE & CO.'S TRACTION ENGINE

This is one of the very few traction engines built with upright boiler, but it has been on the market many years and has been widely used with great success as a general road locomotive.

D. June & Co.'s traction engine.

The engine is mounted on the water tank. The weight of the boiler comes on the hind wheels, and makes this type of engine superior for pulling. It is claimed that it has no equal on the market as a puller. The upright type of boiler has the advantage that the crown sheet is never exposed and it is claimed flues will last longer than in horizontal type. It works equally well whether it stands level or not, an advantage that no other type has.

This type gets up steam more quickly than any other—it is said, from cold water, in twenty minutes. The steam is superheated in a way to economize fuel and water. By being mounted on the tank, the engine does not get hot as it would if mounted on the boiler, and the corresponding straining of parts is avoided. A patent water spark arrester is used which is an absolute protection.

The engine is geared to the traction by a chain, which can easily be repaired as the links wear. The friction clutch works inside flywheel. Engine has a new reversible eccentric, and differential gear, with usual fittings.

NICHOLS & SHEPARD TRACTION ENGINE

The builders of this engine lay special stress upon the care with which the boiler and similar parts are constructed. The important seams are double riveted, and the flue sheet is half inch steel, drilled instead of punched for the flues, and fitted with seamless steel flues, all of the best steel.

The boiler is the direct flue locomotive type. The crown sheet slopes backward to allow it to be covered with water in descending hills. Boiler has round-bottom firebox. Axle passes around below the boiler, and springs are provided.

The engine is mounted on a long heater, which is attached to the side of the boiler. The locomotive link reverse is used, with a plain slide valve.

Cross-head pump and injector are used, and improved pop safety valve. Cylinder is jacketed, and cross-head guides are rigid with cylinder, so that perfect alignment is always secured.

Engines are built to burn coal or wood. A straw burner is provided with firebrick arch. Compound engines are also built.

Nichols & Shepard traction engine.

THE HUBER TRACTION ENGINE

The Huber boiler is of the return flue type, and the gates are in the large central tube. This does away with the low-hanging firebox, and enables the engine to cross streams and straddle stumps as the low firebox type cannot do. The cylindrical shape of the boiler also adds considerably to its strength. The water tank is carried in front, and swings around so as to open the smoke box, so that repairs may be made on the fire tubes at this end easily in the open air.

The huber traction engine.

With water front return flue boilers the workman has to crawl through entire length of central flue. As there is no firebox, the boiler is mounted above the axle, not by bolting a plate to the side of the firebox. The boiler is made fast to the axle, which is mounted on wheels with spring cushion gear, the springs being placed in the wheel itself, between the two bearings of the wheel or the hub on trunnions, which form the spindle for the hub. The wheel revolves on the trunnion instead of on the axle, and there is no wear on the axle. The traction gear has a spring connection so that in starting a load there is little danger of breakage. The compensating gear is all spur. The intermediate gear has a ten-inch bearing, with an eccentric in the center for adjusting the gear above and below. There is a spring draw bar and elastic steering device. An improved friction clutch works on inside of flywheel. Engine has a special governor adapted to varying work over rough roads, etc.

A single eccentric reverse gear is used, with arm and wood slide block (Woolf); and there is a variable exhaust, by which a strong draft may be quickly created by shutting off one of two exhaust nozzles. When both exhausts are open, back pressure is almost entirely relieved.

The steam is carried in a pipe down through the middle of the central flue, so that superheating is secured, which it is claimed makes a saving of over 8 per cent in fuel and water. The stack is double walled with air space between the walls.

A special straw-burning engine is constructed with a firebox extension in front, and straw passes over the end of a grate in such a way as to get perfect combustion. This make of engine is peculiarly adapted to burning straw successfully.

A. W. STEVENS' TRACTION ENGINE

This engine has locomotive pattern boiler, with sloping crown sheet, and especially high offset over firebox, doubling steam space that will give dry steam at all times. A large size steam pipe passes from dome in rear through boiler to engine in front, superheating steam and avoiding condensation from exposure. Grate is a rocking one, easily cleaned and requiring little attention, and firedoor is of a pattern that remains air-tight and need seldom be opened.

The engine is mounted upon the boiler, arranged for rear gear traction attachment. Engine frame, cylinder, guides, etc., are cast in one solid piece.

It has a special patented single eccentric reverse, and Pickering horizontal governor. There is a friction clutch, Marsh steam pump, and injector. Other fittings are complete, and engine is well made throughout.

A. W. Stevens' traction engine.

AULTMAM-TAYLOR TRACTION ENGINE

The Aultman-Taylor Traction Engine is an exceptionally well made engine of the simplest type, and has been on the market over 25 years. There are two general types, the wood and coal burners with locomotive boilers, and return flue boiler style for burning straw. A compound engine is also made with the Woolf single valve gear.

A special feature of this engine is that the rear axle comes behind the firebox instead of between the firebox and the front wheels. This distributes the weight of the engine more evenly. The makers do not believe in springs for the rear axle, since they have a tendency to wear the gear convex or round, and really accomplish much less than they are supposed to.

Aultman-Taylor traction engine.

Another special point is the bevel traction gear. The engine is mounted on the boiler well toward the front, and the flywheel is near the stack (in the locomotive type). By bevel gears and a long shaft the power is conducted to the differential gear in connection with the rear wheels. The makers claim that lost motion can be taken up in a bevel gear much better than in a spur gear. Besides, the spur gear is noisy and not nearly so durable. Much less friction is claimed for this type of gear.

The governor is the Pickering; cross-head pump is used, with U. S. injector, heater, and other fittings complete. A band friction clutch is used, said to be very durable. Diamond special spark arrester is used except in straw burners. The platform and front bolster are provided with springs. The makers especially recommend their compound engine, claiming a gain of about 25 per cent. The use of automatic band cutters and feeders, automatic weighers and baggers, and pneumatic stackers with threshing machine outfits make additional demands on an engine that is best met by the compound type. With large outfits, making large demands, the compound engine gives the required power without undue weight.

AVERY TRACTION ENGINE

The Avery is an engine with a return flue boiler and full water front, and also is arranged with a firebox

Avery traction engine.

besides. There is no doubt that it effects the greatest economy of fuel possible, and is adaptable equally for wood, coal, or straw. The boiler is so built that a man may readily crawl through the large central flue and get at the front ends of the return tubes to repair them.

The side gear is used with a crank disc instead of arm. The reverse is the Grime, a single eccentric with device for shifting for reverse. The friction clutch has unusually long shoes, working inside the flywheel, with ample clearance when lever is off. A specialty is made of extra wide traction wheels for soft country. The traction gear is of the spur variety. There is also a double speed device offered as an extra.

The water tank is carried in front, and lubricator, steering wheel (on same side as band wheel for convenience in lining up with separator), reverse lever, friction clutch, etc., are all right at the hand of the engineer.

The traction gear is of the spur variety, adjusted to be evenly distributed to both traction wheels through the compensating gear, and to get the best possible pull in case of need.

For pulling qualities and economy of fuel, this engine is especially recommended.

BUFFALO PITTS TRACTION ENGINE

The Buffalo Pitts Engine is built either single cylinder or double cylinder. The boiler is of the direct flue locomotive type, with full water bottom firebox. The straw burners are provided with a firebrick arch in the firebox. Boilers are fully jacketed.

The single and double cylinder engines differ only in this one particular, the double cylinder having the advantage of never being on a dead center and starting with perfect smoothness and gently, seldom throwing off belt. The frame has bored guides, in same piece with cylinder, effecting perfect alignment.

The compensating gear is of the bevel type, half shrouded and so close together that sand and grit are kept out. Three pinions are used, which it is claimed prevent rocking caused by two or four pinions.

Cross-head has shoes unusually long and wide. The engine frame is of the box pattern, and is also used as

Buffalo Pitts traction engine.

a heater, feed water for either injector or steam pump passing through it. Valve is of the plain locomotive slide type.

The friction clutch has hinged arms working into flywheel with but slight beveling on flywheel inner surface, and being susceptible of easy release. It is a specially patented device. The Woolf single eccentric reverse gear is used. Engine is fully provided with all modern fittings and appliances in addition to those mentioned. It was the only traction engine exhibited at Pan-American Exposition which won gold medal or highest award. It claims extra high grade of workmanship and durability.

THE REEVES TRACTION ENGINES

These engines are made in two styles, simple double cylinder and cross compound. The double cylinder and cross compound style have been very successfully adapted to traction engine purposes with certain

The Reeves traction engine.

advantages that no other style of traction engine has. With two cylinders and two pistons placed side by side, with crank pins at right angles on the shaft, there can be no dead centers, at which an engine will be completely stuck. Then sudden starting is liable to throw off the main belt. With a double cylinder engine the starting is always gradual and easy, and never fails.

The same is equally true of the cross compound, which has the advantage of using the steam expansively in the low pressure cylinder. In case of need the live steam may be introduced into the low pressure cylinder, enormously increasing the pulling power of the engine for an emergency, though the capacity of the boiler does not permit long use of both cylinders in this way.

The engine is placed on top of the firebox portion of the boiler, and the weight is nicely balanced so that it comes on both sides alike.

The gearing is attached to the axle and countershaft which extend across the engine. The compensating gear is strong and well covered from dirt. The gearing is the gear type, axle turning with the drivers. There is an independent pump; also injector, and all attachments. The band wheel being on the steering wheel or right side of the engine, makes it easy to line up to a threshing machine. Engine frame is of the Corliss pattern; boiler of locomotive type, and extra strongly built.

THE RUMELY TRACTION ENGINE

The most striking peculiarity is that the engine is mounted on the boiler differently from most side crank traction engines, the cylinder being forward and the shaft at the rear. This brings the gearing nearer the traction wheels and reduces its weight and complication.

The Rumely traction engine.

The boiler is of the round bottom firebox type, with dome in front and an ash pan in lower part of firebox, and is unusually well built and firmly riveted.

The traction wheels are usually high, and the flywheel is between one wheel and the boiler.

The engine frame is of the girder pattern, with overhanging cylinder attached to one end.

The boiler is of the direct flue locomotive type, fitted for straw, wood, or coal. Beam axle of the engine is behind the firebox, and is a single solid steel shaft. Front axle is elliptical, and so stronger than any other type.

A double cylinder engine is now being built as well as the single cylinder. The governor regulates the double cylinder engine more closely than single cylinder types, and in the Rumely is very close to the cut-off where a special simple reverse is used with the double cylinder engine.

Engine is supplied with cross-head pump and injector, Arnold shifting eccentric reverse gear, friction clutch, and large cylindrical water tank on the side. It also has the usual engine and boiler fittings.

PORT HURON TRACTION ENGINE

The Port Huron traction engine is of the direct flue locomotive type, built either simple or compound, and of medium weight and excellent proportions for general purpose use. The compound engine (tandem Woolf cylinders) is especially recommended and pushed as more economical than the simple cylinder engine. As live steam can be admitted to the low pressure cylinder, so turning the compound into a simple cylinder engine with two cylinders, enormous power can be obtained at a moment's notice to help out at a difficult point.

Two injectors are furnished with this engine, and the use of the injector is recommended, contrary to the general belief that a pump is more economical. The company contends that the long exhaust pipe causes more back pressure on the cylinder than would be represented by the saving of heat in the heater. However, a cross-head pump and special condensing heater will be furnished if desired.

On the simple engine a piston valve is used, the seat of the valve completely surrounding it and the ports being circular openings, the result, it is claimed, being a balanced valve.

The valve reverse gear is of the Woolf pattern, the engine frame of the girder type, Waters governor, with special patent speed changer, specially balanced crank disc, patent straw burner arrangement for straw burning engines, special patent spark extinguisher, special patent gear lock, and special patents on front axle, drive wheel and loco cab.

The usual fittings are supplied.

Port Huron traction engine.

MINNEAPOLIS TRACTION ENGINE

The Minneapolis traction engine is built both simple and compound. All sizes and styles have the return

Minneapolis traction engine.

flue boiler, for wood, coal or straw. Both axles extend
entirely and straight under the boiler, giving complete
support without strain. The cylinder, steam chest and
guides form one piece, and are mounted above a heater,
secured firmly to the boiler; valve single simple D pattern. Special throttle of the butterfly pattern, large crank

pin turned by special device after it is driven in, so insuring perfect adjustment; special patent exhaust nozzle made adjustable and so as always to throw steam in center of stack; friction clutch with three adjustable shoes. Boiler is supplied with a superheater pipe. Woolf valve and reverse gear. Special heavy brass boxes and stuffing-boxes. Sight feed lubricator and needle feed oiler; Gardner spring governor. Complete with usual fittings. This is a simply constructed but very well made engine.

INDEX

FARM CONVENIENCES
AND HOW
TO MAKE THEM

by Byron D. Halsted

PREFACE

SKILL in the construction and use of simple laborsaving devices is of vast importance to the farmer, and any aid to the development of this manual dexterity is always very welcome.

The volume, herewith presented, abounds in valuable hints and suggestions for the easy and rapid construction of a large number of home-made contrivances within the reach of all. It is an every-day hand-book of farm work, and contains the best ideas gathered from the experience of a score of practical men in all departments of farm labor. Every one of the two hundred and forty pages, and two hundred and twelve engravings, teaches a valuable lesson in rural economy. "FARM CONVENIENCES" is a manual of what to do, and how to do it quickly and readily.

CONTENTS

713

A CONVENIENT BIN FOR OATS

THE usual receptacle for oats, corn, or mill feed, or other grain for domestic animals, is a common bin or box about four feet in hight. It is difficult to get the grain out of such a place when the quantity is half or more exhausted. To obviate this inconvenience, there may be affixed, about one foot from the bottom on one side of the bin, a board, (*B*) figure 1. This is nailed so as to project into the bin at an angle sufficient to allow the filling of a measure between the lower edge of board *B* and top edge of the opening at *M*. The opposite lower side of the bin is covered with boards, as indicated by the dotted line at *R*, for the purpose of placing the contents within easy reach. The top can be completed with hinged cover as well as the delivery space. By using a bin of this form, the last bushel is as easily removed as the first one.

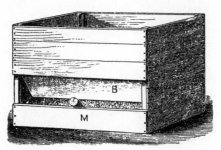

FIG. 1. A bin for oats or other feed.

FASTENINGS FOR COWS

Although stanchions are really the safest fastening for cows, yet some persons object to them because the animals are held in a too confined position, and one which is supposed to be painful, or otherwise objectionable to the cows. Most owners of valuable cows consider safety to be the first requisite in their management, and the question as to what the cow would like as of minor importance. Stanchions have the valuable recommendation that one always finds his cows in the morning just where they were left at night, if they have been properly secured. Nevertheless, for those who dislike stanchions, there are other safe ways of fastening cows. For some

FIG. 2. Fastening by sliding king on a polf.

years we used the method shown in figures 2 and 3. In the first a strong smooth pole was inserted through the floor and "stepped" into the beam beneath and into the floor above. It was also fastened by an iron strap bolted through the front of the trough. A steel ring to which a steel chain was attached was made to slide up and down upon the post, and a leather neck strap, or, in some cases, a leather head stall, was attached to the chain by a suitable ring or loop. The ring could not fall so low as the floor, being held by the edge of the feed-trough, and the cow's feet could not, therefore, be entangled in the chain by getting over it. This is the chief danger in the use of neck straps and chains, but it may be avoided in this way. Another plan is to have an iron rod bolted to the feed-trough, upon which the ring may slide. This is equally secure, and gives more room for movement to the cow. With these ring-ties it is best to have short stalls

FIG. 3. Fastening attached to feed-trough.

to prevent the cows interfering with each other, else one of them may step on to another as it is lying down. The teats are sometimes injured even when stanchions are used, but the danger of this is greater with chain ties.

MOVABLE NESTS FOR HENS

Hens, as a general thing, are remarkably self-willed and obstinate. Perhaps an exception may be made as regards the Brahmas, which are very docile and easily managed. On account of this general peculiarity of fowls, many people who possess a somewhat similar disposition, find no success in keeping them. Their hens will not lay in the nests provided for them, or after sitting

Fig. 3. A movable hen's nest.

a few days upon a nest of eggs, leave them and never return. The consequences are, either no eggs at all, or nests hidden where they cannot be reached; no chickens, and time and labor lost. This may all be avoided if the owners will only study the habits and instincts of their poultry reasonably. One of the most inveterate habits of hens is that of hiding their nests, or seeking them in retired, shaded places. Those who would have plenty of eggs must make their arrangements accordingly. A very cheap and convenient nest is shown in figure 4. It is made of pieces of board eighteen inches long, nailed endwise to three-sided cleats at the top and bottom. The box need not be more than eighteen or twenty inches in length. Some corner pieces are nailed at the front to make it firm, and the back should be closed. These nests may be placed in secluded corners, behind sheds, or beneath bushes in the back yard, or behind a barrel or a bundle of straw. The nest *egg* should be of glass or porcelain, and every evening the eggs that have been laid during the day should be removed. A little cut straw mixed with clean earth or sand, will make the best material for the nest. This should be renewed occasionally, for the sake of cloanliness. When a hen has taken possession of one of these nests, it may be removed at night to the hatching-house, without disturbing her. Before the nests are used, they should be thoroughly well lime-washed around the joints, to keep away lice.

HOW TO GET RID OF STRAW

Many farmers in "the West," and some in what we call "the East," are troubled as to what they shall do with the piles of straw which lie about their fields. Upon the same farms with these nearly useless straw piles, many head of stock are kept, and many more might be kept, which could be made useful in reducing the straw to a condition in which it would serve as manure. If the already urgent necessity for manure upon the western and southern fields were realized, there would be little hesitation in taking measures to remove the difficulty. The chief obstacle is, that these involve either personal or hired labor; the first is objectionable to many, and the second cannot be had for want of the money necessary to pay for it. The least laborious method of using this straw and making it serve the double purpose of a shelter for stock and a fertilizer for the field upon which it has been grown, is as follows: Some poles are set in the ground, and rails or other poles are laid upon them so as to form a sloping roof. This is made near or around the place chosen for thrashing the grain. The straw from the thrashing-machine is heaped upon the rails, making a long stack, which forms three sides of a square, with the open side towards the south, and leaving a space beneath it in which cattle may be sheltered from storms. In this enclosure some rough troughs or racks may be placed, from which to feed corn. Here the cattle will feed and lie, or will lie at nights under shelter, while feeding during the day upon corn in the field. As the straw that is given them becomes trampled and mixed with the droppings, a further supply is thrown down from the stack.

The accumulation may be removed and spread upon the field to be plowed in when it is so required, and the stakes pulled up and carried to another place, where they may be needed for the same purpose. Such a shelter as this would be very serviceable for the purpose of making manure, even where straw is scarce, as in parts of the Southern States. There pine boughs may be made to serve as a covering, and leaves, pine straw, dry pond muck, swamp muck, "trash" from cotton fields, corn stalks, or pea vines, and any other such materials may be gathered and thrown from time to time beneath the cattle. Cotton-seed meal, straw, and coarse hay would keep stock in excellent order, and although there may be little snow or ice during the winter months in those States, yet the animals will be very much better for even this rude but comfortable shelter. In many other places such a temporary arrangement will be found useful in saving the hauling of straw, stalks, or hay from distant fields, and the carting of manure back again to them. It will be found vastly easier to keep a few young cattle in such a field, and go thither daily to attend to them during the winter when work is not pressing, than to haul many loads of hay or straw to the barn at harvest time, or many loads of manure in the busy weeks of spring.

THE MANAGEMENT OF YOUNG BULLS

Many farmers want a method of disciplining bulls so that they may be made more docile and manageable. To do this it would be advisable to work them occasionally in a one-horse tread power. They should be used to this when young, and thus being made amenable to restraint, there will be no "breaking" needed afterwards and consequently no trouble. We have used a Jersey bull in a tread-power in which he worked with more steadiness than a horse, and twice a week he served a very useful purpose in cutting the fodder for the stock. Nothing more was needed than to lead him by a rope from the nose-ring into the tread-power, and tie him short so that he could not get too far forward. He was very quiet, not at all mischievous, and was a very sure stock bull and besides this, the value of his work was at least equal to the cost of his keep. Where there is no tread-power, a substitute may be found in the arrangement shown in figure 5. Set a post in the barn-yard, bore a hole in the top, and drive a two-inch iron pin into the hole. Take the wheel of a wagon that has an iron axle, and set it upon the top of the post so that it will turn on the pin as on an axle. Fasten a strong pole (such as a binding pole for a hay wagon) by one end to the wheel, and bore two holes in the other end, large enough to take the arms of an ox-bow in them. Fix a light-elastic rod to the wheel, so that the end will be in advance of the end of the larger pole. Yoke the bull to the pole, and tie the nose-ring to the end of the elastic rod, in such a way that a slight pull is exerted upon the ring. Then lead the bull around a few

FIG. 5. Manner of exercising a bull.

times until he gets used to it; he will then travel in the ring alone until he is tired, when he will stop. Two hours of this exercise a day will keep a bull in good temper, good condition and excellent health.

A CONVENIENT ICE-HOOK

A very handy ice-hook may be made as shown m fig. are 6. The handle is firmly fastened and keyed into a socket; at the end are two sharply-pointed spikes, one of which serves to push pieces of ice, and the other to draw them to the shore, or out of the water, to be loaded and removed. It may be made of light iron, horse-shoe bar will be heavy enough, and there is no need to have the points steeled; it will be sufficient

FIG. 6. Ice-hook.

if they are chilled, after they are sharpened, in salt and ice pounded together.

HINTS FOR THE WORKSHOP

A grindstone is very seldom kept in good working order; generally it is "out of true," as it is called, or worn out of a perfectly circular shape. A new stone is frequently hung so that it does not run "true," and the longer it is used, the worse it becomes. When this is the case, it may be brought into a circular shape by turning it down with a worn-out mill-file. It is very difficult to do this perfectly by hand, but it is easily done by the use of the contrivance shown in figure 7. A post, slotted in the upper part, is bolted to the frame. A piece of hard wood, long enough to reach over the frame, is pivoted in the slot. This should be made two inches wider than the stone, and be pivoted, so that an opening can be made

FIG. 7. Trueing a grindstone.

731

in the middle of it, of the same width as the stone. This opening is made with sloping ends, so that a broad mill-file may be wedged into it in the same manner as a plane-iron is set in a plane. At the opposite end of the frame a second post is bolted to it. A long slot, or a series of holes, is made in the lower part of this post, so that it may be raised or lowered at pleasure by sliding it up or down upon the bolt. If a slot is made, a washer is used with the bolt; this will make it easy to set the post at any desired height. It should be placed so that the upper piece of wood may rest upon it, exactly in the same position in which the file will be brought into contact with the stone. A weight is laid upon the upper piece to keep it down, and hold the cutter upon the stone. When the stone is turned around slowly, the uneven parts are cut away, while those which do not project beyond the proper line of the circumference are not touched.

Fig. 8. Holder for tools.

A Grinding Frame to hold tools is shown in figure 8. It is made of light pieces of pine, or hard wood. The tool to be ground is fastened to the cross-piece. A sharp point, a nail, or a screw, is fastened to the narrow end of the frame, and, when in use, the point is stuck into the wall of the shed, which forms a rest.

A NON-PATENTED
BARREL-HEADER

Not long since we saw in operation a useful contrivance for pressing the heads of apple or *egg* barrels into place. Both apples and eggs require to be packed very firmly to enable them to be transported in barrels with safety. Apples loosely packed in a barrel will come to market in a very badly bruised condition, and if the packing around eggs is not very firmly compressed, the eggs and packing change places or get mixed up, and it is the eggs, and not the packing, which then suffers. A barrel of eggs properly packed, with layers of chaff or oats an inch thick between the layers of eggs, and three inches at each end of the barrel, will bear to be com pressed as much as three inches with safety; without this compression, eggs are almost sure to be greatly damaged. A barrel of apples may fill the barrel to about two inches above the chime, and will bear to have the head brought down to its place. When barrels containing these perishable articles are thus packed they may receive very rough usage without injury to the contents. The header referred to consists of a bar of half-inch square iron rod, with a large eye or loop at one end, and at the other end two diverging hooks which grasp the bottom of the barrel. The bar is bent to fit the curve of the barrel. When in use, the hooks are placed beneath the lower chime

FIG. 9. Barrel-header.

734

of the barrel, one end of a short lever is placed in the eye, and the lever rests upon a block, which is set upon the head of a barrel properly placed in position. A strap or cord, with a loop or stirrup at one end, is fastened to the other end of the lever. The foot is placed in the loop or stirrup, and the weight of the body thrown upon it brings the head of the barrel into its place; the hands being free, the hoops can be driven down tightly without the help of an assistant. Without the use of the cord and stirrup, two persons are required to head barrels, but with the aid of these the services of one can be dispensed with.

BUILDING RIBLESS BOATS

A method of building boats, by which ribs are dispensed with, has recently been brought into use for coast, lake, and river crafts. These boats are light, swift, strong, and cheap. They have been found to be remarkably good sea boats, and to stand rough weather without shipping water. By this method of building, fishermen and others who use boats can construct their own at their leisure, and in many cases become independent of the skill of the professional boat builder. The materials needed are clear pine boards, one inch thick, a keel of oak or elm, a stem and stern-post of the same timber, and some galvanized iron nails. For small boats the boards and keel should be the whole length of the boat intended to be built; for boats over sixteen feet in length, splices may be made without injuring the strength, if they are properly put together. The materials having been procured, a frame or a set of tressels are made, and the keel is fitted to them in the usual manner, by means of cleats on each side, and wedges. The stem and stern-post are then fitted to the keel in the usual manner, the joints being made water-tight by means of layers of freshly-tarred brown paper laid between the pieces, or by the use of a coating of thick white lead and oil Previously to being fitted together, the sides of the keel, stem, and stern-post are deeply grooved to receive the first strip of planking. The boards are then ripped into strips one inch, or an inch and a half wide, according to the desired strength of the boat. For rough work, such as fishing with nets, or dredging, an inch and a half would be a proper width for the strips. The ripping may be done

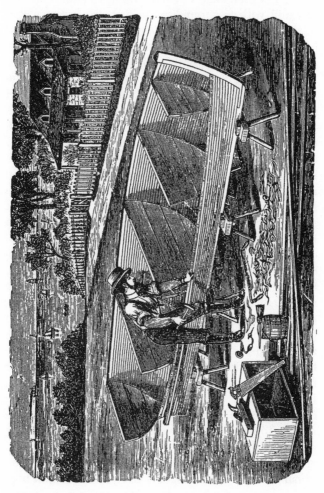

Fig. 10. Building a ribless boat.

with one of the hand circular sawing machines, or at a saw-mill, with great rapidity. The first strip is then nailed to the keel, a coating of tar or white lead having first been given to the groove in the keel already prepared for it. The broad side of the strip is laid next to the keel. A set of molds, corresponding to the lines or form of the boat, are cut out of inch boards, and tacked to the keel in the manner shown in figure 10, with the help of cleats upon each side. Then one strip after another is nailed to each preceding one, and the shell of the boat is built up of these strips. Each strip is trimmed down at the ends in a proper manner, with a drawn knife, or a plane, and as each one is nailed to the preceding one, some of the tar or white lead is brushed over it, to make the joint tight and close. A sufficient number of nails is used to hold the strips firmly together, and the heads are driven down level with the surface of each strip. The work proceeds in this manner, forming the strips as each is fitted, bending them to the shape of the molds, and nailing one alternately upon each side, so that the molds are not displaced by the spring of the timber. When the sides of the boat are completed, the fender and gunwales are fitted, and bolted to them to strengthen them, and cleats are bolted inside for the seats to rest upon. The molds are now removed, and the boat consists of a solid shell an inch and a half thick, with not a nail visible excepting on the top strip, and conforming exactly in shape to the model. To give extra strength, short pieces of the strips are nailed diagonally across the inside, from side to side, and across the keel. In this manner a great deal of additional stiffness and strength is given to the boat. A boat of this kind is easily repaired when injured, by cutting out the broken part and inserting pieces of the strips.

For a larger boat, which requires a deck, the strips are wider and thicker, or a diagonal lining may be put into it; knees are bolted to the sides, and the beams to the knees, the deck being laid upon the beams. The method is applicable to boats of all sizes and for all purposes, and its cheapness and convenience are rapidly bringing it into favor. If the material is ready for use, two men can finish a large boat in two weeks, and a small one in one week. These boats being very light and buoyant, considerable ballast will be necessary to make them steady enough in case sails are used.

TO MEND A BROKEN TUG

No one should go from home with a buggy or a wagon without a small coil of copper wire and a "*multum in parvo*" pocket-knife. This knife, as its name implies, has many parts in a little space, and, among other useful things, has a contrivance for boring holes in leather straps. In case a strap or a leather trace breaks, while one is on a journey, and at a distance from any house, one would be in an awkward "fix" if without any means of repairing damages. With the copper wire and an implement for boring some holes, repairs can be made in a very few minutes. The ends of the broken strap or tug may be laid over each other or spliced; a few holes bored in the manner shown in figure 11, and some stitches of wire passed through in the way known among the ladies as "back stitching." The ends of the wire are twisted together, and the job will be finished almost as quickly as this may be read. If it is a chain that breaks, the next links may be brought together and wire wound around them in place of the broken link, which will make the chain serviceable until home is reached. In fact, the uses of a piece of wire are almost endless. Nothing holds a button upon one's working clothes so securely as a piece of wire, and once put on in this manner, there is never any call upon the women of the house at inconvenient times for thread and needle to replace it. The wire will pierce the cloth without any help, and nothing more is needed than to pass it through each hole of the button and twist the ends to secure them, cutting them off close with a knife. There is scarcely any little thing that will be found of so great use about a farm, or a workshop, or in a mill, or even in a house, as a small stock of soft copper wire.

FIG. 11. Repairing tug.

BUSINESS HABITS

There is probably not one farmer in ten thousand who keeps a set of accounts from which he can at any moment learn the cost of anything he may have produced, or even the cost of his real property. A very few farmers who have been brought up to business habits keep such accounts, and are able to tell how their affairs progress, what each crop, each kind of stock, or each animal has cost, and what each produces. Knowing these points, a farmer can, to a very great extent, properly decide what crops he will grow, and what kind of stock he will keep. He will thus be able to apply his labor and money where it will do the most good. He can weed out his stock and retain only such animals as may be kept with profit. For the want, of such knowledge, farmers continue, year after year, to feed cows that are unprofitable, and frequently sell for less than her value one that is the best of the herd, because she is not known to be any better than the rest. Feed is also wasted upon ill-bred stock, the keep of which costs three or four times that of well-bred animals, which, as has been proved by figures that cannot be mistaken, pay a large profit on their keeping. For want of knowing what they cost, poor crops are raised year by year at an actual loss, provided the farmer's labor, at the rates current for common labor, were charged against them. To learn that he has been working for fifty cents a day, during a number of years, while he has been paying his help twice as much, would open the eyes of many a farmer who has actually been doing this, and it would convince him that there is some value in figures and book accounts. It is not generally understood that a man who raises twenty bushels of corn per acre, pays twice as much for his plowing and harrowing, twice as much for labor, and twice as great interest upon the cost of his farm, as a neighbor who raises forty bushels per acre. Nor is it understood that when he raises a pig that makes one hundred and fifty pounds of pork in a year, that his pork costs him twice as much, or the corn he feeds brings him but half as much as that of his neighbor, whose pig weighs three hundred pounds at a year old. If all these things were clearly set down in figures upon a page in an account book, and were studied, there would be not only a sudden awakening to the unprofitableness of such farming, but an immediate remedy would be sought. For no person could resist evidence of this kind if it were once brought plainly home to him. If storekeepers, merchants, or manufacturers kept no accounts, they could not possibly carry on their business, and it is only because the farmer's business is one of the most safe that he can still go on working in the dark, and throwing away opportunities of bettering his condition and increasing his profits.

HAY-RACKS

We here illustrate two kinds of hay-racks, which have been found more convenient in use than some of the old

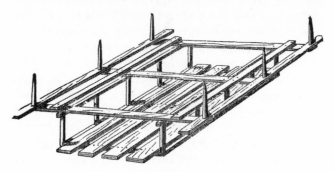

Fɪɢ. 12. Hay-rack.

Kinds. That shown in figure 12 consists of a frame made of scantlings, mortised together, and fitting upon the

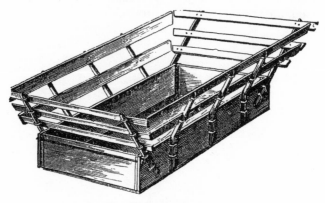

Fɪɢ. 13. Rack for grain.

744

FIG. 15. Extricating a mirred cow.

wagon after the box has been removed. Cross-pieces, which project over the wheels, are bolted to the frame, and to these one or two side-boards are bolted. A few short, sharpened stakes are fixed into the sides of the frame, which help to hold the load, and prevent it from slipping off from the rack during the loading. A strong rack of this kind may be made to carry a very large load of hay. We have seen over thirty hundred-weight loaded upon one of them, and more might have been easily added to the load. The plan of building this rack is readily seen by studying the engraving. At figure 13 is shown a rack made to fit upon a wagon body. When grain is hauled, much is sometimes lost through the rack, by shelling. This is almost always the case in hauling ripe oats, and always in drawing buckwheat. To avoid this loss, we have used a strong wagon-box of rough planks, fitted with

FIG. 14. Support for rack.

iron sockets, bolted securely to the sides. Into these sockets were fitted head and tail racks, as shown in the engraving. For the sides we procured natural crooks, shown in figure 14.

HOW TO EXTRICATE A
MIRED ANIMAL

An animal mired in a swamp gets into a worse pre-
dicament the longer it struggles. The effort to extricate
it should be made in an effective manner, so that the animal may not be encouraged to exhaust itself in repeated
exertions, which are useless, and only sink it deeper in
the mire. The usual method is to fasten a rope around
the animal's horns or neck, and while this is pulled by
some of the assistants, others place rails beneath the
body of the animal for the purpose of lifting it out of the
hole. This plan is sometimes effective, but it often is not,
and at best it is a slow, clumsy, and laborious method.
The materials needed for the method here referred to
are all that are required for a much better one, which is
illustrated in Figure 15. This is very simple, and two men
can operate it, and, at a pinch, even one man alone may
succeed with it. A strong stake or an iron bar is driven
into the solid ground at a distance of twenty-five feet
or more from the mired animal. Two short rails, about
nine feet long, are tied together near the ends, so that
they can be spread apart in the form of a pair of shears,
for hoisting. A long rope is fastened around the horns or
neck of the animal, with such a knot that the loop cannot be drawn tight enough to do any injury. The rope is
cast over the ends of the rails as they are set up upon
the edge of the solid ground, and carried to the stake
or crow-bar beyond. The end of the rope is fastened to
a stout hand-spike, leaving about a foot of the end of it
free. This end is laid against the bar or stake, and the

other end is moved around it so that the rope is wound upon it, drawing it up and with it drawing the animal out of the mire. The rope being held up by the tied rails, tends to lift the animal and make its extrication very easy.

HOW TO SAVE AND
KEEP MANURE

There is no question more frequently or seriously considered by the farmer, than how he shall get, keep, and spend an adequate supply of manure; nor is there anything about the farm which is of greater importance to its successful management than the manure heap There are few farmers now left who pretend to ignore this feed for the land; and few localities, even in the newer Western States, where manure now is thought to be a nuisance. We have gradually come to the inevitable final end of our "virgin farms," and have now either to save what is left of their wonderful natural fertility, or to restore them slowly and laboriously to a profitable condition. We have reached the end of our tether, and are obliged to confess that we have trespassed over the line which bounds the territory of the locust, or have improved the face of the country so much that, the protecting timber being removed, the water supply is becoming precarious, and springs, brooks, and rivers no longer flow as they did heretofore. To some extent the tide of emigration, which has flowed westward so many years, is now eddying or even ebbing, and the cheap, worn lands of the East are finding purchasers, who undertake to bring them back to their former condition. At the same time Eastern farmers are discovering more and more certainly that they must increase their crops, and make one acre produce as much as two have hereto fore done. The only way in which either of these classes can succeed, is by keeping sufficient stock to manure their farms liberally; to feed these animals so skillfully and well that they shall pay for their feed with a profit, and in addition leave a supply of rich manure, with which the soil can be kept in a productive state, and to save and use the manure with such care that no particle of it be lost. It is not every farmer who can procure all the manure he needs; but very many can save what they have, with far greater economy than they now do; and this, although it may seem a question secondary to that of getting manure, is really of primary importance; for by using what one has to better purpose, he opens a way to increase his supply. We have found this to be the case in our own experience, and by strict attention to saving and preserving every particle of manure in its best condition, we have succeeded in so enlarging our supply of fodder that the number of stock that could be fed was largely increased each year, and very soon it was necessary to go out and buy animals to consume the surplus. To bring a farm into improved condition, there is no cheaper or more effective method than this.

The ordinary management of manure, in open barnyards, where it is washed by rains, dried by the sun's scorching heat, and wasted by every wind that blows, is the worst that is possible. In this way half or more of the value of the manure is lost. By figuring up what it would cost to purchase a quantity of manure equal to what is thus lost, the costliness of this common method would be discovered, and the question how much could be afforded to take care of the manure would be settled. When properly littered, one cow or ox will make a ton of manure every month, if the liquid as well as the solid portion is saved. Ten head would thus make one hundred and twenty tons, or sixty two-horse wagon loads in a year. A pair of horses will make as much manure as one cow, or twelve tons in the year. A hundred sheep, if yarded every night and well littered, will make one hundred tons of manure in the year, and ten pigs will work up a wagon load in a month, if supplied with sufficient coarse material. The stock of a one hundred acre farm, which should consist of at least ten cows, ten head of steers, heifers, and calves, a pair of horses, one hundred sheep, and ten pigs, would then make, in the aggregate, three hundred and sixteen tons of manure every year, or sufficient to give twelve tons per acre every fourth year. If this were well cared for, it would be, in effect, equal to double the quantity of ordinary yard manure; and if a plenty of swamp muck could be procured, at least six hundred tons of the best manure could be made upon a one hundred acre farm. If this were the rule instead of a rare exception, or only a possibility what a change would appear upon the face of the country, and what an addition would be made to the wealth of the nation!

GRINDING TOOLS

The useful effect of many tools depends greatly upon the exact grinding of their edges to a proper bevel. A cold chisel, for instance, requires an edge of a certain bevel to cut hard metal, and one of a different angle for softer metal; the harder the work to be cut, the greater

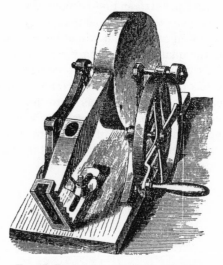

FIG. 16. Device for grinding mill-picks.

should be the angle formed by the edge, and the softer the material, the more acute the edge. The same rule is to be observed in wood-cutting tools. But there are no tools which require more exact and careful grinding than mill-picks, and the first business of a miller is to know how to grind his picks. Upon this depends the dress of the stones, and the quality of work turned out by them.

Figure 16 represents a small grindstone for sharpening picks, which is run by means of friction wheels covered with leather, and provided with a gauge for setting the pick at a variable angle to the stone. This gauge, shown in the engraving, is so serviceable as to be well worth a place in any farm workshop. It consists of a series of steps raised upon a slotted plank, which *is* screwed upon the frame of the grindstone. By means of the slot and a set screw, seen below the pick, the gauge can be set for tools of different lengths, and each step causes the tool set in it to be ground at a different angle.

A METHOD OF
HANGING HOGS

An easy method of hanging a hog or a beef, is by the use of the tripod shown in figure 17. It is made of

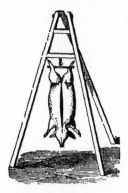

FIG. 17. Tripod set up.

three by three oak scantling, six feet long, connected at one end, in the manner shown, by means of an iron bar one inch thick, passed through a hole bored in each piece. The two outside pieces are fastened together by two cross-pieces, bolted to them, so that they are spread at the bottom sufficiently, which would be about three feet. A hook is fastened to the lower cross-piece, upon which the hog is suspended. To hang the hog the frame is laid upon the ground with the hog between the outside legs, the third leg being drawn backwards. The hog is hooked by the gambrel stick to the cross-piece, the frame is lifted up, and the hinder leg is spread out so as to support it, as shown in figure 17. The frame may be lowered easily when the hog has to be taken down, and as the frames are cheaply made, and occupy little room, it will be well to have several of them. They may be made to serve other useful purposes.

RELIEF FOR BOG-SPAVIN
AND THOROUGH-PIN

Bog-spavin, and thorough-pin, which are in reality the same disease, differing in position only, and that very slightly, may be considered as incurable. But like many chronic disorders, they may be very much relieved by proper methods. They are caused by an inflammatory condition of the synovial membrane of the hock joint, and are chiefly located in the vicinity of the junction of the bones of the leg, or the capsule between the tibia and the astragalus. This inflammation may be primarily caused by sudden shocks, or by continued strains from hard work, and the troubles are common among those horses which are of a lymphatic constitution, soft boned, or hereditarily subject to scrofulous or inflammatory conditions. They are also found lower down the leg, in which case they are the result of inflammation of the sheath of the tendons. They do not always cause lameness, except when the horse is first brought from the stable, and after a short time the stiffness may pass away. At other times there is great heat and tenderness in the parts, and the animal is decidedly lame. The best treatment is by cold applications and pressure upon the part. Blistering, which is sometimes resorted to generally increases the trouble, and may cause a permanent thickening of the tissues, and a stiff joint. Pressure is best applied by a sort of truss, or strap, provided with a single pad in case of spavin or wind-gall, or double pads in case of thorough-pin, which is simply a spavin or wind-gall, so placed that the liquid which is gathered in the sac or puff may be pressed between the tendons or joint, and mad to appear on the opposite side of the leg. In this case it is obviously necessary to apply the

pressure upon both sides of the leg, and a double pad strap will be needed, of the form shown in figure 18. A common broad leather strap, lined with flannel, or chamois leather, to prevent chafing, is used; pads of soft leather, stuffed with wool, are sewn to the strap, and the exact spots where the pressure is to bear, disks of several thicknesses of soft leather or rubber are affixed. The pads must necessarily be made to fit each individual case, as success will depend upon their properly fitting the limb. The pads should be worn continually until the swelling disappears, and meanwhile, at least twice daily, the parts should be bathed for some time with cold water, and cloths wetted with cold water, with which a small quantity of ether has been mixed, should be bound around the parts, and the pads buckled over them so tightly as to exert a considerable pressure. Absolute rest is necessary while the animal is under this treatment.

Fig. 18. Spavin pad.

TOOL-BOXES FOR
WAGONS, ETC

To go from home with a wagon without taking a few tools, is to risk a break-down from some unforeseen accident, without the means of repairing it, and perhaps a consequent serious or costly delay. Those who do business regularly upon the roads, as those who haul lumber, wood, coal, or ores of different kinds, should especially be provided with a set of tools, as a regular appurtenance to the wagon, and the careful farmer in going to market or the mill, or even to and fro upon the farm, should be equally well provided. We have found by experience that a break-down generally happens in the worst possible place, and where it is most difficult to help one's self. The loss of so simple a thing as a nut or a bolt may wreck a loaded wagon, or render it impossible to continue the journey, or some breakage by a sudden jerk upon a rough road may do the same. It is safe to be provided for any event, and the comfort of knowing that he is thus provided greatly lightens a man's labor. At one time, when we had several wagons and teams at work upon the road, we provided the foreman's wagon with a box such as is here described, and it was in frequent use, saving a considerable outlay that would otherwise have been necessary for repairs, besides much loss of valuable time. It was a box about eighteen inches long, sixteen inches wide, and six inches deep, divided into several compartments. It was supplied with a spare king-bolt, a hammer-strap,

Fig. 19. Wagon box.

wrench, some staples, bolts, nuts, screws, a screw-driver, a hammer, cold-chisel, wood-chisel, punch, pincers, a hoof-pick, copper rivets, a roll of copper wire, a knife heavy and strong enough to cut down a small sapling, a roll of narrow hoop-iron, some cut and wrought nails, and such other things as experience proved to be convenient to have. The shape of the box is shown in figure 19. The middle of the top is fixed, and on each side of it is a lid hinged to it, and which is fastened by a hasp and staple, and a padlock or a spring key. The box is suspended to the wagon reach, beneath the box or load, by two strong leather straps with common buckles. Being only six inches deep, it is not in the way of anything, and is readily accessible when wanted.

MAKING A HINGE

A gate with a broken hinge is a very forlorn object, and one that declares to every passer-by, "here lives a poor farmer." If there is one thing more than another worthy of note and a cause of congratulation in this one hundredth year of the existence of the United States, it is the infinite number of small conveniences with which we are supplied, every one of which adds to the sum of our daily comfort. More than this, the majority of these little things, which are in use all over the world, are the inventions and productions of Americans. So plentifully are we supplied with these small conveniences, that we cannot turn our eyes in any direction without coming across some of them. It is these small matters which enable us to have so many neat and pleasant things about our homes, at so little cost of money, time, or labor. One of the greatest of the small conveniences around the farm, or the mechanic's rural home, is the small forge. To make a gate-hinge with the help of this portable forge is a very easy thing. We take a piece of half-inch square bar-iron, as long as may be needed, and heating one end, round it for an inch or two; then, heating the other end, flatten it out gradually to a point for the same length, and bend it over a mandrel, or the nose

of an anvil, into the shape shown in figure 20. We then cut off a piece of round half-inch bar, about two inches long, and drive it into the

FIG. 20.

loop, tightening the loop around it as much as possible. The loop-end is then brought to a welding heat, and the joint closed around the pin, and neatly worked smooth with the hammer. Another piece of square iron is then taken, and worked at each end the same as the first one, the loop, however, is worked open upon a piece of cold 5/8-inch round bar, so that it will be large enough to work easily upon the pin of the first piece. A thread may now be cut upon the round ends, or they may be riveted over a piece of iron plate, or a large washer, when they are driven through the gate-post and the heel-post of the gate. It is best, however, to have a screw-thread

FIG. 21.

and a nut, using a washer under each nut, to prevent the wood from being crushed. The whole then appears as at figure 21, and is a hinge that cannot easily be broken or worn out. In boring the

holes for a hinge of this kind, a bit or an auger of only half-an-inch diameter should be used, so that the edges of the iron should cut their own way into the wood, and when the hinges are driven, a piece of hard wood should be laid upon the ends that are struck, so that they will not be battered by the hammer. Care must be exercised to have them driven in squarely, so that the gate may swing without binding on the hinges. For lighter hinges, the same sized iron may be used, but the ends should be hammered out to a point, and the edges should be

FIG. 22.

notched or bearded with a cold-chisel, as shown at figure 22. These may be driven into a post very readily, if a hole smaller than the

iron be bored to lead the way, and when driven in, will not be easily drawn out. When it is necessary to draw a hinge out of a post or gate, that has become rusted in, or that has been very tightly driven, it may easily be done by boring a hole above it, or on one side of it, or beneath it, a little larger than the iron, and then forcing it into the hole by means of a wooden wedge driven close to it. It will then be loose, and may easily be taken out without difficulty.

SHELTER FOR THE HEAD

Many a severe headache, and a restless night after an exhausting day's work in the harvest field, might be prevented by the use of some simple precautions. The sun beats down upon the head and neck with great force, when the thermometer marks ninety degrees and over in the shade, and the scorching effect of a heat of one hundred and twenty degrees in the direct sunshine is both uncomfortable and dangerous to the health. The head should be protected in such cases by wearing a straw hat, or one of some open material, with a broad brim, and by placing a leaf of cabbage or lettuce, or a

FIG. 23. Neck-protector.

wetted cambric handkerchief in the crown of it. The very sensitive back of the head and neck is best protected by means of a white handkerchief fastened by one border to the hat-band, figure 23, and the rest made to hang down loosely over the neck and shoulders. The neck is thus shaded from the sun's rays, and the loosely flapping handkerchief causes a constant current of air to pass around and cool the neck and head. We have found this to be a most comfortable thing to wear, and its value as a protector to the base of the brain and the spinal marrow is so well known in hot countries, that the use of a similar protection is made inoperative in armies when on the march.

HOW TO LEVEL WITH SQUARE AND PLUMB-LINE

The common carpenter's square and a plumb-line may be made to serve as a substitute for the spirit level for many purposes on the farm or elsewhere, when a level is not at hand. The manner of getting the square in position to level a wall, for instance, is shown in figure 24.

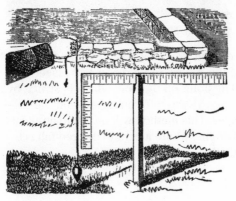

FIG. 24. Manner of levelling a wall.

A piece of board, three feet in length, having one end sharpened, is driven into the ground for a rest; a notch is made in the top of the stick large enough to hold the square firmly in position, as shown in the engraving, A line and weight, held near the short arm, and parallel to it, will leave the long arm of the square level. By sighting over the top of the square, any irregularities in the object to be levelled are readily discovered. A method to find the number of feet in a descent in the ground is illustrated by figure 25. The square is placed as before

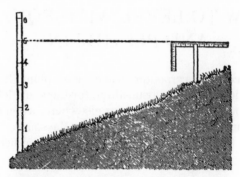

FIG. 25. Measuring a slope with a square.

directed; then a sight is taken over and along the upper edge of the square to a pole or rod placed at a desired point. The point on the pole which is struck by the line of sight shows the difference between the levels of the two places. This method will be found applicable in laying out drains, where a certain desired fall is to be given to the ditch.

KEEP THE CATTLE
UNDER COVER

Even now, in some of the newer regions of the West, the easiest way to get rid of the manure is considered the best. The English farmers have long been obliged to feed farm animals largely for the fertilizers they yield, and this has proved that covered yards are the most economical. These covers are not so expensive as might be supposed at first thought. Substantial sheds, large enough to accommodate a hundred head of cattle, may be built at a cost all the way from $1,000 to $1,500, according to the locality and price of labor and lumber. The roof may be made with three ridge poles resting upon outside walls, and two rows of pillars. There should be ample provision for ventilation and the escape of the water falling upon the roof. The original cost will not be many dollars per head, and the interest on this will represent the yearly cost. If this should be placed at two dollars for each animal, it will be seen that this outlay is more than repaid by the increased value of the housed manure over that made in the open yard, and exposed to the sun and drenching rains. The saving in food consequent upon the warm protection of the animals has been carefully estimated to be at least one-tenth the whole amount consumed. In the saving alone, the covered yard gives a handsome return upon the investment.

WATERING PLACES FOR STOCK ON LEVEL LAND

It is frequently the case that there are underdrains of living water passing through level fields, in which there is no water available for stock. In such a case, a simple

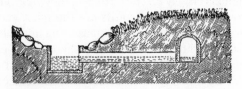

FIG. 26. Taking water from underdrain.

plan for bringing the water to the surface is shown in figures 26 and 27, in which is indicated an underdrain of stone or tile; a pipe of two-inch bore of wood or tile,

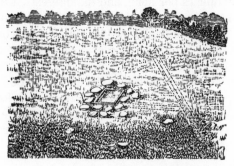

FIG. 27. The watering place.

and about 15 feet in length, is laid level with the bottom of the drain, and connecting with a box one foot or more square sunk into the ground. If the soil be soft, the box is surrounded with stones as shown. A low place or small hollow at some point along the drain is selected for the watering box, or, should the land be nearly level, then with plow and scraper an artificial hollow is soon made at any point desired. Two fields may be thus easily watered by making the box two feet in length, and placing it so that the fence will divide it.

A SHAVING-HORSE

The shingle-horse, shown in figure 28, is made of a plank ten feet long, six inches wide, and an inch and a

FIG. 28. Shaving-horse for shingles.

half thick. A slot is cut through this plank, and a lever, made of a natural crook, is hinged into it. A wooden spring is fixed behind the lever, and is fastened to it by a cord. This pulls back the lever when the foot is removed from the step beneath. The horse may have four legs, but two will be sufficient, if the rear end is made to rest upon the ground. Figure 29 is made of a plank, six

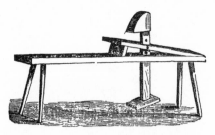

FIG. 29. Horse for general use.

feet long, ten inches wide, and two inches thick. Four legs, two feet long, are fixed in inch and a half holes, as shown below. A bench, eighteen inches long, eight and a half inches wide, and an inch and a half thick, is fixed upon the horse. A slot, eight by one and a half inches, is cut through the bench and the plank, and the lever, two feet eight inches long, is fixed in this by means of a pin passing through the bench. Some extra holes are made in the lever, by which the height of the head above the bench may be changed to suit different sizes of work. A head is put upon the lever, six inches square each way, but bevelled off at the front. The foot-board, five by ten inches, is fastened to the bottom of the lever by a strong pin.

A MILKING-STOOL

The front of the stool (figure 30) is hollowed to receive the pail, which is kept in its place by a wire, fixed as shown in the engraving. The front leg has a projecting rest upon which the bottom of the pail is placed to keep it from the ground, and also from breaking away the

Fig. 30. A milking-stool.

wire by its weight. The milker may either sit astride of this stool, or sideways upon it.

HOW TO TREAT THRUSH

Thrush is a disease of the horse's hoof, quite common in this country. It results oftener from neglect in the stables than from any other cause. The symptoms are fetid odor and morbid exudation from the frog, accompanied with softening of the same. A case recently came under our observation. A young carriage horse, used mostly on the road, and kept in the stable through the year, showed lameness in the left fore foot one morning after standing idle in the stable all the previous day. On removing the shoe, and examining the hoof, a fetid odor was observed. The stable was examined, when the sawdust used for bedding was found to be saturated with urine. The stable was cleaned immediately. Dry sawdust was placed in the stall, and a few sods packed in the space where the horse usually rested his fore feet. The lameness diminished without medical treatment, and in ten days disappeared altogether. A bedding of sawdust or earth, covered with straw or leaves, promotes the comfort of the horse, but it needs watching and systematic renewing. The limit of the absorbing power of the driest soil, or sawdust, is soon reached. If a horse is kept most of the time in the stable, his bedding soon becomes wet, and unfit for his use. It is all the better for the compost heap, and for the horse, to have frequent renewals of absorbments of some kind, that fermentation may not be in progress under his hoofs. The proper place for this fermentation is in the compost heap. Too often the care of the horse is left to a servant without experience in the stable, and the result is permanent disease in the hoofs and legs of the horse. This is most certainly one of the cases in which "an ounce of prevention is worth a pound of cure."

A WESTERN LOCUST TRAP

A great many devices have been used for the destruction of the locusts in those Western States where they have done so much mischief for a few years past. Whether the locusts are to remain as a permanent pest to the Western farmers, or not, remains to be proved. It is certain, however, that through some effects of the climate, the attacks of parasitic enemies, their consumption by birds and other animals, and by the efforts of the farmers themselves, the locusts have of late been greatly reduced in numbers, and their depredations have become almost inconsiderable. Many methods have been adopted for their destruction. Rolling the ground; plowing furrows, and making pits in them in which the insects are caught; burning them in long piles of dry grass; catching them in large sacks, and upon frames smeared with gas tar, and upon large sheet-iron pans containing burning fuel; all these have been tried with more or less success, as well as the negative means of diverting them from their course by means of thick smoke from smothered fires of prairie hay. A most effective method is one invented by a woman in Minnesota. This consists of a large strip of sheet-iron, figure 31, from ten to thirty feet long, turned up a few inches at the ends and one side; a wire is fixed to each end, or at proper places in the front, by which it can be drawn over the ground by a pair of horses or oxen. A light chain or rope is fixed so as to drag upon the ground a foot in advance of the front of the sheet-iron, by which the locusts are disturbed and made to jump, and as the machine is moving on at the same time, they drop upon it. A thick coat of

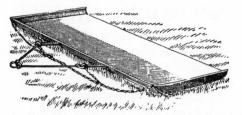

FIG. 31. Trap for catching locusts.

gas-tar is smeared over the surface of the iron, in which the locusts are imbedded and stick fast. The vigorous kicking of the trapped insects helps to keep the mass stirred up, and present a sticky surface. When the trap is full, the locusts are raked off into a pile, and set on fire and consumed. This machine can be drawn over young wheat without injury, as it is not heavy enough to break it down, and being flexible, conforms to the surface of the ground it is passing over. The engraving shows the manner of preparing the sheet-iron for this purpose. The season when the locusts have formerly damaged the newly sprouted wheat is in the spring, and it will be useful for many Western farmers to know of this cheap and effective method, which is not patented, and for which they may thank a farmer's wife of more than usual ingenuity and habits of observation.

SPREADING MANURE

The winter is a good season for spreading manure. It is immaterial whether the ground is covered with snow or not, or whether it is frozen or soft, provided it is not too soft to draw loads over, and that the ground is not upon a steep hill-side, from which the manure may be washed by heavy rains or by sudden thaws. We have spread manure upon our fields several winters, and always with advantage, not only in saving labor and time, but also to the crops grown after it, more especially to oats and potatoes. In spreading the manure, it is the best to drop it in heaps, leaving it to be spread by a man as soon as possible afterwards. This may be done most readily by using a manure hook, with which the manure is drawn out of the sled or wagon-box. Sloping wagon-beds are used for hauling various heavy materials, and why should they not be used for this, the heaviest and most bulky load a farmer has to handle? A wagon, having the box raised (figure 32), so that the forward wheels could

Fig. 32. Wagon with raised box.

774

pass beneath it, would be very convenient on a farm. It could be turned in its own length, and handled with vastly' greater facility than the ordinary farm wagon, which needs a large yard to be turned in, Such a wagon could be unloaded with great ease and very rapidly by the use of the hook, and in case it was desired to spread the load broadcast from the wagon, that could be done perfectly well. But to do this keeps the horses idle the greater part of the time, and is an unprofitable practice. Two teams hauling will keep one man busy in the yard helping to load, and another in the field spreading; the work will then go on without loss of time. In dropping the heaps, they may be left in rows, one rod apart, and one rod apart in the row each load being divided into eight heaps. This will give twenty loads per acre. If ten loads only are to be spread, the rows should be one rod apart, and the heaps two rods apart in the rows. In spreading the manure, it should be done evenly, and the heaps should not be made to overlap. If there is one heap to the square rod, it should be thrown eight feet

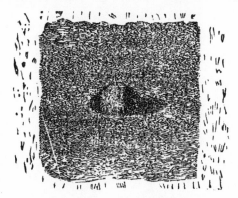

FIG. 33. Manner of spreading.

each way from the centre, covering a square of sixteen and one half feet, as shown in figure 33. One heap then is made to join up to another, and the whole ground is equally manured. There is more in this point than is generally supposed by farmers, many of whom are careless and wasteful in this respect, giving too much in some places, and too little in others. The consequence is uneven growth over the field, rusted grain, or perhaps laid straw in some places, and in others a half-starved crop. Another important point in spreading is, to break up the lumps, and scatter the fine manure. Unless this is done, the field cannot be evenly fertilized. There is work about this, which would tempt some hired men to neglect it, but it should not only be insisted upon, but looked to, and its performance insured.

PUTTING AWAY TOOLS

The wearing out of farm implements is, as a rule, due more to neglect than to use. If tools can be well taken care of, it will pay to buy those made of the best steel, and finished in the best manner; but in common hands, and with common care, such are of little advantage. Iron and steel parts should be cleaned with dry sand and a cob, or scraped with a piece of soft iron, washed and oiled if necessary, and in a day or two cleaned off with the corn-cob and dry sand. Finally, paint the iron part with rosin and beeswax, in the proportion of four of rosin to one of wax, melted together and applied hot. This is good for the iron or steel parts of every sort of tool.

Wood work should be painted with good, boiled, linseed oil, white lead and turpentine, colored of any desired tint; red is probably the best color. Keep the cattle away until the paint is dry and hard, or they will lick, with death as the result. If it is not desired to use paint on hand tools, the boiled oil, with turpentine and "liquid drier," does just as well. Many prefer to saturate the wood-work of farm implements with crude petroleum. This cannot be used with color, but is applied by itself, 80 long as any is absorbed by the pores of the wood.

SELF-CLOSING DOORS

A self-opening, rolling door is shown in figure 34. A half-inch rope, attached to a staple driven into the upper edge of the door, passes parallel with the track,

FIG. 34. Self-closing sliding door.

and beyond the boundary of the door when open, over a small grooved pulley and thence downward; a weight is attached to its end. The door is shown closed, and the weight drawn up. As the door is a self-fastening one, when the fastening is disengaged the weight will draw the door open- By a string or wire connected to the fastening, the door may be opened while standing at any part of the building, or if one end be attached to a post outside, near the carriage way, the door may be opened without leaving the vehicle, a desirable plan, especially during inclement weather. The weight and pulleys should be located inside the building, but are shown outside to make the plan more readily understood. By attaching the rope to the opposite side of the door, it may be made self-closing instead of self-opening, as thought most convenient. The manner of closing a swing-door, as in figure 35, is so clearly shown as to need no description.

FIG. 35. Swing-doork.

VENTILATORS FOR FODDER STACKS

The perfect curing of fodder corn is difficult, even with the best appliances; as usually done, the curing

FIG. 36. Frame.

i very imperfect. The fodder corn crop is one that merits not only the best preparation of the ground and the best culture, but it is worthy of special care in harvesting and curing. The French farmers are giving much attention to this crop, and by good culture are raising most extraordinary and very profitable yields. Seventy tons per acre is not unfrequently grown by the best farmers. We do not average more than eight tons per acre, yet with us the corn crop may be grown under the most favorable circumstances. In a few instances, a yield of thirty tons per acre has been reached by one farmer, but this is the highest within our knowledge. One of the most prominent defects in ordinary American agriculture is, the neglect with which this easily grown and very valuable crop is treated; and one of the most promising improvements in our advancing system of culture is, the attention now being given to fodder corn. A drawback under which we labor is the difficulty of curing such heavy and succulent herbage; this, however, will by and by be removed, both by the adoption of the French system of ensilage, and by better methods of drying the fodder. On the whole, the system of ensilage offers by far the greatest advantages;

the fodder being preserved in a fresh and succulent condition, and the labor of preparing the silos, cutting the stalks, and properly protecting them from the atmosphere, being actually no more than that of drying the crop in the usual manner, storing it in stacks, and cutting it afterwards for use when it is needed. It is impossible, however, that even the best improvements can be introduced otherwise than slowly and with caution; the old system, although it may be less effective and profitable than the new, will be long retained by many; and even in the old methods improvements are being made from season to season by the ingenuity of farmers. We recently saw a very simple but useful arrangement for the ventilation of stacks, and mows in barns, which is applicable to the curing of corn fodder. It consists of a frame, fig-

FIG. 37. Ventilator.

ure 36, made of strips of wood, put together with small carriage bolts. The strips may be made of chestnut, pine, or hemlock, the first being the most durable and best, two inches wide and one inch thick. The illustration shows how these strips are put together. The length of the section shown may be three or four feet. In figure 37 is seen the manner in which the sections are put together. A small stack may have a column of these ventilators in the centre; a large one may have three or four of them; in a mow in the barn, there may be as many as are needful, two or three, or more, as the case may be. When made in this shape, they are so portable, and easy to use, that the greatest objections against ventilators are removed. In stacking fodder corn, it is safest to make the stacks small. Three of these sections, placed together in one column, are sufficient for a stack containing three tons, and which would be about fourteen feet high. The sheaves should be small, and the stack somewhat open at the bottom, so as to freely admit currents of air. The top of the stack should be well protected to keep out the rain; a hay cap fastened over the top would be very effective for this. If a quantity of dry straw could be thrown in between the bundles, and on the top of each layer of them, the perfect curing of the fodder would be then secured.

CORN-MARKER FOR UNEVEN GROUND

The corn-marker, shown in figure 38, is so constructed that it will readily accommodate itself to uneven ground. It consists of two pieces of plank, these form the middle set of runners. Upon these pin two straight pieces of two by four scantling with each end projecting over the runner six inches; through these ends are bored holes for a four-inch rod. Two other pieces of plank, like the former, are procured, and one end of two other pieces of scantling arc pinned to each runner; then these beams are connected to the middle pair by the bolts, as seen in the

FIG. 38. Flexible corn-marker.

engraving, so that, while one runnel is on high ground, the other may be in the land furrow. In turning around, the two outside runners may be turned up against the seat.

A HOME-MADE HARROW

The harrow, figure 39, is a square one. The teeth are set twelve inches from centre to centre, each way. There are four beams in each half, and five teeth in each beam. These beams are four feet eight inches long, mortised into the front piece, which is three feet seven inches in length. The rear ends of the beam are secured by a piece of timber, two by one and a half inches, halved on to the beams and then bolted. The harrow is made of two and a half by two and a half-inch scantling, using locust wood, because of its great durability and firmness. There is nothing particularly new about this harrow, except that it is larger than common, and the novel way of hitching to it by which it is kept steady. The teeth can be made to cut six inches or one inch apart. The manner of hitching is shown in the engraving. The draw-bar is made of three-eighths by one and three-quarter iron, three feet

Fig. 39. An excellent harrow.

four inches in length. The chain is attached to this by a hook at one end, the other being fastened to the harrow by a staple. The chain is about two feet long. The entire cost is about twelve dollars.

CLEARING LAND BY BLASTING

The explosive used is dynamite or giant powder. It is a mixture of nitro-glycerine with some absorbents, by which this dangerously explosive liquid is made into a perfectly safe solid substance, of a consistence and appearance not unlike light-brown sugar. It is not possible to explode dynamite by ordinary accident, nor even by the application of a lighted match. A quantity of it placed upon a stump and fired with a lighted match, burns away very much as a piece of camphor or resin would do, with little flame but much smoke, and boils and bubbles until only a crust is left. There is not the

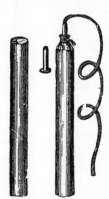

FIG. 40.   FIG. 41.

least danger, therefore of igniting the powder dangerously, until properly placed for the blast. In this respect it has a very great advantage over ordinary blasting powder, which may be exploded by a spark. The powder, as it is manufactured, is made into cartridges about eight inches in length, and of any required diameter. The cartridges are wrapped in strong parchment paper, covered with paraffine, and the true form is shown at figure 40. They are fired by a cap (also in figure 40), which is inserted into the end of the cartridge. The fuse, which is of the common kind, is inserted into the open end of the cap, which is pinched close upon it with a small pair of pliers, so as

FIG. 42. The stump before the explosion.

to hold it firmly. The cartridge is then opened at one end, the cap with the fuse attached inserted, and the paper tied tightly around the fuse, with a piece of twine. The cartridge ready for firing is shown at figure 41.

Our first operation was upon a green white-oak stump, thirty inches in diameter, with roots deeply bedded in the ground. To have cut and dug out this stump with axe and spades would have been a hard day's work for two or more good men. The shape of the stump is shown at figure 42. A hole was punched beneath the stump, as shown in the figure, with an iron bar (figure 43), so as to reach the centre of it. Two of the cartridges were placed beneath the stump, and were tamped with some earth; a pail of water was then poured into the hole, which had the effect of consolidating the earth around the charge. The fuse was then fired. The result

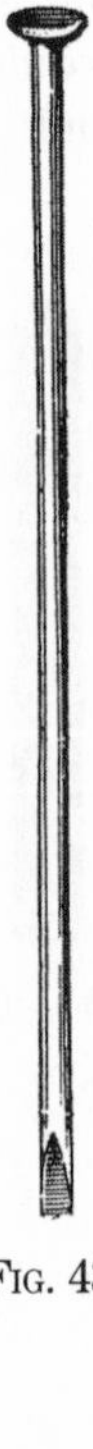

FIG. 43.

was to split the stump into numerous fragments, and to throw it entirely out of the ground, leaving only a few shreds of roots loose in the soil. The result is shown in figure 44, on the next page; the fragments of the stump

FIG. 44. The effect of blasting the stump.

in the engraving were thrown to a distance of thirty to fifty feet, and many smaller ones were carried over one hundred feet. The quantity of powder used was less than two pounds. A portion of the useful effect produced by the explosion, consisted in the tearing of the stump into such pieces as could easily be sawed up into fire-wood; by which much after-labor in breaking it up, when taking it out in the usual manner, was saved. This test was perfectly successful, and proved not only the thorough effectiveness of this method, but its economy in cost and in time. Several other stumps were taken out in the same manner; the time occupied with each being from five to ten minutes. Smaller stumps were thrown out with single cartridges, and in not one case was anything left in the ground that might not be turned out with the plow, or that would interfere with the plowing of the ground. The explosive was then tried upon a fast rock, of about one hundred and fifty cubic feet, weighing about ten tons. The shape of the rock before the explosion is shown in figure 45. A hole was made, with the bar, in the ground

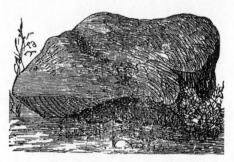

Fig. 45. The ruck as it was.

beneath the rock, and three cartridges were inserted and exploded. To have produced this result (shown in figure 46), by the ordinary method, that was here done in a few minutes by one man, would have occupied, at least, two men, with drills, sledge, etc., two or three days. The application of this method is seen to be of great value where the saving of time is an object. An acre of stumps or rocks may be cleared in one day by one or two men, and the material left ready for use as fire-wood, or as stones for fences or buildings. The cost in money is also reduced in some cases very considerably, and almost

Fig. 46. The rock after blasting.

absolute safety to the careful operator is insured. It would be generally advisable to secure the services of an expert, and that the parties who have work of this character to be performed, should jointly engage such a man, who could either do the whole work, or do it in part, and instruct a foreman or skilful workman sufficiently in a day to perform the remainder. The most favorable seasons for operating upon stumps and rocks are fall and spring, when the ground is saturated with water. It should be explained that this explosive is not injured by water, although a long-continued exposure to it would affect some qualities of it.

PREVENTABLE LOSSES ON THE FARM

It is a "penny wise and pound foolish" system, to breed from scrub stock. There is not a farmer in this region who has not access to a pedigreed Shorthorn bull, by a payment of a small fee of two to five dollars, and yet we find only one animal in ten with Shorthorn blood. It is a common practice to breed to a yearling, and as he is almost sure to become breechy, to sell him for what he will bring the second summer. Many farmers neglect castrating their calves until they are a year old. We think ten per cent, are thus permanently injured, must be classed as stags, and sold at a reduced price. Fully half the calves so stunted never recover.

With many, the starving process continues through the entire year. They are first fed an insufficient quantity of skim milk; then in July or August, just at the season when flies are at their worst, and pastures driest, they are weaned, and turned out to shift for themselves, and left on the pastures until snows fall, long after the fields yield them a good support. They are wintered without grain, spring finds them poor and hide-bound, and the best grazing season is over before they are fairly thrifty.

The keeping of old cows long past their prime is another thing which largely reduces the profits of the farmer. We have found quite a large per cent. of cows, whose wrinkled horns and generally run-down condition show that they have long since passed the point of profit. A few years ago, these cows would have sold at full prices for beef, now they will sell only for Bologna at two cents per pound. Thus cows have, in a majority of cases, been kept, not because they were favorites, or even because they were profitable, but from sheer carelessness and want of forethought. Another fruitful cause of loss to the farmer is, attempting to winter more stock than he has feed for. Instead of estimating his resources in the fall, and knowing that he has enough feed even for a hard winter, he gives the matter no thought, and March finds him with the choice of two evils, either to sell stock, or buy feed. If he chooses the former, he will often sell for much less than the animals would have brought four months earlier, and if the latter, will usually pay a much higher price for feed than if it had been bought in autumn. Too often he scrimps the feed, hoping for an early spring, and so soon as he can see the grass showing a shade of green around the fence rows, or in some sheltered ravine, turns his stock out to make their own living. This brings one of the most potent causes of unprofitable cattle raising; namely, short pastures. The farmer who is overstocked in winter, is almost sure to turn his cattle on his pastures too early in the spring, and this generally results in short pasture all summer, and consequently the stock do not thrive as they ought, and in addition, the land which should be greatly benefited and enriched, is injured, for the development of the roots in the soil must correspond to that of the tops, and if the latter are constantly cropped short, the roots must be small. The benefit of shade is lost, and the land is trampled by the cattle in their wanderings to fill themselves, so that it is in a worse condition than if a crop of grain had been grown on it. From all those causes combined, there is a large aggregate of loss, and it is the exception to find a farm on which one or more of them does not exist, and yet without exception they may be classed as "preventable," if thought and practical common sense are brought to bear in the management.

A CRADLE FOR DRAWING
A BOAT

When it is necessary to draw a boat out of the water, a cradle should be used. This is very easily made out of some short boards and a piece of plank. The boards are cut so that when three thicknesses are bolted together, the joints shall be broken and not come opposite each other, as shown in figure 47. The cradle should be made

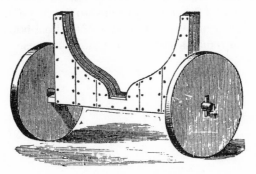

Fig. 47. Cradle for a boat.

to fit the boat tightly, midway between stem and stern, so that when it rests upon it, the boat will be evenly balanced and firmly held. The cradle is mounted upon two wheels, which may be made of hard wood plank. A piece of two-inch plank may be sawn out for the axle, and the upper part of the cradle firmly bolted to it. Such a cradle as this may be made light or heavy, and if desired may be furnished with iron wheels, so that it will sink in the water. It can then be run clown under the boat, and that be drawn upon it. By hauling upon the ring-bolt in its stern, the boat can be drawn up out of the water, and easily moved on land.

When it is desired to lift a boat out of the water, and suspend it in a boat-house, all that is necessary to be done is to fix two strong hooks, or rings, in the top of the house, and a ring-bolt at each end of the boat. A pair of double-sheaved blocks is provided for each end of the boat. The blocks are hooked to the rings in the house and to those in the boat, which is then drawn up, one end at a time, alternately, until high enough. If two persons are in the boat, both ends may be hauled up at once. The loose end of the rope is fastened to the ring of the boat, or to a ring or a cleat at the side of the boat-house. Then the boat remains suspended in the boat-house.

FEED-RACK FOR SHEEP

The rack, figure 48, is made of poles for the bottom and top, and cross-bars fitted into them. The bottom bar

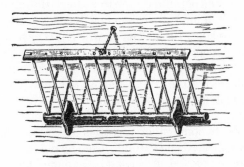

Fig. 48. Feed-rack for sheep.

slides loosely in brackets, which are fixed to the wall of the shed, and the upper bar is secured by a cord, which passes over a small pulley in a hole in the wall above the rack; a weight being attached to the outside end of the rope, serves to keep the rack always against the wall. When the hay is put in the rack is drawn down, and, when filled, is pushed back against the wall, holding the hay closely, and being kept in place by the weight. This prevents the hay from being pulled out too freely by the sheep or cattle. It is recommended that the grain-trough be placed beneath a rack of this kind, so that the chaff which falls from it may be caught in the trough and saved for use, instead of being trampled under foot.

HOW TO MANAGE NIGHT-SOIL

The fertilizing properties of night-soil are well known. The principal reason why this valuable material is neglected and permitted to go to waste, is the difficulty of handling it. If improperly handled, it is disagreeable and difficult to apply to the uses to which it is best adapted. There are many cases in which it could be made use of very conveniently, if rightly managed. In country towns and villages it is difficult to dispose of it, and it becomes a serious nuisance to householders, and a detriment to the public health, when it ought to be turned to profitable uses. In some other countries this refuse matter is eagerly collected and carefully used by the farmers. The methods employed in England, Germany, and France might very well be adopted by us, and a large quantity of fertilizing material be gathered. By the methods there in use, the night-soil is easily handled and prepared for distribution upon the land, or for mixing in composts. Arrangements are made with persons in towns and villages who wish to have the soil removed, and the time being fixed (this is always in the night, from which circumstance the name given to the material is derived), wagons with tight boxes, or carts, are sent to the place. Carts are mostly used, as indeed they are in Europe for most of the farm work. The carts, or wagons, carry out a quantity of earth, chopped straw, ashes, or such other absorbent as may be conveniently procured, and some sheaves of long straw, or else the ashes or other absorbent used, which is frequently the sweepings and scrapings of streets, is prepared Upon the ground or near by. This material is then disposed in the form of a bank enclosing a space of sufficient size to hold the night-soil, as shown in figure 49. A reserve heap is kept to be

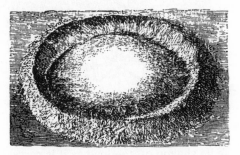

Fig. 49. Preparing night soil.

mixed with the night-soil as it is emptied into the place prepared for it. Wheelbarrows with tight boxes are generally used to convey the soil from the cesspool. When the whole has been removed from the cesspool, the cut straw is mixed in and the banks of earth are turned over upon the pile, which can then be handled with shovels or forks, and is ready to be loaded into the wagon. Some of the long straw is laid in the bottom of the wagon-box, and the mixed mass is thrown upon it, layer after layer alternately with straw, until the top of the wagon-box is reached. It is most convenient to have a rack, or flaring side-boards, to confine the upper part of the load, but this is not necessary if the loading is properly done. The manner of loading the top is as follows: a bundle of straw is spread so that half of it projects over the side or end of the load. A quantity of the mixed stuff is forked on to the straw, the loose projecting ends of which are turned back on to the load when more is laid upon it. The doubled straw holds the loose stuff together, which

Fig. 50. Manner of loading night-soil.

might else be shaken off the load as it is carried home. In this manner the load is built up until it is completed, when it appears as shown in figure 50. Loads thus made are carried many miles without losing anything on the journey, and the mass, which would seem to have no coherence, is kept solidly together. Carts are sometimes loaded to a hight of two or three feet above the side-boards, and are made to carry a load for three horses. By this management this material is no more disagreeable than ordinary manure, and the work of moving it is rendered quite easy.

THE USE OF LIME IN BLASTING

There are some forces, apparently insignificant, which act with irresistible power through short distances. The expansion of water in freezing is a force of this kind. The increase in bulk in changing from the liquid to the solid state of ice is only about one-tenth, yet it exercises a power sufficient to break iron vessels and rend the hardest rocks. Every one who has slaked a lump of quicklime by gradually pouring water upon it, has observed that the first effect of the contact between the water and lime is to cause a swelling of the lump. It generally expands and takes up considerable more room than before. This expansive force has recently been successfully applied to coal mining in England. Powdered quicklime is strongly compressed into cartridges about three inches in diameter, and each has running through it a perforated iron tube, through which water can be forced. These cartridges were used in a coal mine in place of the usual blasting charge, water Was forced into them, and the expansion of the lime threw down a mass of coal weighing about ten tons, with little of the small coal made with the usual blast. The exemption from danger and the avoidance of smoke, have caused coal mine owners to regard this new method with favor. Some of our ingenious reapers may find a useful hint in this.

A WATER AND FEED TROUGH

A supply of water in the cow-stable is a great convenience; a simple arrangement for furnishing it to the cows in their stalls may be made as follows: Sheets of galvanized iron are bent to form a trough, and fitted into the floor joists under the feed-box, as indicated in figure 51, making a trough three inches deep and sixteen inches wide. The flanges on each side are nailed to the joists, and the sheets of iron riveted together at the ends, and made water-tight by cement. The trough runs the entire length of the feed floor, and is supplied with water from a pipe, pump, or hose; a pipe at the other end carries away the surplus water and prevents overflow, and another pipe with a faucet is provided for emptying the trough. The feed-box is built over the water trough, a part of

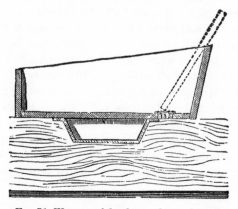

Fig. 51. Water and feed trough combined.

its floor being a trap—indicated by dotted lines in the engraving—by which admission to the water is gained.

Before opening this trap, the manger is swept clean; and if there were no other advantage than this compulsory cleansing of the mangers after each feeding, it would be sufficient to pay for the cost of constructing such a watering arrangement.

THE CONSTRUCTION
OF STALLS

It is rare, even in these days of progress, to see a well-arranged stall in a farmer's barn. No horse stall should be less than six feet in width, nor of a length less than nine feet. This affords room for the animal to lie down and rise comfortably without bruising hips and limbs, and also for the attendant to pass in and out. The partition between stalls should be of sufficient hight to prevent playing, biting, and kicking. Hacks of iron are neat and serviceable. The horse eats its food from the ground, and because many first pull out a greater portion of the hay from the rack, we shall dispense with the rack as commonly used, and substitute a single manger which serves for both hay and grain.

Whatever may be the foundation of the stall, whether of brick, stone, cement, clay, or wood, it should have inclination enough to carry off all fluid. Over this place a flooring composed of strips of plank, four inches in width by two inches in thickness, with an inch intervening between each strip. This need not extend more than half the length of the stall, the upper portion being compact. The essential point is that the horse shall stand with an equal weight upon all the extremities. This custom of confining a horse to a sloping stall, in one position sometimes for days, is a cruel one, and very detrimental to the limbs and feet, as it brings about, sooner or later, serious affections in these parts. A loose box is far preferable to the stall, wherever practicable. Every stable or barn should be provided with one at least, in case of sickness or accident. By the arrangement of a floor as just described, the bedding is kept dry and the animal clean and comfortable. Litter should be always kept beneath the animal; it gives an air of comfort to the place and invites to repose of body and limbs by day and night. Stalls for both horses and cattle should be of sufficient hight as also all door and passage ways about a barn. Formerly, it was the custom to build in such a way that no horse, and not even a man of respectable hight could enter a door-way with out danger of knocking his skull, and inflicting serious injury. There are stalls in country barns so low that a horse cannot throw up his head without receiving a blow against the beams above. Animals undoubtedly acquire the trick of pulling back, or of making a sudden spring when passing a door-way, from having been obliged to run the gauntlet of some narrow, low, ill-contrived passage-way. The man who should now be guilty of building in this way would deserve to have his own brains knocked, every time he passes in and out, as a gentle reminder of his folly. All barn-doors should be high, wide, and, when practicable, always slide.

The common mode of securing cattle in the barn, especially milch cows, by placing their necks between stanchions, is not to be advocated, especially when they are confined in this way for many hours at a time without relief, as is often necessary in the winter season. A simple chain about the neck with a ring upon an upright post affords perfect security, while it gives the animal freedom of movement to head and limbs—and conduces to its comfort in various other ways. Animals should not be overcrowded, as is too often the case in large dairy establishments—a fact which will make itself evident sooner or later in the sanitary qualities of the milk, if in no other manner. We cannot deny the fact, if we would, that everything, however trifling, that contributes to the welfare of our domestic animals is a gain to the owner of them pecuniarily, and what touches a man's pocket is generally considered to be worth looking after, at all times and in all places.

HOG-KILLING
IMPLEMENTS—RINGING

The stout table on which the dead porkers are lain to be scraped and dressed after being scalded, is made with its top curving about four inches in a width of four feet, and consisting of strips of oak plank, as represented in figure 52. This curved top conforms to the form of the

Fig. 52. A dressing table.

carcass, and holds it in any desired position better than a flat surface. For scrapers, old-fashioned iron candle-sticks are used; the curved and sufficiently sharp edges at either end serving as well as a scraper made for the purpose, and its small end has an advantage over the latter for working about the eyes and other sharp depressions, A cleaver for use in cutting up the pork is shown in figure 53; it has a thirteen-inch blade, three inches

FIG. 53. Handy meat cleaver.

wide at the widest part, and one-quarter inch thick at the back. This is a convenient implement, easily and cheaply made by a good blacksmith, if it cannot be had at the stores; any mechanic can put on the wooden handle. In figure 54 is represented a home-made hog-ringing

FIG. 54. Hog-ringer and key.

apparatus. The blacksmith makes an instrument resembling a horse-shoe nail, of good iron, about three inches long, three-sixteenths of an inch wide, and one-thirty-second of an inch thick, tapering to a point; the "head" is merely the broad flat end curled up. Just before using, this needle-like instrument has its corners rubbed off on a file; it then is easily pushed through the septum of the pig's nose. A key with its tongue broken off and a slot filed in the end, is used to curl up the projecting end, and the ringing is done. The "rings" cost about seventy-five cents a hundred, and are effective and easily applied.

HOW TO MIX CEMENT

The article to be used is the Rosendale cement. This is nearly as good as the imported Portland cement, and much cheaper. The cement is made from what is known as hydraulic lime-stone—that is a rock which contains, besides ordinary lime-stone, some clay, silica, and magnesia. Pure lime-stone contains only lime and carbonic acid, in the proportions of fifty-six parts of the former to forty-four of the latter in one hundred. When this stone is burned, the carbonic acid is driven off by the heat, and pure or quick-lime is left. When this is brought in contact with water, the two combine, forming hydrate of lime; during the combination, heat is given out; the operation is called slaking. When the water is just sufficient to form the combination, a fine, dry powder is produced, which we call dry slaked-lime. When the water is in excess, the surplus is mixed mechanically with the lime, and forms what is called the milk of lime, or cream of lime, according to its consistence; it is this pasty substance which we mix with sand, to form building mortar. But when we have clay mixed in a certain proportion, either naturally or artificially, with the lime-stone, and this stone or mixture is burned in the same manner as ordinary lime-stone, we get what is known as hydraulic lime, because it combines with a much larger proportion of water than pure lime, and in combining with it, instead of falling to powder, like ordinary lime, it hardens into stone again. This hardening takes place even under water; the hydraulic lime combines with just so much water as is required to "set" or harden, and leaves the remainder. It possesses this property, also, when mixed, with sand in proper proportions, and when so mixed, the cement will adhere very firmly to the surface of any stone to which it may be applied. This property is made available in constructing works of concrete, which consists of broken stone mixed with such a quantity of cement, that, when it is packed closely, the surfaces of all the pieces of stone are brought into contact with the cement, and the spaces between the fragments of stone are filled with it. That there may be no more cement used than is actually needed, the mixture is rammed down solidly, until the fragments of stone are brought into close contact with

FIG. 55. Box for mixing cement.

each other. The composition of the impure or hydraulic lime-stone, which behaves in this useful manner, is, in the case of some of the Kingston stone, as follows: Carbonic acid, 34.20 per cent.; lime, 25.50; magnesia, 12.35; silica, 15.37; alumina (clay), 9.13; and peroxide of iron (which is useless or worse), 2 25. On account of this difference in character between lime and cement, a different treatment is necessary for each, and each is put to different uses. The cement makes a much harder and more solid combination with sand, and is therefore chosen when great strength is required. Its rapid setting,

FIG. 56. Side of cement box.

when mixed with water, also requires that it be used as soon as it is mixed, and renders a rapid mixture necessary. The cement and sand should, therefore, be mixed together dry, and very thoroughly. Four parts of sand to one part of cement are the proportions generally used. These may be mingled in a box of suitable character, and the mass is so spread as to have a hollow in the centre, into which water is poured. The sides of the heap are gradually worked into the water, with a common hoe, in such a way as to prevent the water from spreading about, and as it is absorbed more water is poured in, until the whole is brought to a thin semi-liquid condition. A box very suitable for this operation is shown in figure 55. This is made of pieces of plank, prepared as follows: The side pieces are shown at figure 56. The end pieces are made with tenons, which fit in mortises in the side pieces, and the frame thus made is held together by keys driven into the holes seen in the tenons. The bottom planks are fastened together with cleats, so placed as to receive the frame and fit snugly. Iron bolts are put through holes in the cleats, and through the holes in figure 56, and by means of nuts with washers under them, the whole box is brought firmly together. Such a box, after having been used for this purpose, will be found very useful for mixing feed in the barn, or for many other purposes, and may, therefore, be well made at the first. When the mortar is mixed, the broken stone may be thrown into it, beginning at one side, and the whole is

FIG. 57. Machine for making cement.

worked up thoroughly with the hoe, so that every piece of stone is coated with the cement. A machine, that is easily made, may be used for this mixing, and is also very useful for mixing ordinary mortar for building or plastering. It is shown in figure 57. It consists of a box set upon feet, with a smaller box attached at the rear end, having an opening at the bottom where the mortar is seen escaping, and a shaft, having broad, flat arms on it, placed at a somewhat acute angle with the line of the shaft, so that they will operate as a screw to force the mass along the spout and out of it at the opening. A crank handle is fitted to this shaft, and if a fly-wheel can be borrowed from a feed-cutter, or a corn-sheller, and attached to the shaft as shown, so much the better. The materials to be mixed are thrown into the box, and by turning the handle, the whole will be thoroughly incorporated with great rapidity and ease.

RINGING AND HANDLING BULLS

Now that more attention is given to improving farm stock, a bull is kept upon nearly every large farm. The high-bred bulls are spirited animals, and are exceedingly dangerous if the utmost caution is not exercised in managing them. Experienced breeders are not unfrequently caught unawares, and unceremoniously lifted over the fence, or forced to escape ingloriously from one of their playful animals, or even seriously injured by the vicious ones. It should be made a rule, wherever a bull is kept, to have him ringed, before he is a year old, and brought under subjection and discipline at an early age, while he can be safely and easily handled. Some time ago we assisted at the ringing of a yearling bull, which severely taxed the utmost exertions of six persons with ropes and stanchions to hold him. A slip of the foot might have caused the loss of a life, or some serious injuries. To avoid such dangerous struggles, a strong frame, similar to that in figure 58, in which to confine the bull, may be used. The frame consists of four or six stout posts

FIG. 58. Stall for bull.

811

set deeply in the ground, with side-bars bolted to it, forming a stall in which the bull can be confined so that he cannot turn around. The frame may be placed in the barn-yard or a stable, and may be made to serve as a stall. At the front, a breast-bar should be bolted, and the upper side-bars should project beyond this for eighteen or twenty inches.

Fig. 59. Strap.

The forward posts project above the side-bars some inches. The ends of these posts, and the side-bars, are bored with one-inch holes, and at the rear of the frame there should be tenons or iron straps to receive a strong cross-bar, to prevent the animal from escaping should the fastenings become broken or loosened. The bull, led into the frame, is placed with his head over the breast-bar, and the horns are tied with ropes an inch in diameter to the holes in the bars and posts. He is then secured, and his head is elevated so that the trochar and cannula can be readily used to pierce the cartilage of the nose, and the ring inserted and screwed together. Before the ring is used, it should be tested to ascertain that it is sound and safe.

When the ring is inserted, the straps shown in figure 59 should be used, for the purpose of holding it up and out of the way, so as not to interfere with the feeding of the animal until the nose has healed and become calloused. The straps may be left upon the head permanently, if desired, when the front strap will offer a

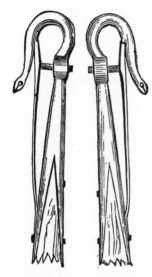

FIG. 60. Staves.   FIG. 61.

convenient means of catching him by the staff, when necessary to do so in the field. The staff is a matter of the greatest importance. This should be made of the toughest ash or hickory, and not less than five feet long. With a staff of this length, the herdsman can check the wildest bull, and by resting the butt-end of it upon the ground, can throw the animal's head up, and prevent him from approaching too near. The hook of the staff is shown of two kinds in figures 60 and 61. One is furnished with a spring, by which it is closed. A metal bar attached to the spring and passing through a hole in the staff, prevents the ring from slipping along the spring. The other is provided with a screw by which if is closed.

SLED FOR REMOVING
CORN-SHOCKS

A sled used for moving corn-shocks from a field which
is to be sown with winter grain is shown in figure 62. It
is simply a sled of the most ordinary construction, and
which any farmer can build. It is made of two joists or
planks of hemlock, though oak might be better; say three
inches thick, a foot wide, and fourteen to sixteen feet
long, rounded at one end and connected by three strong
cross-pieces, being in form just such a sled as a farmer
boy would make to use in the snow, with the addition of
cross braces before and behind. The under edge of the
runners should be rounded off to the extent of one and
a half to two inches, to turn more easily. There should
be also short standards before and behind. The runners
may be four to five feet apart, according to the length of
the corn. A side view of the runner with the standards
is given in figure 62, and a top view of the complete sled

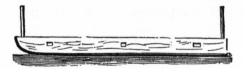

Fɪɢ. 62. Side of bled.

in figure 63. First, cut off the corn and put it in shocks
in the usual way, making the shock smaller than usual.
Let it stand thus a few days to dry, then a pair of horses
are hitched to the sled, which is driven alongside the
shock. The shock is pushed over on to the sled, and so
one shock after another until the sled is full. The load is

814

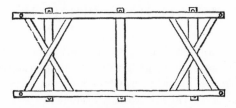

FIG. 63. Top of sled.

then driven to an adjoining field, where the shocks are set up on end again, and about four of them made into one and tied at the top, or reared against a fence.

The particular advantages of this plan are: First, that by use of the sled and method of loading and unloading the shocks, all actual lifting of the corn is avoided, and the labor and expense reduced more than one-half. Second, by permitting it to dry a few days, its weight, is greatly reduced, and the handling much lighter. Third, the corn being partially dried, it can be put together in larger shocks the second time, and will keep better. By this method one man can clear two acres or more in a day, according to the weight of the crop.

A TAGGING TROUGH

Sheep should be tagged in early spring, and a table for this purpose is shown in figure 64. The sheep is placed on this table feet upwards, in which position it is perfectly helpless, and will not struggle. Then the soiled wool about the hind parts, the belly, or the legs is clipped off with great ease, less than a minute being needed to tag a sheep. Half time will serve for some shearers to do this. In large flocks these tables will be necessary, and those who have small ones will find them very useful.

LIME AND LIME KILNS

The periodical use of lime as a fertilizer is necessary to good culture. In the best cultivated parts of the country, lime is used once in every rotation of five crops, the usual rotation being two years, grass, corn, oats, wheat, or rye, seeded to grass or clover again. The lime is applied to the land when it is plowed for the fall grain, and is harrowed in before the seed is drilled, or it is harrowed in with the seed, sown broadcast. The quantity used is from forty to fifty bushels per acre. The effect of lime is both mechanical and chemical; it opens and loosens heavy clays, and

FIG. 64. Trough for tagging sheep.

consolidates light, loose, sandy, or peaty vegetable soils; it has the effect of liberating potash from the soil, and of decomposing inert organic matter, and reducing it to an available condition. But while it is beneficial, it cannot be used alone without exhausting the soil of its fertile

817

properties. This is evident from what has been said of its character; at least this is true, so far as regards its effects beyond affording directly to the crops any lime that they may appropriate from the supply thus given. All the benefits received beyond this is a direct draft on the natural stores of the soil. It is therefore necessary, to good agriculture, that either a thrifty clover sod should he plowed under, at least once in the rotation, or that a liberal dressing of manure be given, or both of these. In those localities where the benefits to be derived from the skilful use of lime are best known and appreciated, this method is practised; a heavy sod being plowed under, after having been pastured one year, for the corn, and a good coating of manure being given when the land is plowed for fall grain. Under such treatment, the soil is able to maintain itself and return profitable crops. It is not where this course is pursued that complaints are

Fig. 65. Improved lime kiln with elevated track.

prevalent of the unprofitableness of farming. The use of lime is spreading gradually into the Western States, where the competition of the still farther and fresher western fields is being severely felt. The experience of Eastern farmers is now being repeated in what were once the Western States, and every appliance of scientific and thorough agriculture is found to be needed to maintain those Western farmers in the close contest for a living. This kiln, figure 65, is intended to stand upon level ground, and is furnished with a sloping track, upon which self-dumping cars containing fuel or lime may be drawn up by horse-power with a rope and pulleys. The body of the kiln may be twenty feet square at the bottom, and thirty feet high, with a flue above the stack of ten to twenty feet. The stack may be built of stone or brick, but should be lined with fire-brick or refractory sandstone. The arch is protected by the shed under the track. At B, B, are two bearing bars of east-iron, three by two inches thick, which support the draw-bars, C. These are made of one and a half inch round wrought iron, having rings at the outer end, and of which there are four to the foot across the throat of the kiln, which is four or five feet in diameter. The rings serve to admit a crow-bar, by which the bars, or some of them, are drawn out to let down the charge of lime. The open space, D, is intended for the insertion of the bar to loosen or break the lime, should the throat become gorged. A cast-iron frame, with an aperture of three by twenty-four inches, is built into this opening. It also serves to kindle the kiln, and is closed by an iron door. The car should be made of wood, and lined with sheet-iron; it is hinged to the front axle, and hooked to the draft-rope, so that when the fore-wheels strike the block, E, at the mouth of the kiln, the car tips and dumps its load. The iron door, F, which closes the kiln, is raised or lowered by means of the rope and ring,

G, which passes over a pulley fixed upon the side of the flue, A covered shed will be needed to protect the top of the stack, and a gallery should be made around it, for a passage-way for the workmen. This kind of kiln is suited only for the use of coal as fuel; when wood is used for burning the lime, common pits or temporary kilns are to be constructed.

FALL FALLOWING

The old practice of summer fallowing, or working the soil for one year without a crop, for the purpose of gaining a double crop the second season, is now, very properly, obsolete. While some may question the propriety of this opinion, there can be no doubt as to the value of fall fallowing. The constant turning and working of the ground during the fall months cost nothing but time and labor, at a season when these cannot be otherwise employed, and so, in reality, cost nothing. But the benefits to the soil are very considerable. Especially is this the case with heavy clay soils, and less, in a descending ratio, through the gradations from heavy clay down to light loams—at least it is so considered by many; and it is reasonable to suppose that if the atmospheric effects upon the particles of a clay soil serve, to some extent, to dissolve the mineral particles, they may easily do the same service for a sandy soil, and help to set loose some of the potash contained in the granitic or feld-spathic particles of such a soil. The mechanical effects of the fall working are certainly more useful upon clay than a light loam; but there are other purposes to serve than merely to disintegrate the soil, and mellow and loosen it. There are weeds to destroy, and the forwarding of the spring work by the preparation of the ground for early sowing. These services are as useful for a light soil as a heavy one, and as it is reasonable to look for some advantage from the working in the way of gain in fertility on light as well as heavy soils, it is advisable that owners of either kind should avail themselves of whatever benefits the practice affords. Fall fallowing consists in plowing and working the soil with the cultivator or the harrow. This may be done at such intervals as may be convenient, or which will help to start some weeds into growth, when these may be destroyed by the harrow or cultivator. Heavy soils should be left in rough ridges at the last plowing, with as deep furrows between them as possible, in order to expose the largest surface to the effects of frost and thaw. Light soils may be left in a less rough condition, but the last plowing should be so done as to throw the furrows on edge, and not flat, leaving the field somewhat ridged. A very little work in the spring will put the ground into excellent order for the early crops, and for spring wheat, especially, this better condition of the soil will be of the greatest benefit. When thus treated in the fall, the soil is remarkably mellow, and is dry enough to work much earlier than the compact stubble land which remains as it was left after the harvest. As to the time for doing this work, the sooner it is begun, and the oftener it is repeated, the better. It is not too late to finish when the ground is frozen or there is an inch of snow on the ground

UNLOADING CORN

Every little help that will ease the troublesome labor of transferring the corn crop from the field to the crib is

FIG. 66. Board for unloading.

gratefully accepted. We have used both of the contrivances here shown (figures 66 and 67), to help in getting the ears out of the wagon-box. At the start it is difficult to shovel up the corn, and until the bottom of the wagon-box is reached, the shovel or scoop cannot be made to enter the load. But if a piece of wide board is placed in a sloping position, resting upon the tail-board of the wagon (figure 66), the shovel can be used with ease at

FIG. 67. Unloading arrangement.

the commencement of the unloading. Another plan is to make the box two feet longer than usual, and place the tail-board two feet from the end, figure 67. When the tail-board is lifted, the ears slide down into this recess, from which they can be scooped with ease.

STONE BOATS

For moving plows, harrows, etc., to and from the fields, and for many other purposes, a stone boat is far

Fig. 68. Plank stone boat.

better than a sled or wagon, and is many times cheaper than either. Two plans of construction are illustrated. The boat shown in figure 68 is of plank, six feet in length, one foot at one end being sawed at the angle shown. Three planks, each one foot in width, will make it of about the right proportion. A railing two by three inches is pinned upon three sides, while a plank is firmly pinned at the front end, through which the draw-bolt passes. That shown in figure 69 has some advantage over the

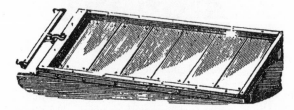

Fig. 69. Stone boat with runners.

former, a cheaper quality of wood and of shorter length can be used, and when one set of runners is worn out, others can be readily attached without destroying the frame. Oak or maple plank should be used for the best boats, and when runners are used, the toughest wood at hand should be selected. Don't think because it is only a stone boat it is not worthy of being taken care of.

A DUMP-CART

The dump-cart, figure 70, is a handy contrivance, a good deal used in some parts of this State, and is simply an ordinary ox-cart, the tongue shortened and fastened by a king bolt to the forward axle of a wagon, as shown in the engraving, It can be turned very short, as the wheels have a clear swing up to the cart-tongue, and is very convenient for hauling anything that is to be dumped: such as stones, earth, wood, manure, etc. The seat of an old mowing machine is fastened to the cart-tongue, on which the driver sits. Horses or oxen may be used.

Fɪɢ. 70. Improved dump-cart.

TO PREVENT WASHING
OF HILL-SIDES

Much damage is done by the washing of hill-sides into deep gullies by heavy rains. Where sloping ground is cultivated this is unavoidable, unless something is done to prevent it. In some cases deep plowing and loosening the sub-soil will go far to prevent washing, as it enables the water to sink into the ground, and pass away without damage, by slow filtration. But where the subsoil is not very porous, and when the rain falls copiously and suddenly, the water saturates the surface soil in a few minutes, and the surplus then flows down the slope, cutting the softened earth into many channels, which by and by run together. Then the large body of water possesses a force which the soil cannot resist, and carries the earth down with it, often doing serious and irreparable damage in an hour or less. Of the many plans which have been suggested and tried to prevent this washing, the most successful is the terracing of the slope. This is done by plowing, with a swivel plow, around the hill, or back and forth on the slope, commencing at the bottom and throwing the earth downwards in such a manner that a flat terrace is formed, which has a small slope backwards from the front of the hill. When this terrace has been formed, the plowing is commenced ten or twelve feet above, and another terrace is made in the same manner. This is continued to the top of the slope. If thought desirable, the inner furrows on each terrace may be made to form a water channel, and this may be connected with the channel on the next slope lower down, in some safe manner, either by a shute of boards

827

or of stone, to prevent washing of the soil at these points where the fall will be considerable. This, however, is a side issue, which does not necessarily belong to the main work. The arrangement of the hill-side is shown in figure 71, in which the original outline of the hill, and

Fig. 71. Profile of a terraced hill.

the arrangement of the terraces, which are cut out of it, are given. When a heavy rain falls upon the terraced hill, the effect will be to throw the water backwards from the outer slope, into the channels at the rear of the terraces; and there, as well as upon the broad surface of the terraces, there is abundant means of escape by sinking into the soil. If not, and the amount of water is too great to be thus disposed of, it may be carried down the slope, by arranging the furrows as drains in the way previously indicated. Hill-sides of this character should be kept in grass, when the slope is too steep for comfortable plowing, after it has been thus arranged; or it may be planted with fruit trees, vines, or timber, upon the slopes, leaving the terraces to be cultivated, or the slopes may be kept in grass, and the terraces cultivated. But in whatever manner the ground may be disposed of, it would be preferable to leaving it to be gullied by rains, barren, useless, and objectionable in every way.

A LOG MINK-TRAP

A mink-trap is made by boring a two-inch or two and a half inch hole in a log, four or five inches deep, and into

Fɪɢ. 72. Mink-trap.

the edges of this hole drive three sharpened nails, so that they will project half an inch or so inside, as shown in figure 72. The bait being at the bottom, the mink pushes his head in to get it, but on attempting to withdraw it is caught by the nails. Musk-rat is good bait for them, and a highly praised bait is made by cutting an eel into small bits, which are placed in a bottle and hung in the sun, and after a time become an oily and very odorous mass. A few drops of this are used. The above simple mink-trap may be made by using any block of wood, or a stump of a tree, large or small, and the same plan may be made use of to trap skunks, or, by using a small hole and some straightened fish-hooks, it will serve to catch rats or weasels, enemies of the rural poultry yard, which may be thinned off by the use of this trap.

PLOWING FROM THE INSIDE
OF THE FIELD

There is but one reason why plowing should not be done from the inside of the field, and that is, the imaginary difficulty in "coming out right." There are several points in favor of this method: When a field is plowed, beginning at the outside, there is always a dead furrow running from each corner to the centre; besides this, the team is obliged to run out, and turn upon the plowed land at every corner, making a broad strip which is much injured by the treading, especially if the land is clayey and rather moist. By beginning at the middle, all this is avoided; the horses turn upon unplowed land, and the soil at each plowing is thrown towards the centre of the field, as it should be. There is no difficulty in finding the centre of the field from which to begin the plowing. Suppose we have a rectangular field like the one shown in figure 73; any person who can measure by pacing, is able to find the middle of the ends, *A D* and *B C;* the points *K* and *L.* From *K*, pace towards *L*, a distance equal to

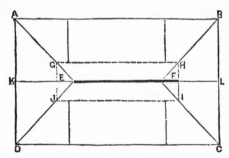

FIG. 73. Plan for rectangular field.

830

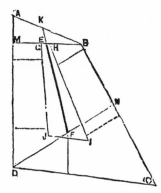

FIG. 74. Plan for irregular field.

one-half *A D*, which gives the point *E*. Also the same distance from *L*, towards *K*, giving *F*, and the work of fixing the central point is done. Run a furrow from *A* and *D* to *E*, and from *B* and *0* to *F*; these define the corners and assist in the turning of the plow. The plowing then begins by back-furrowing from *E* to *F*; plowing on the ends as soon as possible. After the work has progressed for a time, as far as indicated, for example, by the dotted lines, *G*, *H*, *I*, *J*, pace from the furrow to the outside (see dotted lines), at or near each end of the furrow, as a correction, and, if necessary, gauge the plow until the furrow on all sides is equally distant from the boundary. When the field is of irregular shape, it is not difficult to begin in the centre and plow outward —in fact, this system is of most importance here, because all the short turning in the middle of the field, incident to the irregularity of the field, comes on unplowed ground.

In figure 74 we have a piece of very irregular shape. From a point on *A D*, at right angles to *B*, pace the distance to *B*, and place a stake at the middle point, *E*. In the same way, determine the point *F* on the line *N D*. In a line with *E*, *F*, measure from *K* a distance equal to *M E* (one-half the perpendicular distance across the end of field), and also in like manner determine the point *F*—which gives the central line, *E F*. The plow should be run from the four corners, as in the first case, to make the corner lines. The plowman will use his judgment, and plow only upon the lower portion at first, until the plowed land takes the shape *G*, *H*, *I*, *J*, when the correction is made. From this time on the furrow runs parallel with the boundary, and the work continues smoothly to the end.

A WIRE-FENCE TIGHTENER

Having occasion recently to tighten some wires in a trellis, we made use of the following contrivance. Into a

FIG. 75. Wire tightener.

small piece of wood a few inches long we put two screws about three inches apart, and near to one end one other screw, leaving the heads projecting about half an inch. By placing the wire between the two screws, and turning the piece of wood around, the wire was drawn tight; and by engaging the head of the single screw upon it, the tension was maintained. The operation of the contrivance is shown at 1, and the method of arranging the screws or pins appears as 2. By using a strong piece of wood two feet long, and strong iron bolts, fastened with nuts upon the back side, this device may be used to tighten fence wires.

PLANTING CORN—A MARKER

What would be thought of a mechanic who should rip his boards from a log with the old-fashioned whip saw and plane them or match them by hand, or who should work out his nails on the anvil one at a time by hand labor? He would hardly earn enough to find himself in bread alone. Yet in an equally old-fashioned, costly, and unprofitable way do thousands of farmers plant and cultivate their corn crops. The ground is plowed, harrowed

Fig. 76 Runner and tooth for marker.

and marked out both ways, either with the plow, or sometimes by a quicker method, with a corn marker. The seed is dropped by hand and covered by hand with a hoe; the crop is hoed by hand or plowed in the old method, leaving the ground ridged and deeply furrowed, so that in a dry season the corn suffers for want of moisture. All this costs go much that the farmer's labor brings him about fifty cents a day, upon which he lives, grumbling that "farming does not pay." This method would be ruinous in the West where corn is a staple crop, and that it is not so in the East is simply because it is not grown to a large extent. But there is no crop that may be grown so cheaply and easily in the East that produces so much feed as corn. Fifty bushels of corn and four tons of fodder per acre contain more dry nutriment than thirty tons of turnips or mangels, and may be grown with less labor and less cost, if only the best methods are employed. Now, with the excellent implements and machines that are in use for planting and cultivating corn, no farmer can afford to work this crop in the old-fashioned method. There is no longer any need to plant in squares, for the crop may be kept perfectly clean when planted in drills, if the proper implements are used. There are several corn planters by which the seed may be dropped

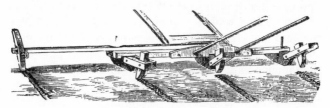

FIG. 77. The marker at work.

and covered at the same time in single or double drills, at the rate of eight to twenty acres per day. By using the Thomas harrow a few days after planting, every young weed will be killed, and the crust, which so often gathers upon the surface, will be broken up and the surface mellowed. The harrow may be used without damage until the corn is several inches high. Then anyone of the many excellent horse hoes may be used by which the weeds may be cut out of the rows close to the corn until the crop is so high that farther working is useless. This method of cultivation may cost two dollars per acre, or less, as the ground may have been kept free from weeds in previous years, while on the old-fashioned system it may cost ten dollars per acre, or more, as the weeds may have been allowed to get further ahead.

Nevertheless, there are farmers who will still work on the hand-to-mouth plan, and will still mark out their crops by a marker and drop the seed by hand. For these it will be convenient to have at least a good marker. It will mark uneven as well as level ground; it can be set to any width between rows; any farmer or smart boy can make it, and the inventor, who is a farmer in Canada, does not propose to patent it. The marker is made of two by four scantling, one piece being eight feet long. In this five holes are bored, one for each of the runners, one and one-eighth inch in diameter. The runners are also of two by four timber, and eighteen inches long. Holes one and one-eighth inch in diameter are bored through the runners, in which are placed hard wood pins fourteen inches long. These are driven in from the bottom, the ends being left broad, so that they may not pass through the holes, and projecting an inch and a half. This is shown in figure 76. The small pin which passes through the larger one serves to connect the runner with the principal timber, and by shifting the large pin from one hole to another, the runners may be brought from four feet to one foot, or even six inches apart, and made to mark rows of widths increasing by spaces of six inches up to four feet. When one of the markers meets with an obstruction it is lifted by it, as seen in figure 77, and passes over it. A guide marker is fixed by a hinge to one of the outside runners, and carries a scraper which is held in place by a pin, by moving which the distance of the next row may be regulated. A pair of light shafts may be attached to the marker, and a pair of handles by which it may be guided.

FEED TROUGH AND HALTER

The trough rests on the floor and is four feet long. *A*, *A*, are inch auger holes; a rope, four feet long, is put through them and tied. Another rope, *D*, has a ring

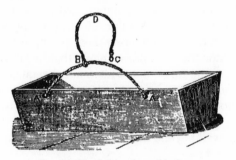

FIG. 78. Feed though and halter.

spliced on one end, and a "snap hook" on the other. The longer rope passes through the ring, *B*, and when the rope, *D*, is put over the neck of the cow, the "snap," *C*, hooks into the ring. This allows the animal to stand or lie down with comfort.

THE HORSE-SHOE AND ITS APPLICATION

Any excess of growth at the toe renders the pasterns more oblique, and, as a consequence, throws undue weight upon the "back sinews," whereas, too great height of heels has a similar effect upon the joints of the extremities, by rendering them too upright. Taking as our guide the foot of the animal that has never been brought to the forge, and which, in consequence, must be considered as a correct model, let the external wall of the hoof be reduced by means of the rasp to a level with the firm unpared sole. If there is no growth of the external wall beyond this level, then there is nothing to be removed.

In the selection of a shoe for the healthy foot, we must bear in mind the object in view, which is to protect the parts from excessive wear. This protection is to be found in a metallic rim of proper size and shape, securely adjusted. Almost every shoe in common use meets this end more or less satisfactorily, and we have already remarked that the proper preparation of the foot that has been previously shod is of vastly mere importance than the particular kind of shoe to be adopted. At the same time, there are faults in the shoe most commonly employed, which had their origin in its particular adaptation to the foot after this had undergone more or less severe mutilation at the hands of the farrier, and which have been retained more through custom than through actual necessity, as we have reason to hope. The most prominent of these faults consists in extreme narrowness of rim with a concavity upon the upper or foot surface, in order to prevent the sole from sustaining least weight or pressure, which it is perfectly unfitted to do after being pared down to a point of sensitiveness. In a state of nature we know that every portion of the foot comes to the ground and sustains its share of weight, and in the shod state it should do the same, as far as practicable. Hence, the shoe should be constructed with its upper surface perfectly flat, and with a breadth sufficient to protect a portion of the sole, and to sustain weight. It should be bevelled upon the ground surface, in imitation of the concavity of the sole, and not upon its upper surface, where the space thus formed serves as a lodging place for small stones and other foreign bodies. In shape it should follow the exact outline of the outer wall, being narrowed at the heels, but continued of the same thickness throughout. The lateral projection at the quarters, and the posterior one at the heels are unsightly, of no benefit, and should never be allowed where speed is required.

HOW TO MAKE A
FISHING SCOW

Boat-building should be done during the winter, when in-door work is more agreeable, and leisure is more ample, than in the summer. A boy who can handle tools, may make a very handsome boat or scow, such as is

Fig. 79. View of fishng scow.

shown at figure 79, at a cost of five dollars or less, in the following manner. Procure five three-quarter or half-inch clear pine boards, twelve feet in length and eight inches wide; four boards ten feet long, one inch thick, and one foot wide, and three strips ten feet long, one and a quarter-inch thick, and three inches wide Plane all these smoothly on both sides, and have them all free from loose knots or shakes. Cut two of the one-inch boards sloping at each end to a straight line for two feet, and then slightly rounding the middle of the board. Cut two pieces of the one and a quarter-inch strips into lengths of two feet ten inches, and nail them to the ends of the side-boards, as shown in figure 80. If strips of soft brown paper are dipped into tar and placed between the joints, they will be made closer and more water-tight. Cut the eight-inch boards into three feet lengths, and nail them across the bottom, as shown in figure 80; where

FIG. 80. Putting on the bottom.

the bevel ends, the two bottom boards must be bevelled slightly upon one of their edges, so as to make a close joint. Then take two of the one and a quarter-inch strips, and make cuts in each on one side with the saw, one inch deep, as follows: measuring from one end, mark with a pencil across the strip three feet ix inches from the end; then mark again across the strip one inch and a half from the first mark, and score between these marks with an x. Then measure three inches and make another mark, and then an inch and a half and make still another mark, and score as before between these last two with an x. Then do precisely the same on the same side of the strip, measuring from the other end. Then on the edges of the board score with gauge or make a line with a pencil exactly one inch from the marked side. Then make the cuts on the pencil lines down to the score on the edge, just one inch deep, but no more. Cut away the wood in the places that were marked with an x, leaving four slots one inch and a half wide, one inch deep, and with three inches between them upon each strip. Nail these strips with the cut side inwards, to the upper edge of the side-board, on the outside of the boat, as seen in figure 81. The spaces left in the gunwales are for the rowlocks. The strips should be well nailed near the rowlocks, and if a quarter-inch, flat-headed, counter-sunk carriage-bolt were used on each side of them, it would be very much better than so many nails. A thin washer, or burr, should be used beneath the nut of each bolt. The row-lock pins should be made of hard maple or oak, in the shape shown at *a*, figure 81. They are one inch thick, one

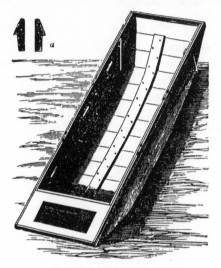

Fig. 81. Interior of boat.

and a half inch wide at the lower part, which fits into the slot, with a shoulder of half an inch, and the top is bevelled off neatly as shown. The seats, of which there are two, are made ten inches wide. The cleats for the seats, one inch thick, one and a half inch wide, and ten inches long, are nailed three inches below the upper edge of the side-board. The middle seat goes exactly in the centre of the boat, with each edge four feet seven inches from the end of the boat. The end seats are placed with the backs two feet from the ends of the boat, leaving eight inches between each seat and the edge of the row-lock nearest to it. There are cleats for three seats, but only two seats are used at once. When one seat is used, the rower sits in the centre, and he can use either of the rowlocks, the boat being double-bowed. When two seats are used, one person only rows at one time, but either can row without changing seats, and one always faces to the direction in which the boat moves. This arrangement of seats is very convenient. Eighteen inches of each end is closed in, and makes a locker for holding fish-lines, hooks, or the "painter," which is a light rope for tying up the boat when not in use. This may be fastened to a ring-bolt or a hole bored in one of the locker covers. The long bottom-board, seen in figure 81, eight inches wide and half an inch thick, is nailed as shown, by wrought nails driven from the outside and clinched on the inside. The seat cleats are nailed in the same manner, as are also the side strips. Every nail is counter-sunk and the hole filled up with putty. The seams are puttied or filled with a strip of cotton sheeting pushed in with the blade of a dinner knife. If the joints are made as well as they may be, this is not needed, but two coats of paint will make all tight. The inside should be painted lead-color, made by mixing lampblack with white paint to a proper shade. The outside may be painted white or a light-green, with the gunwale of a light-blue. A few days will be required to harden the paint before using the boat. None but seasoned boards should be used.

CROWS AND SCARECROWS

Probably there is no point upon which a gathering of half a dozen farmers will have more positive opinions than as to the relations of the crow to agriculture. It is likely that five of these will regard the bird as totally bad, while the minority of one will claim that he is all good. As usual, the truth lies between the extremes. There is no doubt that the crow loves corn, and knows that at the base of the tender shoot there is a soft, sweet kernel. But the black-coated bird is not altogether a vegetarian. The days in which he can pull young corn are few, but the larger part of the year he is really the friend of the farmer. One of the worst insect pests with which the farmer, fruit-grower, or other cultivator has to contend is, the "White Grub," the larva of the "May Beetle," "June Bug," or "Dor-Bug." It is as well established as any fact can be, that the crow is able to detect this grub while it is at work upon the roots of grass in meadows and lawns, and will find and grub it out. For this service alone the crow should be everywhere not only spared, but encouraged. We are too apt to judge by appearances; when a crow is seen busy in a field, it is assumed that it is doing mischief, and by a constant warfare against, not only crows, but skunks, owls, and others that are hastily assumed to be wholly bad, the injurious insects, mice, etc., that do the farmer real harm have greatly increased. Shortly after corn is planted, the crows appear, and are destructive to young corn. Some assert that the crow pulls up the corn plant merely to get at the grub which would destroy it if the bird did not. How true this may be we do not know, but as the corn is destroyed in either

case, it may be as well to let it go without help from
the crow. The first impulse of the farmer, when he finds
his corn pulled up, is to shoot the crow. This we pro-
test against. Even admitting that the crow does mischief
for a short time, it is too useful for the rest of the year
to be thus cut down in active life. Let him live for the
good he has done and may do. It is vastly better to keep
the crown from pulling the young corn, for two or three
weeks, and allow them all the rest of the year to destroy
bugs and beetles in astonishing numbers. The corn may
be protected by means of "scarecrows," of which there
are several very effective kinds. Crows are very keen,
and are not easily fooled; they quickly understand the
ordinary "dummy," or straw man, which soon fails to
be of service in the corn-field. It has no life, no motion,
and makes no noise, and the crow soon learns this and
comes and sits upon its outstretched arm, or pulls the
corn vigorously at its feet. A dead crow, hung by a swinging cord to a long slender pole, is recommended as far
better than a straw man—as it, in its apparent struggles
to get away, appeals impressively to the living crow's
serve of caution. But the crow may not be at hand to be
thus employed, and if it were, the farmer cannot afford
to kill it. Better than a dead crow is a glass bottle with
the bottom knocked out, which may be done with an
iron rod. The bottle is suspended to an elastic pole by
a cord tied around its neck; the end of the cord should
extend downward into the bottle, and have a nail fastened to it and within the bottle, to serve as a clapper.
If a piece of bright tin be attached to the cord extending
below the bottomless end of the bottle, all the better.
A slight breeze will cause the tin to whirl, and, in the
motion, cast bright reflections rapidly in all directions,

while the nail keeps up a rattling against the inside of the bottle. An artificial "bird," to be hung in the same manner, may be made from a piece of cork—one used in a pickle-jar—into which a number of large goose or chicken feathers are fastened so as to roughly imitate a dilapidated bird. A rough head may be carved and put on, to make the deception more complete. As this "bird" catches the wind, it will "fly" here and there in a peculiar manner not at all enticing to the corn-loving crows.

FLOOD FENCE

The weak point of a fence is where it crosses a stream; a sudden freshet washes away loose rails, and a gap is

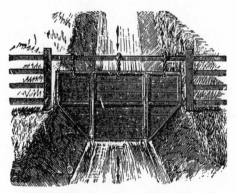

Fig. 82. A flood gate.

left through which trespassing cattle soon find a passage. Many devices have been used. The one shown in figure 82 is self-acting: when water rises high enough, it opens, and when the flood falls it closes again. It may be made of rails, bars, or fence strips.

CLEARING SLOUGH LAND

In clearing up land that is covered with tussocks of coarse grass and a tough sod, and digging out ditches to drain such land, much useless labor may be given that could be spared by skilful work. The spade is commonly used for this purpose, but, as in digging *dry* ground, this slow tool may be replaced to very great advantage by the plow and the horse-shovel. In working in swamps these more effective tools may be made available in many cases. To cut off the tussocks with grub-hoes, while they are tough in the summer time, is very hard and slow work; but if a common horse-scraper is used they can be torn up, or cut off, with the greatest ease. The scraper should be furnished with a sharp steel-cutting blade in the front, which may be riveted on, or fastened with bolts, so that it may be taken off and ground sharp. If there are wet and soft places the scraper may be drawn by a chain of sufficient length to keep the horse upon dry ground, as shown in figure 83. This plan has been

FIG. 83. The horse-shovel at work.

tried by the writer with success, and with a great saving of time and expense; the digging of a pond twenty feet wide along the edge of a swamp, was performed with one man, a boy, a team, and a horse-shovel, as quickly as ten men could have done it with spades. In cutting tough swamp, the plow may be used to break up the surface when the horse-shovel will remove the muck very fast. If the swamp is wet, and water flows in the excavation, the digging may still be done with the horse-scraper by adding to the length of the handles and using planks upon each side for the man to stand upon, and planks upon the inner side of the excavation for the scraper to slide upon with its load of muck. The muck may be thrown in heaps on the side of the pond or ditches, and it will be found convenient to leave it upon one side instead of in a continuous heap, as this will greatly facilitate its final disposal in whatever way that may be.

HOW TO DRESS A BEEF

There is a way of slaughtering that is not butchering, and it may be done painlessly by taking the right course. The barn floor or a clean grass-plot in a convenient spot

Fɪɢ. 84. The proper place to strike.

will be a suitable place for the work. To fasten the animal, put a strong rope around the horns, and secure the head in such a way that it cannot be moved to any great distance, and in a position to allow a direct blow to be easily given. The eyes may be blinded by tying a cloth around the head so that there will be no dodging to avoid the stroke. The place for the stunning blow is the centre of the forehead, between the eyes and a little above them. The right place is shown at a, figure 84. The best method is to fire a ball from a rifle in the exact spot, and this may be done safely when the animal is blinded, by holding the weapon near to the head, so that a miss cannot be made; otherwise a blow with the back of an axe made when the striker is on the right side of the animal, and the head is fastened down near the ground, will be equally effective. So soon as the animal falls, the throat is divided with a cut from a long, sharp knife; no jack-knife should be used, but a long, deep, sweeping stroke which reaches to the vertebræ as the head is held back This divides all the blood-vessels, and death is almost instantaneous, but at any rate painless. When the carcass has been freed from blood, it should be turned on its back, and the skin divided from the throat up the brisket, along the belly to the legs, and up the legs to the knees, where the joints should be severed, taking care, however, to cut off the hind feet below the hock joints about two or three inches. The skin is then stripped from the legs and belly, and as near to the back as may be by turning the carcass. The belly is then opened, and the intestines taken out; the brisket is cut through, and the lungs and gullet removed. It is now necessary to raise the carcass. This is done on the rack, the forward legs of which are placed on each side of the carcass, and the gambrels are placed upon the hooks shown in figure 85. The legs of the rack are then raised as far as

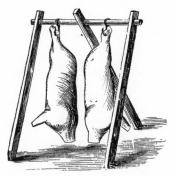

Fig. 85. Rack for a carcass of beef.

possible, and as the carcass is lifted, the hinder leg is brought up to hold what is gained until the carcass is clear of the ground; the hide is then wholly removed, the carcass washed and scraped from anything adhering, and then divided down through the backbone, leaving the sides hanging. As a matter of safety from dogs or other dishonest animals, it is well to have the work done in the barn, laying down a quantity of straw to protect the floor, if thought necessary, as the beef should remain at least twelve hours to cool and set.

A FARM CART

While there are different kinds of farm carts, we have not yet hit upon *the cart—i. e.*, one that meets with general approbation. The writer, having much work for a cart, has designed one which is intended to do all the work of the farm more easily than a wagon or any other cart. For the carriage of manure, of fodder-corn, green clover, or other soiling fodder, for hauling roots and such work, a cart is needed with a low body, that can be turned around in its own length to back, or even turn in a manure cellar or on a barn floor. All this can be done with this cart, and when hay, straw, or green fodder is to be loaded, the fore and hind racks may be put on, and greatly increase its capacity. With four-inch wheels, this cart can be drawn, when loaded, over plowed ground or muddy roads, and scarcely sink below the surface. The cart body consists of a frame eight feet long, three and a half feet wide, and fourteen inches deep, thus holding, when heaped, about a cubic yard and a half of manure, or as near as possible one ton. The frame is made of three by four timber for the top, and two by three for

FIG. 86. Axle fastening.

the bottom, sides, and cross-bars, and is covered with basswood or willow boards on the bottom, the front, and the sides near the wheels. The rear end is closed when desired by a sliding tail-board. The axles are fixed to the frame, as shown in figure *86*, and pass through the middle side posts under the upper slide bar and a wide iron strap, which embraces the top of the frame, and passes under the bottom, as shown in the engraving, being screwed by bolts to the timbers. The wheel is the same size as ar ordinary wagon wheel, viz., four feet; this brings the bottom of the cart body to within one foot of the ground, and in loading, the lift is only a little more than two feet from the ground. The saving of labor and the effect of work are thus greatly increased, a man being able to load twice as much with the same force, into a cart of this kind, as into a wagon-box four feet high. The rear end of the cart may be provided with a roller, fitted into the rear posts, which serves to ease the unloading of the cart when it is tipped, the rear end then easily moving over the ground as the cart is drawn over the field when unloading manure. But as the cart body is so low there will rarely be any need for tipping the cart. To enlarge its capacity, there are movable racks fitted before and behind, as shown in figure 87. The cost of two of these carts is not more than

Fig. 87. The cart with movable racks.

that of a wagon, and may be less, if economy is exercised in making them. The shafts may be bolted to the sides and so arranged that the cart can be tipped over when the load requires it.

BRACES FOR A GATE POST

On the side of the post, and near the surface of the ground, spike an inverted bracket, made of a two-inch plank of white oak, or other hard wood. The bracket

FIG. 88. Bracing a gate post.

should be not less than six inches wide, and a foot long. There should be two of these braces, one on the gate side of the post when the gate is shut, and one on the gate side when open. Under the bracket place a flat stone firmly settled in the ground, on which the bottom of the bracket is to rest; a piece of plank, as long as it lasts, will do instead of the stone.

The hang of the gate can then be exactly adjusted by putting a thin stone or piece of wood between the bottom of the bracket and the flat stone or plank. This is a simple and effective method of supporting a post, where there is no other convenient way of bracing, and even in almost all cases, it gives additional firmness. If the lower end of the post is of good size, and is well put in, this method of bracing will hold a very heavy gate.

WHIPPLE-TREES FOR
PLOWING CORN SAFELY

We have found it beneficial to cultivate our corn crop until the rows become impassable for a horse, or until it was four feet or more high. But to do this with the

FIG. 89. Whipple-tree.

wide whipple-tree, the ends of which project beyond the traces, and break down the stalks, is impossible. It may, however, be done by using a whipple-tree specially provided for it. This is made as follows: a piece of oak timber, two inches thick, three wide, and twenty inches long, is rounded at the corners, and deeply grooved at the ends, so that the trace-chains may be entirely imbedded in the grooves. A small hole is bored through each end, into which a small carriage bolt is inserted, being made to pass through a link of the trace-chain, and it is then fastened beneath with a nut. The trace-chains should be covered with leather where they, will rub against the corn, and a flap of leather should be left to cover the front corners of the whipple-tree, as shown in figure 89. A ring or an open link is fastened at the part of the chain which is attached to the clevis, and one at each end by which it is hooked to the traces. With this arrangement one may cultivate his corn without injury, and the same method may be applied to the whipple-trees, for plowing or cultivating amongst trees in the orchard or garden.

WHAT TREES TO PLANT FOR FUEL AND TIMBER

The attention of our people in the older States is being very properly turned to planting rocky ridges and worn-out pastures with forest trees. This work is done by those who have no expectation of cutting the timber themselves, but with a view to improve their property for future sale, or for their heirs. These old pastures now are worth $10, or less, per acre. Forty or fifty years hence, covered with heavy timber, they would be worth three hundred dollars, or more, per acre. Two elements may safely enter into this calculation of the profit of tree planting the steady growth of the trees, and the constant increase in the price of fuel and timber. There is great difference in the price of the varieties of wood, but still more in the rapidity of their growth. Hickory grows more rapidly than white oak, and in most markets is worth a quarter more for fuel. Chestnut grows about three times as fast as the white oak, and for many purposes makes quite as good timber. It is in great demand by ship-builders, and cabinet-makers. The chestnut, the tulip tree, and the hickory attain a good size for timber in twenty to twenty-five years, and the spruce and pine need about fifty years. The maples grow quite rapidly, and are highly prized, both for fuel and for cabinet purposes. On light sandy land, the white pine will grow rapidly, and cannot fail to be a good investment for the next generation. As a rule, the more rapid growing trees, if the wood is valuable, will pay better than the oaks.

TO STEADY PORTABLE MILLS

Figure 90 shows a contrivance for steadying portable mills, which has been used for several years. It is an

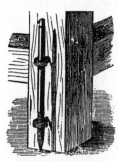

FIG. 90. Leg of mill.

iron rod of suitable size, about a foot long, fastened by iron brackets to the leg, or post of the mill. Three or four inches of the rod is a screw, and fits one of the brackets through which it runs, and can be turned up or down. The lower end of the rod is pointed, and the upper end squared, that it may be turned with a wrench. The rod is fastened firmly to the side of the post (one on each of the forward posts), and turned down so that the point shall enter the floor sufficiently to hold it firmly.

SPLITTING RAILS AND POSTS

Autumn is the best season for cutting timber, as many farmers have learned by experience. The seasoning process is much more perfect, because there is no layer of growing sap wood. Insects do not work in autumn cut timber, as in that cut in the spring or summer, and the wood does not "powder post." It is best to split the logs into rails or posts at once, and not wait until the timber has become seasoned in the log. The logs will split easier, the rails will season quicker, and be more durable. The splitting of rails is a work that requires good judgment, otherwise much timber will be wasted. Some persons will make rails that are large at one end, and gradually tapering to a sliver at the other, and are worthless for fencing purposes. Set the wedge at the top end of the log, after first "checking" with the axe, by driving with the beetle, so as to divide the log into two equal parts. Now drive in two wedges, as shown in figure 91, both at the same time. Next use a wooden wedge or "glut," either in the end of the log, or on the top a little back from the end. After halving the log, quarter it, and then proceed on the principle that a rail should be about three by three inches. The size of the log will determine the number of rails to be made. For example, in figure 92, six rails are made by first halving the quarter, then splitting off the inner part half-way

Fɪɢ. 91. Position of wedge.

from the centre, and afterwards halving the outer part. Should the logs be larger, twelve rails are made from each quarter, as shown in figure 93,—or forty-eight rails from the log. In splitting logs into posts, a broad and smooth side is to be sought. Suppose we have the same sized log as the one split into forty-eight rails, or twelve rails per quarter, figure 93—the splitting would be, in each case, from the centre to outside with cross splitting midway. The number of posts would be determined by the size of the posts desired. If the logs are of the size of the quarter, shown in figure 92, there is no cross splitting, unless a small piece for a stake is taken from the centre. When the logs are only large enough for four posts, and a broad surface is desired, as in bar posts, they may be split by first "slabbing," and afterwards splitting through the centre; all the split surfaces

FIG. 92.

FIG. 93.

to be parallel. If still smaller, three posts can be made, by splitting off two slabs on opposite sides, as in the case above, and not divide the heart, and finally when the log will make only two, it can be halved.

A MIXTURE OF GRASSES

It is a well-known fact that mixed crops are more productive than those sown singly. Thus one acre sown to oats and barley, or oats and peas, will yield as much, or nearly as much, as two acres sown singly to either crop. So in grass lands, Clover and Timothy, mixed, will produce nearly twice as much as if the ground were seeded to one of these alone. It is also a well-known fact that our grass lands are not so productive as we could wish, and the reason of this may be, and probably is, that we have but one or two kinds of herbage in them. If we examine an old, thick, luxuriant sod, in a pasture or a meadow, it will be found to consist of a variety of grasses and other plants, each of which seems to vie with the other in occupying the soil for itself. This is the result of natural seeding, and gives us a lesson which we may well profit by. There is another reason why grasses should be mixed; this is that the periods of greatest vigor of different varieties occur at different times. We can therefore secure a succession of herbage for a long season by sowing a variety of grass seeds.

To give examples, we might mention that a mixture of Orchard Grass, Red Clover, Timothy, and Kentucky Blue-Grass will produce a pasture which will be in good condition for grazing from April, when the first mentioned grass is in fine condition, up to October, when the last is in its most vigorous state; the Clover and Timothy serving to fill up the interval. With one of these alone there would be but one month of good herbage, and that coarse, if given the whole field to itself. In like manner, a quantity of Rye Grass added to a meadow would help to furnish a quick growing herbage which rapidly and constantly recuperates after cutting or eating down.

The fact is, that we make much less of our advantages in regard to our meadows and pastures than we might. On the average, seven acres of pasture are required to keep one cow through the pasturing season, when by the best management one acre, or at the most two, ought to be sufficient. This is due in great measure to the prevalent fashion of seeding down with but one variety of grass, with clover added sometimes, a fashion which, hereafter, experience teaches us should be more honored in the breach than in the observance.

HITCHING A CRIB-BITER

Those persons who have a horse that is a crib-biter and windsucker, and which practices his vice when hitched to a post in the street, is recommended to try a hitching-rod, such as shown in figure 94 It consists of a 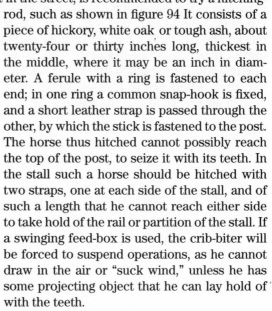 piece of hickory, white oak or tough ash, about twenty-four or thirty inches long, thickest in the middle, where it may be an inch in diameter. A ferule with a ring is fastened to each end; in one ring a common snap-hook is fixed, and a short leather strap is passed through the other, by which the stick is fastened to the post. The horse thus hitched cannot possibly reach the top of the post, to seize it with its teeth. In the stall such a horse should be hitched with two straps, one at each side of the stall, and of such a length that he cannot reach either side to take hold of the rail or partition of the stall. If a swinging feed-box is used, the crib-biter will be forced to suspend operations, as he cannot draw in the air or "suck wind," unless he has some projecting object that he can lay hold of with the teeth.

Fɪɢ. 94.

HOW TO INCREASE VEGETABLE MATTER IN THE SOIL

The amount of vegetable matter in the soil may be increased by various methods; one is by large applications of barn-yard manure, say fifty cords to the acre. But this would be very expensive, and is out of the question in common farming. It may be done by putting on peat or muck, when these are near to the fields. But this involves a considerable outlay for labor in digging the peat, and a still larger expense in carting it, whether it first pass through the yards and stables, or be carted to the fields for composting or spreading upon the surface to be plowed in. On some farms this may be the cheaper method of supplying vegetable matter to the soil. But on others the most economical method is the raising of clover, to be fed off upon the land, or to be turned in. If a ton of clover may be worth nine dollars, as a fertilizer, the growing of the plant is a cheap method of improving the land. Two tons for the first crop and a ton for the second is not an uncommon yield for land in good heart. The roots of clover also add largely to the vegetable matter in the soil. The first crop may be pastured, waiting until the crop is in blossom, and then turning in cattle enough to feed it off in three or four weeks. They should be kept constantly upon the field, that the whole crop may be returned to the soil. This will, of course, help the second crop, which maybe turned in with the plow soon after it is in blossom. If the equivalent of three tons of dried clover hay, and one ton of roots have been grown to the acre, about thirty-six dollars' worth of manure have been added to the soil, and it has been distributed more evenly than would have been possible by any mechanical process. There has been no expense for carting and spreading peat, or for composting. On the contrary, there has been the equivalent of two tons of clover-hay consumed upon the field, worth, as fodder, twenty-four dollars. This will more than pay the cost of seed, of plowing twice and other labor. This is generally admitted to be the cheapest method of increasing the vegetable matter and the fertility of soils in common farming. And this, it will be seen, requires some little capital.

OPEN LINKS

An open link, shown in figure 95, is made of three-eighth inch iron rod, and when used to connect a broken chain, is simply closed by a blow from a hammer or a stone. There being no rivet, the link is not weakened in any way. Figure 96 shows another link, made of malleable cast-iron, in two parts, which are fastened together by a rivet in the centre. A few of these links may be carried in the pocket, and are ready for instant use in case of an emergency. The last-mentioned links are kept

Fɪɢ. 95. Common link.   Fɪɢ. 96.

for sale at the hardware stores; the first named may be made in a short time by a blacksmith, or any farmer who has a workshop and a portable forge.

CARE OF THE ROOT CROPS

Sugar beets and mangels, if early sowed, will need little care. They ought not to stand too thick, however, and it would certainly pay to go through the rows, thinning out all superfluous plants, whether beets or weeds, leaving the plants six to eight inches apart. If the leaves are not so large as to forbid horse-hoeing, this should be done and the crop "laid-by." No root crop should ever be left after horse-hoeing, without a man going through it immediately after, to lift and straighten up any plants which may have been trodden Upon, covered with earth, or injured in any way. Rutabagas, and any turnips in drills, need the same general culture. One of the great advantages of the introduction of roots into the rotation is that, when properly treated, no weeds ripen seeds. Even red sorrel and snapdragon succumb to two or three years' cropping with mangels or Swedes. This advantage is often lost by careless cultivators, and nothing offers surer evidence of heedless farming. The crop itself may be very fine, but if kept clear of weeds it would be enough better to pay for the trouble, and the weeds would then be where they will make no more trouble forever.

Turnips may be sown as late as the middle of August, but the land should be in good heart, and good tilth. Swedish turnips (rutabagas) sowed as late as the first of August, will usually make a crop delicious for the table, and, though small, bring a good price. Thus they are often used to follow early potatoes by market gardeners, though by them usually regarded as a farm crop.

TRAP FOR SHEEP-KILLING DOGS

In many places the losses by dogs are so great as to prevent the keeping of sheep altogether; thus this profitable and agreeable industry is made impracticable over the greater portion of the country; unless such precautions are taken as will add greatly to its trouble and cost. With small flocks only, this extra cost and trouble are too onerous, and it is only where sheep are kept in large flocks that it will pay to employ shepherds to constantly watch them, or take other necessary precautions. In several of the States—West Virginia and Tennessee more particularly—very stringent laws have recently been enacted for the protection of sheep against dogs, which will go far to encourage the raising of flocks. In other States, where the influence of the owners of dogs is of more weight than that of sheep-owners, these latter are obliged to look out for themselves, and protect their sheep as they may be able. For such the contrivance here described and illustrated, may be useful. It is made as follows: In the meadow or field, where sheep are pastured during the day, a small pen, eight feet square, is made, and fenced strongly with pickets or boards. This pen is divided into two parts (*A*, *B*, figure 97) by a cross-fence. The pen is wholly covered over on the top with strong lath. Two gates (*a*, *b*) are made so that they will swing open of their own accord, and remain so, unless held closed or fastened. The gate, *a*, is furnished with a latch, by which it is fastened when closed. This gate is intended to admit the dog into the part of the pen, *A*, when he is attracted to it by a sheep confined for the

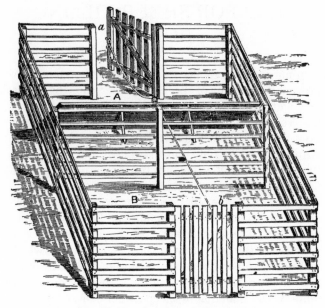

FIG. 97. Trap for sheep-killing dogs.

purpose in the other part of the pen *(B)*. In the part of
the pen, *A*, is a heavy board, reaching across it. One
edge of this board rests upon the ground against two
pegs, which keep it from slipping backwards. The other
edge is kept up by means of two shaky slender supports.
A rope is fastened to the upper edge of this board, and to
the gates, so that one half of it, when the board is propped
up, allows the gate, *a*, to swing open, and the other half
holds the gate, *b*, shut, and thus keeps the sheep confined. The trap is now set. A dog, prowling in search of
mutton, finds the sheep, and seeks an entry into the pen.
He finds the open gate, and rushes in, over the board

set upon its edge, and knocks this down. This closes
the gate, *a*, which is at once latched and fastened. The
gate, *b*, is allowed to swing open, and the sheep is liberated, and, of course, proceeds homeward without delay,
while the dog is imprisoned. We need not suggest any
method of dealing with the prisoner, as there are many,
more or less effective, which will suggest themselves.
We think it would be an improvement upon this plan, if
the sheep be confined in the pen, where it can be seen by
the dogs, and an additional apartment, if not more than
one, made, in which other dogs could be trapped. Sheep-
hunting dogs usually go in couples, and if only one dog
were trapped, the sheep escaping from the pen would
be caught by the other before it could reach home. With
two or three traps all the dogs could be caught, and in
a short time the locality would be rid of them, or, being
identified, their owners could be made responsible for
their trespasses. It would be necessary to have the pen
made very strong, so that the dogs should not tear their
way out of the trap, or into the pen in which the sheep
is confined. Stout wire-netting would make a safe fence.
So far as regards what are called dog-laws, it would be
well if these should provide, amongst other things, that
every dog must wear a collar, bearing its owner's name;
that the owner of any dog which is caught in pursuit of
sheep upon the property of any person other than the
owner of the dog, should be held liable for damages for
the trespass, and that any dog caught trespassing, and
being without a collar bearing its owner's name, should
be destroyed by the person capturing it. As any citizen
has as much right to keep a dog as another has to keep
a sheep, without being taxed for it, and can only be held
liable for what damage his dog may do, it does not seem

just that any tax should be levied upon dogs. The only just claim that can be made by a sheep-owner is that he shall be protected in the enjoyment of his property, and that the person by or through whom he is injured should recompense him. In the case of irresponsible owners of dogs, from whom no recovery can be made, the dogs should be destroyed by a proper officer. If the right of persons to keep dogs, when they wish to do so, without being taxed, is recognized in this manner, much of the opposition to the enactment of what are called "dog-laws," would be removed, and the protection of sheep made much less difficult, and productive of much greater profit.

HOW TO USE A FILE
PROPERLY

The file is very frequently used in such an imperfect manner as to greatly reduce its value as a mechanical tool. The chief difficulty in using a file is in keeping it in a perfectly horizontal position as it is moved over the work, and in maintaining an equable pressure upon the work meanwhile. Perhaps the most difficult work in filing, and that which is most frequently ill-done, is in sharpening saws. The bearing of the file upon the work is very narrow, and unable to guide its direction, and unless the file is held very carefully the direction varies continually, so that the saw tooth is filed rounding instead of flat, or sloping instead of horizontal, or at exact right angles with the line of the saw, as it should be in a mill-saw or a rip-saw. When the file is held as shown in figure 98 (a very common manner of holding

FIG. 98. Improper use of the file.

it), it is almost impossible to do good work upon a saw. When the file is pushed on to the tooth, the weight or pressure of the right hand is exerted upon the longer portion of the tool, making it act as if it were the longer arm of a lever, and thus depresses that portion below the horizontal, as at *a*. When pushed forward, the pressure is then exerted upon the longer portion of the file, which is carried from the horizontal in the contrary direction. The work is thus made round. Or if the pressure of the left hand is guarded against, that of the right hand is seldom altogether controlled, and the work is left sloping, as in figure 99; the position at the commencement being shown at *a*, and that at the finish of the stroke at *b*. This is a very common error with sawyers in mills, as well as with many good carpenters in filing their rip-saws.

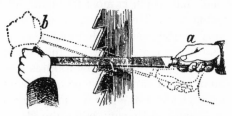

FIG. 99. Another wrong way of filing.

To avoid either form of this error, the file should not be held with the ball of the thumb pressing upon the

FIG. 100. Proper. method of filing.

handle of the file, as in figures 98 and 99; but the end of the file should be taken lightly between the thumb and fore-finger, as in figure 100. There is no uneven pressure in this case, and the direction of the file may easily be kept perfectly level. In filing the base of the tooth, or the under portion of any work which cannot be turned over, the end of the file should be supported upon the ends of the fingers, as in figure 101, or be held by the end of the thumb, in an easy gentle manner. If held lightly, and not grasped too firmly, the arm or wrist will not be tired so soon as when it is held rigidly; and the motion of the file will be more even and regular.

FIG. 101. Filing underneath.

When the arm is wearied by working in one direction, it may be rested by reversing the position of the file, taking the handle in the left hand, grasping the end between

FIG. 102. To rest the hand in filing.

the fingers and thumb of the right hand, and drawing the file towards the body, instead of thrusting it away from it. The file is then held as in figure 102. This is an excellent position in which to hold the file when finishing off a saw tooth, or when touching it up at noon.

A MITRE-BOX

A mitre-box of an improved form is shown in figure 103. The greatly increased use of moulding in house

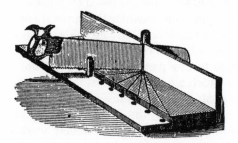

FIG. 103. Mitre-box.

building renders a mitre-box very necessary in the workshop. In the one here described, a bevel of any angle may be cut. At the rear of the box is a slotted post, which works in a socket, so that it will turn readily in any direction. From the post, lines are laid out upon the bottom at various angles. At the termination of each line is a round hole, into which a pin may be fitted. The pin is used as a guide for the saw in cutting a mitre-joint, as shown in the illustration.

THE MANURE HARVEST

In the midst of the harvest of grain, and grass, and tubers, we must not forget the compost heap, in which we garner and store the unsowed crops of a future season. The saying that "anything that grows in one summer will rot before the next," is a safe guide in collecting vegetable matter for the compost heap. When sods, muck, and weeds form a part of the heap, it is not alone the material which we are assiduous in collecting, and put into the heap, that constitutes its whole value. The fermentation induced by the dung and liquid manure, and the action of the lime or ashes added, work upon the earth, adhering to the roots of the weeds, and forming a considerable part of both sods and muck, and develop an admirable quality of plant food. Hence this element of the compost heap, which is generally overlooked as possessing any special value, should never be wanting. It has, moreover, its own offices to perform, in promoting decay, in the formation of humus, and in preserving, locking up, and holding on to valuable ingredients of plant food.

The compost heap should always be laid in even layers, and each layer should go over the entire heap for thus only can final uniformity be had. We do not mean special-purpose composts, but those made for general farm crops. It would be well if every particle of dung, liquid manure, straw, litter, leaves, weeds, etc., could be worked together into uniform fine compost, and there is really no substantial reason why this should not be done. The gardener would plead for certain special composts. It might, perhaps, be well to make a special hen-manure compost for corn in the hill, and taking the general compost as a basis, to make one for turnips, by the addition of a large percentage of bone-dust. All this may be done—establish once the rule to compost everything of manurial value, and we have in prospect an abundance of farm-made fertilizers at all times and for all crops—victory over weeds, a good place for decomposable trash of all kinds, a sacred burial ground for all minor animals and poultry, whose precincts need never be invaded. There will besides be no stagnating pool in the barn-yard, for all liquids will go to the tank, to be pumped over the compost heaps—no nasty, slumpy barn-yard, for everything will be daily gathered for the growing com post heap, and the harvesting of the manure crop, and its increase day by day, ail the year round, will be a source of constant pleasure to master and men.

FASTENING CATTLE
WITH BOWS

Everything connected with this method of fastening cattle in the stable, by means of bows, is so simple in construction, that it is within the reach of every farmer. It requires no outlay, as each one can make all the parts for himself. The bow, figure 104, passes around the animal's neck in the same manner as an ox-bow, and is made of a, good piece of hickory, by bending a strip of the right length, and three-quarter inch in diameter into the bow form. After the bow-piece, *A*, is made of the right size and shape, with one end left with a knob, to prevent the clasp from slipping off, and the other out as shown in front view in figure 104, *G*, and side view at *F*, to fit into the slot, in the clasp, it is carefully bent until its ends are brought together, fastened, and left so for a considerable time, when it will take its form and be ready for use. The clasp is shown at *B*, *D*, and consists

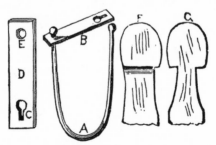

Fig. 104. Bow and clasp.

of a piece of hard-wood—hickory is best—three-quarters of an inch in thickness, and long enough to hold the

top of the bow well together. A round hole is bored in
one end, *E*, through which the bow passes as far as the
knob, the other end is cut with a hole for the passage
of the other end of the bow, and a slot, *C*, into which
its narrow neck springs when the bow is secured about
the creature's neck. A smooth, stout hickory pole, two
and a half inches in diameter, reaching from the floor to
the beam overhead, serves as a stanchion to which to
attach the animal, by means of a small bow, and stationary clasp, figure 105, or an iron ring, *A*. If a little more

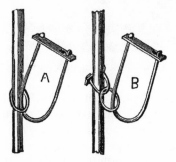

FIG. 105. Bow and attachment.

room is desired for young stock, a link or two of chain,
B, can be inserted between the bow and stanchion ring.
In fastening the cattle, the bow is raised when it passes
around the neck, and the clasp is brought on, and the
end of the bow is sprung in place. When the animal is
to be let loose, the end of the bow is pressed in, the
clasp slips off, the bow falls, and the work is done in
less time than it takes to describe it, and everything is
out of the way. Figure 106 shows a stable arranged for
this method of fastening; aside from the manger, there
is but the stanchion poles, one for each animal. There

FIG. 106. Stable showing bow and stanchions.

is sufficient freedom of movement of the head, but not
an excess; the animal can stand or lit down with perfect
comfort, as the bow moves with ease the whole length
of the stanchion. After a week's practice, the animals
will take their place with their heads by the side of the
stanchion, with a precision that is remarkable. Having
used the method, here illustrated, for several years, the
writer has found it inexpensive, easy in application, and
safe.

THE PRESERVATION OF WOOD

It is known that the decay of timber first begins through the fermentation of sap within the pores, and that it is continued after this by the absorption of water. The neutralization of the acids in the timber by the use of lime, has been made use of to preserve it from decay with success; but the most effective methods have been to saturate the pores with oils or mineral salts. Creosote and petroleum have been used successfully, but few persons are aware of the enormous absorptive capacity of timber for these liquids. Cypress wood, when dry, will absorb thirty-nine pounds, or five gallons of oil per cubic foot, and California redwood and pine absorb twice their weight when perfectly dry. But it is not necessary for perfect preservation that timber should be fully saturated. One gallon per cubic foot, for the most porous woods, will be fully effective, and a coating of one and three-quarters of a pint per square foot for weather boards, or half as much for shingles, will render them perfectly water-proof. In some careful experiments recently made, dry spruce absorbed, during two days' soaking, nearly two per cent, of its weight of water, and but one-tenth as much when treated with oil; dry pine absorbed sixteen per cent. of its weight, and oiled pine absorbed none that could be ascertained by careful weighing. Pine has proved to be the most easily water-proofed of any timber. Those who wish to preserve shingle roofs, will be able to draw their own inferences as to the usefulness of coating them with crude petroleum.

A NEST FOR EGG-EATING HENS

In the winter season hens frequently acquire the habit of eating eggs. Sometimes this vice becomes so confirmed that several hens may be seen waiting for another one to leave her nest, or to even drive her off, so that they may pounce upon the *egg*, the one that drops it being among the first to break it. In this state of affairs there is no remedy, except to find some method of protecting the *egg* from the depredators. The easiest way of doing this is to contrive a nest in which the egg will drop out of reach. Such a nest is shown in figure 107. It consists of a box with two sloping false half-floors; one of these being depressed below the other sufficiently to make a space through which the e*gg* can roll down to the bottom floor.

A door is seen in the side of the nest, through which the eggs can be removed. The sloping half-floors are shown by dotted lines. Upon the back one, close to the back of the nest, a glass or other nest-egg is fastened by a screw or by cement. The false floors may be covered with some coarse carpet or cloth, and the bot-

Fig. 107. Safety-nest.

tom floor with some chaff or moss, upon which the eggs may roll without danger of breaking. If the eggs do not roll down at once, they will be pushed down by the first attempt of a hen to attack them.

PLOWING GEAR FOR A KICKING MULE

Below is presented a plan for hitching a mule which has a habit of kicking when harnessed to a plow, but which goes very well in shafts. Kicking is a vice which sometimes belongs to horses as well as mules, and the following expedient has been found effective in curing it. Take a pair of light shafts from a wagon, or make a

FIG. 108. Plowing gear for kicking animals.

pair, and fit to the end of it a bent strap of iron, as shown in figure 108. When the mule or horse is hitched into the shafts the end may trail on the ground, and the beast may be exercised with the shafts alone. When used to these, the bent bar is fastened to a plow by means of a clevis, and any difficulty there will soon be overcome. This device has been used, not only for plowing, but for drawing a stone boat, railroad cars, and other similar vehicles.

A LEAF FORK

A useful plan for making a fork to gather leaves is shown in figure 109. The fork is made of tough ash, with ten teeth, similar to the fingers of a cradle, three feet long, and slightly turned up. The head into which the butts of the teeth are inserted, is thirty inches long. A light cross-bar of tough wood is fastened to the teeth, about eight inches from the head, by means of copper

FIG. 109. Fork for gathering leaves.

wire and a light screw to each finger. A handle is provided and fixed in its proper place, being flattened somewhat to keep it from turning in the hand. The handle should be braced by two strong wires. With such a fork leaves may be loaded very easily and rapidly.

PREPARATION OF THE WHEAT GROUND

Wheat demands for its perfect development, among other favorable conditions, besides showers and sunshine, depth and richness of soil, thorough tilth, and freedom from excess of moisture. Soil that will yield good clover will bear good wheat. Wheat follows corn very well, but this involves rather late sowing. Where there is a market for new potatoes, which, as they are intended for immediate use, may be freely manured, the potato ground —well plowed and harrowed with a dressing of bone-dust, superphosphate, or, if there is much organic matter in the soil, with a dressing of lime—forms an admirable seed-bed for wheat. One of the best rotations, including winter wheat, is corn on sod, early potatoes, wheat, clover and timothy, the grass to be mowed as long as it is profitable—the manure being applied in the hill for corn, and put on broadcast very liberally for the potatoes. Winter wheat follows none of the usual root crops well, for it ought to be sowed and up before the middle of September, although it often docs well sowed nearly a month later.

When wheat follows clover, a crop of clover-hay is often taken off early, and a second crop allowed to grow, which is turned under about the first of August for wheat. In case we have very dry weather in July, the growth of clover will be meagre. If, however, the clover stubble be top-dressed at once, as soon as the early crop is cut, with a muck and manure compost, or any fine compost, "dragged in" with a smoothing harrow, the second crop will be sure to start well, while none of the manure will be lost. Lime, or ashes, if they can be obtained, are to be spread after plowing under the clover and manure, and thoroughly harrowed in. Forty bushels of ashes to the acre is about right, and where hearths of old charcoal pits are accessible—ashes, charcoal-dust, and baked earth, are all excellent—they form a good substitute for ashes and for lime. Sixty to one hundred bushels of evenly dry-slaked lime is a usual application, which, if it could have been mixed with an equal quantity of soil or sods during the slaking, would be all the better.

The soil, and particularly wheat ground, is not well enough tilled in this country. We plow fourteen to sixteen-inch furrows, and use a skim-plow; this leaves the surface so mellow, and covers the sod so perfectly, that we think it hardly needs harrowing at all, and only smooth it over with a harrow, and let it go. The skim plow is a great advantage, but we should take narrow furrows.

The following practice, on heavy land especially, is excellent: Turn under the first crop of clover as deep as possible, just before it is in full blossom; cross-plow the first or second week in August; then put on seventy-five bushels of lime, or more, and harrow it in lightly. Sow early after a soaking rain, and apply at the time of sowing two hundred and fifty pounds or more of superphosphate to the acre.

HOW TO DRIVE A
HORSE-SHOE NAIL

Most farmers hesitate to attempt to fasten on a loose shoe for fear of injuring the foot by driving the nail in a wrong direction. It is such a saving of time and money to be able to put a shoe upon a horse in a hurried busy time, that every farmer ought to learn how to do it. He may practice upon a piece of soft pine wood in a rough way, when he will find how easy it is, by properly preparing the nails, to make the point come out in exactly the proper place. To prepare the nail it should be laid upon the anvil (which every workshop should have for such work as this), or a smooth iron block, and beaten out straight. The point should then be bevelled, slightly upon one of the flat sides, and the point also bent a very little from the side which is bevelled. It will then be of the shape shown in figure 110. In driving such a nail into a piece of soft wood, or a horse's hoof which is penetrated easily in any direction, if the bevelled side is placed towards the centre of the hoof and away from the crust, the point will be bent outwards, and will come out lower or higher on the crust as the bevel and curve is much or little. A little practice will enable one to cause the point to protrude precisely at the right place. By turning the bevel outwards, in driving the nail, the course will be towards the centre of the foot as shown by the line *b*, in figure 111. The nail is sometimes started in the wrong direction by careless blacksmiths,

Fɪɢ. 110. Nail.

FIG. 111. Driving nails.

and the horse is lamed in consequence. If the mistake is discovered, and an attempt made to draw out the nail, a piece of it may be broken off, and at every concussion of the foot the fragment will penetrate further, until it reaches the sensitive parts, and great suffering will follow. Many a horse is supposed to have navicular disease (because that happens to be one of those obscure affections of the foot which has no outward sign), when the trouble is a fragment of nail broken off by a bungling shoer. We have examined the foot of a horse which was killed because of an incurable lameness, and found a piece of nail thus bedded in the centre of the foot, surrounded with an abscess which had eaten into the bone. The torture suffered by this horse must have been intense, and it was supposed to be a case of navicular disease, while the real cause was unsuspected. In driving nails into the hoof, great caution should be exercised. The hand, or the thumb, should be held over the spot where the point of the nail is expected to come out, and if it does not appear when it should do so, the nail must be withdrawn. Use no split or imperfect nail, and have the point very carefully prepared. The course taken by a nail properly pointed and driven is shown by the lines curved outwards at a, *a*, in figure 111.

SCREW-DRIVERS

To drive a screw with a screw-driver, as it is usually pointed and handled, is a disagreeable task. If the screw goes in with difficulty, the driver slips out of the groove, or it cuts the edges of the groove so that the screw is useless. This is because the point of the tool is not ground properly. It should be ground with an even and long bevel, at least an inch long in small tools, and two inches in large ones. The sides of the bit should be kept straight, and not tapered off nor the corners ground off or founded. There should be no sharp edge ground upon the end of the tool, and the grinding should be lengthwise, or from handle to point, and not crosswise. The edge should be slightly rounded. The degree of roundness given may be such as would make it equal to an arc of a circle ten to twelve inches in diameter; for small tools this may be lessened considerably. The shape of a well-pointed screwdriver is shown in figure 112. Flat handles should be abolished as a nuisance; after an hour's use of a driver with such a handle, the hand will be stiff and sore. The handle should be round. Screw-drivers are

Fɪɢ. 112. Screw-drive.

used more frequently than necessary. We have driven hundreds of screws in all sorts of timber, hard white oak even, with the hammer, just as nails are driven, without the use of a screw-driver, and found them to hold perfectly well. This, of course, can be done only with the sharp taper-pointed screws, and if any one uses the old blunt-pointed kind, he is too far behind the times to be much of a mechanic or farmer either.

TO PREVENT COWS SUCKING THEMSELVES

There are many devices to prevent cows from sucking themselves. A spiked halter is shown in figure 113. A buckle at the upper part, behind the ears, makes it quite easy to detach it. Figure 114 shows how the spikes

FIG. 113. Spiked halter.

FIG. 114. Making the halter.

are secured. The spikes should not be over two inches in length. They are best made of wrought nails, which are sold at the hardware stores. They are placed in an iron vise and the heads flattened as much as possible by pounding with a hammer; they are then driven into a piece of thick leather, and secured by sewing or riveting it upon another piece of leather, as shown at B in figure 114.

ABUSE OF BARN CELLARS

A great change has come over the farm during the last thirty years, in all our thrifty farming districts, in the general use of barn cellars. Formerly such an arrangement of the barn was a novelty, and farmers have slowly learned its great advantages—the greater comfort of cattle, the cheaper cleaning of stables, the more convenient watering of stock, the larger use of peat. muck, and headlands in the compost heap, and the greater value of the manure made under cover. Now the cry is raised of damage to fodder and stock from the barn cellar. Almost any good thing can be perverted and become a nuisance, and it were strange if men who do not read much, and think less, could not abuse the barn cellar, which is the stomach of the farm. The same kind of men not infrequently abuse their own stomachs, and suffer grievously in consequence. "If you make your barn cellar tight, carbonic acid gas and ammonia are thrown off and injure the quality of hay stored in the rooms above, and the health of the cattle in the stables. If you turn your pigs into the cellar to make compost, and keep them from the air and the light, they become diseased, and you put bad meat into your barrel to breed disease in your family." These are not uncommon complaints, circulating in our agricultural journals. Well, suppose we admit these things to be true, what of it? Is there any necessity for having a barn cellar without ventilation? If you leave one end open towards the south, you certainly have ventilation enough—and the gases that are evolved from fermenting manure are not going through two-inch stable plank and the tight siding of the barn when they have the wind to carry them off. If a barn cellar is properly managed, and seasonably furnished with absorbents, the ammonia will be absorbed as fast as it is formed. There will be no odor of ammonia that the nostrils can detect. If the pigs do not do the mixing fast enough, the shovel and the fork, the plow and the harrow, can be added. The making of compost under the barn is nice work for rainy days in winter, and is more likely to pay than any work exposed to the storm. The keeping of pigs under the barn is a question of two sides, and however we may decide it, barn cellars will stand upon their own merits. Any farmer who makes a business of raising pork for the market will find a well-appointed pig-sty, with conveniences for storing and cooking food, a paying investment

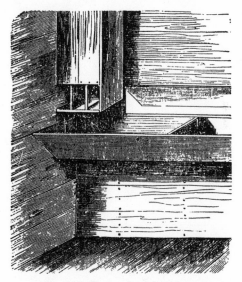

Fig. 115. Hay rack and manger.

If he sees fit to utilize the labor of his pigs by making compost in a well-ventilated barn cellar, their health is not likely to suffer from the wholesome exercise, or that of his family from the use of the flesh. Swine, furnished with a dry sleeping-apartment and plenty of litter for a bed, will keep comfortably clean, and not suffer from overwork in the compost heap. If anything is settled in the experience of the last thirty years, it is the economy of the barn cellar. Our most intelligent farmers, who can command the capital, invest in them. A nice appendage to them is a watering trough fed by a spring or a large cistern in the embankment, to catch all the water, and bring it out by a faucet upon the stable floor above. This works admirably.

HAY-RACK AND MANGER

A cheap and convenient hay-rack and manger is shown in figure 115. The front of the manger should be of oak or other hard wood plank, two inches thick, and one foot wide, the lower edge of which is placed about two and a half feet from the floor; the bottom should be one foot wide. The side of the hay-rack is one foot wide, the front is eighteen inches wide; the top and bottom being of the same width, so that hay will not lodge. The bottom is made from one and a half inch hard board, and is placed one foot above the top of the manger. Two guards, one inch in diameter, and one foot in length, are placed in an upright position across the opening. At the front of the manger is a swinging door, which is shown partly open. This opens into the feed-passage. The manger may have one end partitioned for feeding grain. All corners should be smoothed and rounded off, and to make it durable, attach a thin, flat bar of iron to the upper edge of the manger by screws or rivets.

A BARN BASKET

Figure 116 shows a home-made basket or box for use in the barn or in gathering crops. It is made of two pieces of light board, twelve inches square, for the ends, fastened together by laths sixteen, eighteen, or twenty inches long, for bottom and sides. These are securely nailed. The handle consists of a piece nailed to each end, and connected by a light bar. This box is quickly made, and will be found very handy for gathering many crops in the field, as it may be made to hold exactly one bushel, half a bushel, or any other definite quantity, by changing the size. To hold a bushel, which is two thousand, one hundred and fifty cubic inches, the box may be scant twenty inches long, twelve inches wide, and nine deep, or scant eighteen inches long, twelve inches wide, and eleven inches deep. For half a bushel, scant

Fig. 116. Convenient barn basket.

eighteen inches long, ten inches wide, and six deep; or fifteen inches long, nine inches wide, and eight inches deep. For a peck, ten inches long, nine wide, and six deep; or eight inches square, and scant eight and a half inches deep.

THE TREATMENT OF
KICKING COWS

It is safe to say that a kicking cow is not naturally disposed to this vice, but has been made vicious by some fault of her owner. There are few men who possess sufficient patience and kindness to so manage a cow, from calfhood until she comes to the pail, that she will be kind and gentle under all circumstances. There are nervous, irritable cows, that are impatient of restraint, which are easily and quickly spoiled when they fall into the hands of an owner of a similar disposition. One who is kind and patient, and who has an affection for his animals, is never troubled with kicking cows, unless he has purchased one already made vicious. Unfortunately, few persons are gifted with these rare virtues, and, therefore, there are always cows that have to be watched carefully at milking time. Cows sometimes suffer from cracked teats, or their udders may be tender from some concealed inflammation, and they are restless when milked; so that,

Fig. 117. Cow-fetter.

now and then, in the best regulated dairies, there will be cows that will kick. Many devices have been recommended to prevent such cows from exercising this disagreeable habit. Different methods of securing the legs have been tried. The best plan that we have heard of, or have tried, is shown in figure 117. This fetter is fastened to the cow's near leg, by means of the strap in the centre, the curved portions embracing the front of the leg above and below the hock. It will be perceived that, while the cow can move her leg to some extent, and is not hobbled, as when the legs are tied together, yet she cannot lift it to kick, or to put her foot in the pail. We have seen this "fetter" tried upon a cow that had very sore cracked teats, and that kicked furiously when milked, but with the fetter she was unable to kick or hinder milking.

HOW TO BUILD A
BOAT-HOUSE

Any kind of a house that is large enough may be used, if provided with the needed fittings named below. Where the level of the water is liable to little change, the house

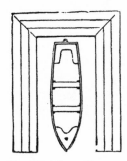

FIG. 118. Plan of house.

need not be raised much above the surface of the water, but the floor may be made so low that one can easily step out of the boat to the floor. Of course there should be a channel made in the centre of the house, deep enough to float the boat when loaded. The plan of the floor is shown in figure 118, with the boat in the centre. The floor

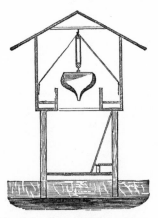

FIG. 119. Section of boat-house.

901

should be protected by a light railing around it (see figure 119), to prevent accidents from slipping when the floor is wet Where the water level changes, the house should be raised on posts, or bents, as may be necessary to keep it above high water. A hanging ladder, that may be drawn up, is provided for use at low water.

WASTE LANDS—MAKE THEM USEFUL

Waste land abounds everywhere. It is fenced, and has the appearance of farm-land, but the owner, if a farmer, would be better off without it than he is with it. No one locality seems to be better or worse than another in this respect, unless it be that the Southern States have the most waste land, and the Eastern States come next in this respect. There are rocky fields, and fields covered with loose stones; swamps and wet ground, and land covered with wretched brush and small timber, and in the South, especially, barren and gullied hill-sides. It is true, that to clear up these lands, and make them fruitful, will cost in labor, if the labor is valued at the current rates, more than the land would bring if offered for sale. But this is not the right way to look at this matter. In reality, it will cost nothing to clear these lands, because their owners may do it by working when they would otherwise be idle. The way to do it is to set about it. To clear an acre or two at a time, of those fields that can be cleared; and to plant with timber, of some valuable sort, that ground which is too rough for the plow, instead of permitting it to grow up with useless brush. In many cases, the worst trouble that farmers suffer is, that they have more land than they can care for, under their present system of management. Hundreds of farms are worked as grain farms, that are not well suited for any other use than dairy farms, and ground is plowed that should be kept in permanent grass. In some cases, the owners of land have discovered their proper vocation, as in the dairy district of Central New York, in the fruit and grain farms of the western part of that State, in the pasture farms of the blue-grass region of Kentucky, and in the corn-growing and pork-raising prairies of the West. If the system of culture in these places were changed, the farmers would he poor instead of being rich, and one sees very little waste land in these localities. There are districts where the surface is hilly, and not so well suited for arable purposes as for pasture, but where, instead of grass and cows, side-hill plows and poor corn fields, washed and gullied by rains, are to be seen. Here are waste lands in plenty; and their owners show every sign of poverty and want of thrift. It is not easy to change these circumstances quickly, but it is easy to begin—just as it is easier to start a stone rolling down a hill, than to throw it down bodily; and when it is once started, it goes slowly at first, and may need help, but it can soon take care of itself, and speedily reaches the bottom. It is just so with such improvements as are here referred to. They are necessarily begun slowly, but when one or two acres of these waste places are reclaimed, the product of these adds to the farmer's resources. He is richer than before by the increased value of these acres, and he is better able to reclaim more. When these in their turn are improved, the means for further improvements are greatly enlarged; the ambition of the man to excel in his vocation is excited, and he speedily becomes a neater, better farmer, and necessarily his circumstances are improved. Thus the rough waste lands, which give a disagreeable appearance to the landscape, and are a stigma upon its character and that of our farmers, in the eyes of our own citizens and of foreigners, might in a short time be improved and a source of profit.

A RAT-GUARD

To keep rats away from anything that is hung up, the following simple method may be used. Procure the bottoms of some old fruit-cans, by melting the solder which holds them upon a hot stove. Bore holes in the centre

FIG. 120. Guard against rats.

of these disks, and string a few of them upon the cord, wire, or rope upon which the articles are hung. When a rat or mouse attempts to pass upon the rope by climbing over the tin disks, they turn and throw the animal upon the floor. This plan, shown in figure 120, will be found very effective.

A CRUPPER-PAD FOR HORSES

Many horsemen desire a method by which to prevent a horse from carrying its tail upon one side, and from clasping the reins beneath the tail. We cannot advise the operation of "nicking," which consists in cutting the skin and muscles upon one side of the tail, and tying it over to the cut side, until the cuts heal, when the skin, being drawn together, pulls the tail permanently over to that side. A different form of the operation causes the tail to be carried up in a style that is supposed to be more grace-

ful, and prevents the horse from clasping the reins when driven. As a preventive of both of these habits, the pad shown in figure 121 is often used by horsemen, instead of the cruel and unnecessary operation of "nicking." This appliance is made of leather, is stuffed with hair or wool, and is about three inches in diameter at the thickest part, gradually tapering toward each end, where it is fastened to the crupper straps. It

Fig. 121. Crupper-pad.   should be drawn up close to the roots of the tail, and by exerting

a pressure beneath it, the tail is carried in a raised position, and is not thrown over to one side. If it is, a few sharp tacks may be driven into the inside of the pad.

A DAM FOR A FISH POND

In making a fish pond, by placing a dam across a stream, it should be borne in mind that success depends upon the proper construction of the dam, whether it

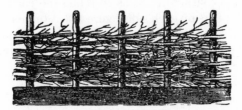

FIG. 122. Stakes and brush.

be large or small. Any defect here will make the whole useless. The main point in the construction of a dam is, to have a complete union between the earth of the bed and that of the dam. This cannot be done by throwing the earth upon an old surface. A new surface must be made, solid and firm, to receive the new earth. In addition, there should be a central core of some strong material, that will serve to strengthen and bind the new construction. In making a dam or embankment to retain or exclude water, the beginning should be to dig a shallow ditch, removing sod or uneven ground, or if the earth is bare, to disturb it thoroughly with the pick, so as to provide binding material to unite with the bottom of the dam. A line of stakes is driven into the ground, and filled with brush woven in, or wattled, as in figure 122. In building the dam, all the sods and vegetable matter should be placed on the outside, where these will root, and bind the surface together; the rest

907

of the earth should be well trodden, or rammed down firmly, and it the soil is puddled by admixture of water in the process of ramming, the work will be better for it. The water-way in the stream should be tightly boarded or planked. Three posts may be driven or set on each bank of the stream, and boards nailed, or planks spiked for a larger structure, so as to retain the earth of the embankments on each side, figure 123. A timber is fitted as a mudsill, to the front and rear posts, and one to the central posts; the latter at such a height as will raise the

Fig. 123. Waste-gate for pond.

water to the desired depth. The spaces between these timbers are boarded and planked, and may be filled in with earth, well rammed, and mixed with straw and fine cedar brush, under the covering. If it is desired to raise the water to a greater depth, loose flash-boards may be fitted with cleats, on the centre of the waste-way, or a wire-gauze fence may be placed there, to prevent the escape of the fish. If freshets are apt to occur, a sufficient number of these waste-ways should be provided to carry off the surplus water, and prevent overflowing and wasting of the dam. The dam of a fish pond should always be made high enough for safety against overflow, and to guard against percolation, and washing away by undermining, it should be made three times as wide as it is high, with slopes of one and a half foot horizontal on each side, to one foot in perpendicular height, if any plants are set upon a dam or embankment, they should be of a small, bushy growth, such as osier willow, elders, etc., but nothing larger, lest the swaying caused by high winds should loosen and destroy the bank.

A WAGON JACK

In figures 124 and 125 is shown a most convenient home-made wagon jack, in constant use for ten years, and has proved most satisfactory. The drawings were made with such care, the measurement being placed upon them, that the engravings tell nearly the whole story. Figure 124 shows the jack when in position to hold the axle, at *a*. When not in use, the lever falls down out of the way, and the affair can be hung up in a handy place. Figure 125 shows the "catch-board," and the dimensions proper for a jack, for an ordinary wagon, buggy, etc. It is so shaped and fastened by a din between

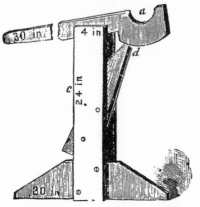

FIG. 124. Michigan wagon jack.   FIG. 125. Catch-board.

the upright parts of the jack, that it is pushed in position, *d.* by the foot at *c*, when the axle is raised; and falls back of its own accord when the lever is raised a trifle to let the wheel down. All the parts are made of inch stuff, the

910

foot board, catch, and lever should be of hard wood; the upright boards between which they are placed can be of pine or other soft wood. Persons who see this simple and convenient wagon jack frequently say, "Why don't you get it patented?" but the inventor thinks that such simple things, which any one can make, ought to be contributed for the common good, and in the same spirit we commend it to any who may be in need of a good wagon jack.

WILL YOU FEED
HAY OR WOOD?

A great deal has been said and written about the proper time of cutting hay. The best time, all things considered, is to cut the grass just after it has come into full bloom, though many think the preferable time is just when it is coming into full blossom. As it is impossible to always mow every field just at the right moment, the general safe rule is, we think, to be all ready to begin at full bloom, and finish before it is entirely past.

There is this important fact to be kept in mind, viz., that as soon as grass of any kind has attained its growth, and is full of juices, it begins to change more and more into woody fibre, and that when fully ripe a large part of the stems or stalks differ very little in composition from dry wood. And every one knows that dry wood is neither easily digested nor nutritious. It stands to reason that a stalk of grass cut when it is full of juice containing sugar, gum, and protein compounds, and cured thus, must be more nutritious than if left standing until a part of these constituents have changed into woody fibre. Feeding hay not cut until it is thoroughly ripe, is giving the animals that which is in part only so much wood. The practical lesson is, make a good ready well in advance, now, and have the barns, mows, stacking arrangements, mowers, scythes, horse and other rakes, forks, wagon racks, in short, all things, in perfect order—and the work planned, so as not to let any hay-field get into the fully ripe condition. Head work beforehand will save hard work and worry, and secure better hay.

A BRACE FOR A KICKING HORSE

Those so unfortunate as to own a kicking horse know something of the patience that it requires to get along with it—and will welcome anything which will prevent the kicking and finally effect a cure. The writer knew a horse, which was so bad a kicker that after various trials, and after passing through many hands, and getting worse all the time, to be perfectly cured in the course of three months by the use of the device here given. This is a simple brace, which acts upon the fact that if the head be kept up, the horse cannot kick. A kicking horse is like a balance, when one end goes up, the other must go down. The brace is shown in figure 126, and consists of a one-half inch iron rod, which may be straight, or, for the looks, bent into a grace-

FIG. 126. The brace.

ful curve. It is forked at both ends; the two divisions of the upper end are fastened to the two rings of the bit, while the lower ends fit upon the lower portion of the collar and hames. The upper ends can best be fastened to the bit by winding with wire, which should be done smoothly, so as not to wear upon the mouth. The lower end is secured by means of a strap fastened to the upper loop, and passing around the collar is buckled through

913

the hole in the lower part of the end of the brace. The brace need not be taken from the bit in unharnessing. Any blacksmith can make such a brace, taking care to have it of the proper length to fit the particular horse. Keep its head at about the height as when "checked up," and the horse will soon be cured.

HOW TO SAVE LIQUID MANURE

In ordinary farm practice, by far the larger part of the liquid manure of the stock kept is lost. No effort is made to save it. There is no barn cellar, no gutter behind the stabled animals, no absorbents. Analysis shows that the liquid manure is quite as valuable as the solid, or even more so. In 1,000 pounds of fresh horse dung there are 4.4 pounds of nitrogen, 3.5 of potash, and 3.5 of phosphoric acid. In horse urine there are 15.5 pounds of nitrogen, and 15.0 of potash. In 1,000 pounds of fresh cattle dung there are 2.9 pounds of nitrogen, 1.0 of potash, 1.7 of phosphoric acid. In the urine, 5.8 pounds of nitrogen, 4.9 of potash. These are the most valuable constituents of manure, and no farmer can afford to have them so generally run to waste. There is very little loss where there is a gutter well supplied with absorbents, and a barn cellar well coated with dried peat, muck, or headlands, to absorb the liquids as fast as they fall. But barn cellars are still in the minority. Mr. Mechi had a very expensive apparatus for distributing the liquid manure over his farm, by means of tanks and pipes, and thought it paid, but failed to convince his contemporaries of the fact. However that may be, it is out of the question to apply liquid manure in this manner, economically, upon the average farm. It takes too much capital, and requires too much labor. By the use of absorbents, it can be done economically on a small or large scale, "with very little waste. Some use a water-tight box, made of thick plank, covering the floor of the stall. This is a very sure way to save everything, and the only objection to it is the expense of the box, and the increased labor of keeping the stalls clean. We used for several years dried salt-marsh sod, cut in blocks eight or ten inches square, taken from the surface of the marsh in ditching. This had an enormous capacity for absorbing liquids, and a layer of these sods would keep a horse or cow comfortably dry for a fortnight. Refuse hay or straw was used on top for purposes of cleanliness. The saturated sod was thrown into the compost heap with other manure, where it made an excellent fertilizer. Later we used sawdust, purchased for the purpose at two cents a bushel, as bedding for a cow kept upon a cemented floor. A bed a foot thick would last nearly a month, when it was thrown out into the compost heap. The sawdust requires a longer time for decomposition, but saves the liquid manure. Our present experiment, covering several months, is with forest leaves, principally hickory, maple, white ash, and elm. A bushel of dried leaves, kept under a shed for the purpose, is added to the bedding of each animal, and the saturated leaves are removed with the solid manure as fast as they accumulate. The leaves become very fine by the constant treading of the animals, and by the heat of their bodies, and the manure pile grows rapidly. It is but a little additional labor to the ordinary task of keeping animals clean in their stalls, to use some good absorbent, and enough of it, to save all the liquid manure. What the absorbent shall be is a question of minor importance. Convenience will generally determine this matter. No labor upon the farm pays better than to save the urine of all farm stock by means of absorbents. These are in great variety, and, in some form, are within the reach of every man that keeps cattle or runs a farm. Stop this leak, and lift your mortgage.

AN OPEN SHED FOR FEEDING

A feeding-trough in a yard, which can be covered to keep out snow or rain, is a desirable thing, and many devices have been contrived for the purpose, most of which are too costly. We give herewith a method of constructing a covered feeding-trough, which may be made very cheaply of the rough materials to be had on every farm. A sufficient number of stout posts are set firmly in the ground, extending about ten feel above the surface. They should be about six feet apart and in a straight line, and a plate fastened to their tops. A pair of rafters supported by braces, as shown in figure 127, is fitted to each post. A light roof of laths is laid, and covered with bark, straw, corn-stalks, or coarse hay. Strips are fastened from one brace to another, and laths or split poles nailed to them, about six inches apart, to make a feed-rack. A feed-trough for grain or roots is built upon each side. For sheep, the shed and rack may be made only eight feet high at the peak, and the eaves four feet from the ground; giving better shelter.

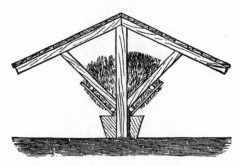

Fig. 127. An open feeding-shed.

917

A SHADE FOR HORSES' EYES

The most frequent cause of weak eyes in horses is a badly-arranged stable. Foul gases irritate and inflame the tender membranes of the eye and head, and horses brought from dark stables into bright sunlight, or onto glittering snow, are dazzled and blinded. The existing weakness or irritation is intensified, and the poor animal suffers unsuspected torments. The remedy is to purify the stable and give it sufficient light, shaded by blinds, from before and behind the horse, or from both sides, avoiding a light from only the front, rear, or one side light. A shade for weak or inflamed eyes may be constructed by fastening wires to the bridle and covering it with oiled cloth in the manner represented in figure 128. Thus a soft, subdued light reaches the eyes, while the horse can still see the ground immediately before him. It will be a timely job to prepare such a shade for use before the snow of winter comes.

Fig. 128. To protect the eyes.

918

TEST ALL SEEDS—IMPORTANT

No one can, by merely looking at them, positively tell whether any particular lots of field, garden, or flower seeds have or have not sufficient vitality of germ to start into vigorous growth. Yet it is a severe loss, often a disastrous one, to go through with all the labor and expense of preparation and planting or sowing, and find too late that the crop is lost because the seeds are defective. All this risk can be saved by a few minutes' time all told, in making a preliminary test, and it should be done before the seed is wanted, and in time to get other seed if necessary. Seeds may not have matured the germ; it may have been destroyed by heat or moisture; minute insects may have, unobserved, punctured or eaten out the vital part of a considerable percentage.

Select from the whole mass of the seed, one hundred, or fifty, or even ten seeds, that will be a fair sample of all. For larger seeds, as wheat, corn, oats, peas, etc., take a thin, tough sod, and scatter the counted seeds upon the earth side. Put upon the seeds another similar sod, earth side down. Set this double sod by the warm side of the house or other building, or of a tight fence, moistening it occasionally as needed. If very cold, cover, or remove to the kitchen or cellar at night. The upper sod can be lifted for observation when desirable. The swelling and starting of the seeds will in a few days, according to the kind, tell what percentage of them will grow—a box of earth will answer instead of sods, both for large and small seeds. Small seeds of vegetables or flowers, and even larger ones, may be put into moist cotton, to be kept slightly moist and placed in the sun or in a light warm

Fig. 129. Home-made roller.

room. For small quantities of valuable flower seeds and the like, half a dozen will suffice for a trial test. With any seed, for field or garden, however good, it is always very desirable and useful to know exactly how many or few are defective, and thus be able to decide how much seed to use on an acre, or other plot.

A FIELD ROLLER

A very good field roller may be easily made in winter, when timber is being cut. Use a butt-log of an oak tree, in the form shown in figure 129. The log need not be a very large one, because the frame, in which it is mounted, enables it to be loaded to any reasonable extent, and the driver may ride upon it, and thus add to the weight. A roller will be found very valuable in the spring when repeated frosts have raised the ground and thrown out the stones.

A PORTABLE SLOP BARREL

A barrel mounted upon wheels, as shown in figure 130, will be found useful for many purposes about the farm, garden, or household. The barrel is supported upon a pair of wheels, the axles of which are fastened to a frame connected with the barrel by means of straps bolted to the sides. The frame may be made of iron bent in the form shown in figure 131, or of crooked timber having a sufficient bend to permit the barrel to be tipped for emptying. A pair of handles are provided, as shown in the engraving. When not in use, the barrel rests upon the ground, and may be raised by bearing down upon the handles. The barrel may be made to rest in notched

FIG. 130. Portable barrel for slops.

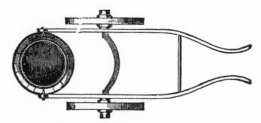

FIG. 131. Plan of frame of barrel.

922

bearings upon the frame, so that by raising the handles, the wheels may be drawn away from the barrel, and the latter left in a convenient place until it needs removal. This contrivance will be useful for feeding slops to pigs, or for removing the waste of the house to the barn-yard.

WHERE AND HOW TO APPLY
FERTILIZERS

It is often difficult to decide—for barn-yard or stable manures, or for any artificial fertilizer—whether to use it in the hill or broadcast it; and whether to apply it on the surface, or bury it deeply. Here is a hint or two. If not strong enough to injure the first tender roots, a little manure near at hand gives the plant a good send-off, like nourishing food to the young calf or other animal; the after-growth is much better if the young animal or plant is not dwarfed by imperfect and insufficient diet. Therefore, drilling innocuous hand fertilizers in with the seed is useful, as is putting some well-rotted manure or leached ashes into hills of corn, potatoes, indeed with all planted seeds. But there arc good reasons for distributing most of the manures or fertilizers all through the soil, and as deeply as the plant roots can possibly penetrate. The growth and vigor of all plants or crops depend chiefly upon a good supply of strong roots that stretch out far, and thus gather food over the widest extent of soil. If a flourishing stalk of corn, gram or grass, be carefully washed, so as to leave all its roots or rootlets attached, there will be found a wonderful mass of hundreds and even thousands of roots to any plant, and they extend off a long distance, frequently several feet—the farther the better, to collect more food and moisture. Put some manure or fertilizer in place two feet away from a corn or potato hill, or from almost any plant, and a large mass of roots will go out in that direction. So, if we mix manures or fertilizers well through the whole soil, they attract these food-seeking roots to a greater

924

distance; and they thus come in contact with more of the food already in the soil, and find more moisture in dry weather. A deeply-stirred soil, with manure at the bottom, develops water-pumping roots below the reach of any ordinary drouth, and the crops keep right on growing—all the more rapidly on account of the helpful sun's rays that would scorch a plant not reaching a deep reservoir of moisture.

A MILL FOR CRUSHING BONES

To save the expense of a purchased bone-mill, one may be made as described below, which will crush them into a condition much more valuable for manure than the whole bones, if not quite as good as if finely ground. Make a circular mould of boards, six feet wide and two feet deep. Hoops of broad band-iron are fitted to the inside of the mould, and secured to it about one inch apart. The mould is then filled with a concrete of Portland cement, sand, and broken stone. Place in the concrete when filling binding pieces of flat bar-iron, to prevent the mass from cracking when in use. In the centre place squares of band-iron, as a lining for a shaft by which the crusher is turned. When the concrete is set and hardened, the frame may be taken apart; and, as in setting the concrete will expand somewhat, the iron bands around the mass will be found to have become a tight solid facing to the wheel. The wheel is then set up

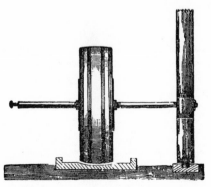

Fig. 132. Home-made bone-mill.

926

on edge, and a square shaft of yellow pine, six inches thick, is wedged into the central space. This shaft is fitted to an upright post by a loose band of iron and a swivel joint, so that the wheel may be made to revolve around it. Any other suitable connection may be used for this purpose. A hollow trough of broken stone and well rammed concrete is then laid in the track of the wheel as it revolves, and the crusher is complete and ready for a pair of horses to be attached to it, figure 132. A crusher of this kind may be put up at a country mill, or as a joint affair by a few farmers uniting their efforts, and thus utilizing a valuable fertilizing material, which is now wasted for want of means to render it available.

LIME AND LIMESTONE

In the first place, limestone, marble, calc-spar, chalk (of rare occurrence in this country), marl, and oyster, and other shells, are all essentially the same in composition, however they may differ in texture, form, and other particulars. They are all different forms of the carbonate of lime; that is, they consist of the alkaline earth, lime, in combination with carbonic acid, and in the case of shells, with animal matter. As a general thing, we only know carbonic acid as a gas. It has a very weak hold of the lime, for if we drop a fragment of limestone into strong vinegar, the acetic acid of the vinegar will unite with the lime (forming acetate of lime), while the carbonic acid, being set free, will be seen to pass off in small bubbles. In this case we free the lime from its carbonic acid, by presenting to it a stronger acid, that of vinegar. But if instead of using another acid to displace the carbonic acid, we place limestone in any of its forms, in a strong fire, the carbonic acid will be driven off by the heat, and there will be left, simply lime. This is called quick lime, or caustic lime, and by chemists oxide of the metal calcium, or calcium oxide. Lime, then, is limestone without its carbonic acid. All the forms of limestone are very little soluble in water; lime itself is more soluble, though but slightly so, requiring at ordinary temperatures about seven hundred times its own weight of water, yet it gives a marked alkaline taste to water in which it is dissolved. Lime in this condition, as quick lime, or when combined with water, "slaked" as it is called, is much employed in agriculture. A small portion of lime is required by plants, but the chief use of lime, when applied to the soil, is to bring the vegetable matters contained in the soil into a condition in which they can be used as plant food. This application of lime as a fertilizer has long been followed by farmers, and in many cases with the most beneficial results. Within a year or so great claims have been made for ground limestone, especially by the makers of mills for grinding it; some of these have asserted that it was superior to burned lime, and superior to nearly all other fertilizers. The question which most interests farmers is, has limestone, however fine it may be, any value as a fertilizer? To this the answer would be both "yes" and "no." Upon a heavy clay soil the carbonate of lime, or limestone in any form, appears to have a beneficial effect; it makes such soils friable and open, so that water and air may penetrate them. While its action upon th6 vegetable matter in the soil is far less prompt and energetic than that of quick-lime, yet its presence, affording a base with which any acid that may be present in the soil may unite, is often beneficial. To extol ground limestone as "the great fertilizer of the age," to even claim that it is equal to lime itself, is a mistake. Both have their uses. It should be borne in mind by inquirers about the value of ground limestone, that many soils already contain more lime in this form than can ever be utilized, and need no addition.

A FARM WHEELBARROW

The wheelbarrow is an indispensable vehicle on the farm and in the garden. Applied to hard uses it needs to be strong and durable. A barrow of the ordinary kind, used on farms, soon becomes weak in the joints and falls to pieces. The movable sides are inconvenient, and the shape necessarily adopted when movable sides are used greatly weakens the structure. It will be noticed at first sight that the wheelbarrow, shown in figure 133, is most strongly supported and braced, that the box, instead of weakening it, greatly strengthens it, and that it is stout and substantial. It is put together at every part by strong bolts, and can be taken apart to pack for transportation, if desired, and a broken part readily replaced.

Fig. 133. Farm wheel barrow.

TO PREVENT THE BALLING OP HORSES

When the snow upon the roads is cohesive and packs firmly, it collects upon the feet of horses, forming a hard, projecting mass, in a manner known as "balling." This often occurs to such an extent as to impede the motion of the horse, while it causes the animal great discomfort, and is sometimes dangerous to the rider or driver. The trouble may be prevented very easily by the use of guttapercha. For this purpose the gutta-percha should be crude, *i.e.*, not mixed with anything or manufactured in any manner, but just as imported. Its application depends upon the property which the gum has of softening and becoming plastic by heat, and hardening again when cold. To apply it, place the gutta-percha in hot water until it becomes soft, and having well cleansed the foot, removing whatever has accumulated between the shoe and hoof, take a piece of the softened gum and press it against the shoe and foot in such a manner as to fill the angle between the shoe and the hoof, taking care to force it into the crack between the two. Thus filling the crevices, and the space next the shoe, where the snow most firmly adheres, the ball of snow has nothing to hold it, and it either does not form, or drops out as soon as it is gathered. When the gutta-percha is applied, and well smoothed off with the wet fingers, it may be hardened at once, to prevent the horse from getting it out of place by stamping, by the application of snow or ice, or more slowly by a wet sponge or cloth. When it is desired to remove the gum, the application of hot water by means of a sponge or cloth will so soften it that it may be taken off. As the softening and hardening may be repeated indefinitely, the same material will last for years. For a horse of medium size, a quarter of a pound is sufficient for all the feet.

TO PREVENT CATTLE THROWING FENCES

To prevent a cow from throwing fences or hooking other cows, make a wooden strip two and a half inches wide and three-quarters of an inch thick, and attach it to the horns by screws; to this is fastened, by a small bolt, a strip of hardwood, three inches wide, half an inch thick, and of a length sufficient to reach downward within an inch of the face, and within two or three inches of the nostrils. In the lower end of this strip are previously driven several sharp nails, which project about one-quarter of an inch. The arrangement is shown in figure 134; the strip, when properly attached, allows the animal to eat and drink with all ease, but when an attempt is made to hook or to throw a fence, the sharpened nails soon cause an abrupt cessation of that kind of mischief.

Fig. 134. Cattle check.

FEED BOXES

In figure 135 a box is shown firmly attached to two posts. It has a hinged cover, *p*, that folds over, and may be fastened down by inserting a wooden pin in the top of the post near *n*. The one given in figure 136 maybe

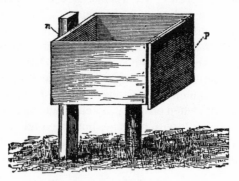

FIG. 135. Covered feed box.

placed under shelter, along the side of a building or fence. One side of the top is hinged to the fence or building, the bottom resting upon a stake, *e*. When not in use, the box may be folded up, the end of the strap, *b*, hooking over the pin, *a*, at the side of the box. A good portable box, to be placed upon the ground, is shown in figure 137. It is simply a common box, with a strip of board, *h*, nailed on one side and projecting about eight inches. When not in use, it is turned bottom up, as shown in figure 138. The projecting strip prevents three sides of the box from settling into the mud or snow. The strip is also a very good handle by which to carry it. Those who now

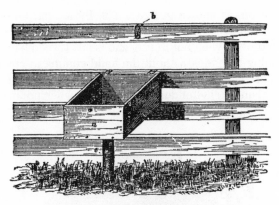

FIG. 136. Hinged feed box.

use portable boxes will find the attaching of this strip a
decided advantage. A very serviceable portable feed box
is made from a section of half a hollow log, with ends

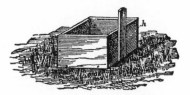

FIG. 137. Portable feed box.

FIG. 138. Feed box inverted.

nailed on, as shown in figure 139. By letting the ends project above the sides four or five inches, it may be turned over when not in use, and easily turned back by grasping the sides without the hand coming in contact with earth or snow. All feed boxes and racks should be placed under shelter during summer, or when not in use.

Fig. 139. Box from hollow log.

A CATTLE TIE

Judging from the numerous stanchions and arrangements for fastening cattle in stalls, illustrated from time to time in the public prints, the perfect cattle-fastening has not yet been invented. We do not claim perfection for the arrangement given in figure 140, but it will be difficult to devise a cheaper one, and we doubt if any better or more satisfactory one is in use. The fastening consists of a three-fourth inch rope, which is run through the partitions of the stalls, one long rope being used for the tier of stalls, although short pieces may be employed if desired. This rope is knotted on either side of each

Fig. 140. Cattle tie.

partition, and a good swivel snap for use with a rope, is tied in the rope in front of the centre of each stall. The rope should pass over, very nearly, the front of the manger—from the side of the cattle—and for cattle of

937

ordinary hight, it ought to be about two feet from the floor. When put in, the rope should be drawn up tightly, as it will soon acquire considerable and sufficient slack from the constant strain from the animals. With this arrangement each cow must be provided with a strap or rope about the neck, the rope or strap being supplied with a free-moving iron ring. When the animal is put in the stalls the snap is fastened in the ring, and if the snap is a good one—none but the best swivel snaps should be used—an animal will rarely get free from it. This fastening, it will be noticed, admits of considerable fore and aft motion, and but slight lateral movement. The cost of this arrangement it is difficult to state accurately, it is so small. The rope for each stall will cost less than five cents; the snaps will cost ten cents when bought by the dozen, and the time of putting these fittings in each stall is less than fifteen minutes. The rope will wear two years at least.

A BEEF RAISER

Two posts are set about fifteen feet high. A deep mortise is cut in the top of each to receive the roller, which is grooved at the points of turning. One end of the roller extends beyond the post, and through this end three two-inch holes are bored. Three light poles are put through these holes, and their ends connected by a light rope. In raising the beef the middle of a stout rope is thrown over the roller; the ends are drawn through the loop, and after the beef is fastened to the loose ends the roller is turned against the loop by means of the "sweep," or lever arms, figure 141. A heavy beef can be easily raised, and may be fastened at any light desired, by tying the end of one of the levers to the post with a short rope.

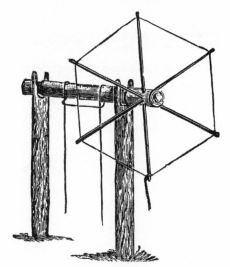

FIG. 141. A beef raiser.

A CEDAR STEM SOIL-STIRRER

A convenient and quickly-made implement for stirring and mixing manure and fertilizers with the soil, may be made as follows: A cedar stem is cut about eight feet long, and the branches cut off several inches from the stem, leaving long spurs on all sides for its whole length, as shown in figure 142. A horse is hitched by a chain to the butt end, and the driver guides the implement by a rope fastened to the rear end of the stem. By means of the guide-rope the implement may be lifted over or around obstacles, and turned at the end of the field. Such an implement is specially useful in mixing fertilizers with the soil, when applied in drills for hoed crops.

FIG. 142. A stirrer made of a cedar stem.

A HINT FOR PIG KILLING

Lay a log chain across the scalding trough, and put the pig upon it. Cross the chain over the animal, as shown in figure 143. A man at each end of the chain can easily turn the pig in the scald, or work it to and fro as desired.

FIG. 143. Scalding a pig.

MENDING BROKEN TOOLS

Farming tools, such as shovels, rakes, forks, etc., that are much used, will often, through carelessness or accident, become broken, and, with most men, that means to be thrown one side, as utterly useless. By exercising a little ingenuity, they could in a short time be fitted up to do service for several years. The head of hand-rakes often becomes broken at the point where the handle enters, and not unfrequently the handle itself is broken

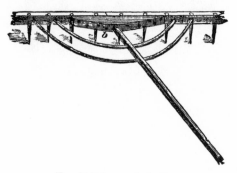

Fig. 144. A mended rake.

off where it enters the head. In either case the break is easily made good by attaching a small piece of wood to the head, by small nails or screws, as shown in figure 144. Should the head be broken where one of the bows passes through, it may be mended in a like manner, *b*. Shovels and spades, owing to the great strain to which they are often subjected, especially by carelessly prying with them, crow-bar fashion, are frequently broken, and usually at the point where the wood enters the blade.

This break, bad as it is, should not consign the broken parts to the rubbish pile, especially if the blade and the handle be otherwise in fair condition. Remove the iron straps or ferule from the handle; firmly rivet a strip of iron, *a*, figure 145, on top of the handle, and a similar one underneath, to the blade and handle, as shown in the engraving. Other broken tools may be made to do good service by proper mending.

FIG. 145. Mending a shovel.

A LARGE FEED-RACK

The width of the rack is seven feet, but it can be any length desired; hight, ten feet; hight of manger, two and a half feet; width., one and one-half foot. Cattle can eat from both sides. The advantage of such a rack, shown in figure 146, is that it will hold a large quantity of feed, and so securely that very little can be wasted by the feeding animals.

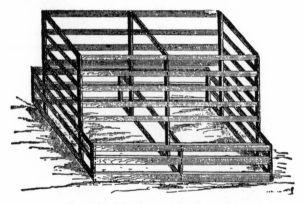

Fig. 146. A large yard fodder-rack.

944

BARN DOOR FASTENING

One of the best barns in the country has its large double doors fastened by a bar of iron, about six feet long,
which is bolted to one of the doors at its middle point.
The ends of the bar are notched, one upon the upper
and the other on the under side, to fit over sockets or
"hooks" that are bolted to the doors. One hook bends

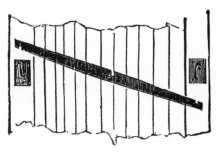

<small>Fɪɢ. 147. Iron bar door fastening.</small>

upward, and the other downward, and the bar moves in
the arc of a circle when the door is being unfastened or
bolted. The construction of this door fastening is shown
in figure 147. A wooden bar may replace the iron one,

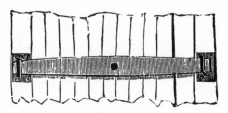

<small>Fɪɢ. 148. Wooden bar door fastening.</small>

945

and may be of a size and length sufficient to make the fastening secure. A wooden bar is shown in figure 148. Such a method of fastening could be used for a single door, provided it needs to be opened and closed only from the inside. By putting a pin in the bar near the end that passes by the door-post, so that it will reach through a slot in the door, such a "latch" might be used for any door.

A "FORK" STABLE SCRAPER

A very handy stable scraper is made of an inch board, five inches wide, and about eight inches longer than the Width of a four-tined fork. Bore a hole for each tine a quarter inch in diameter from the edge of the board to about two inches from the opposite edge, the holes passing out upon the side. The lower part of the board is bevelled behind, thus forming a good scraping edge. After the coarse manure is pitched up, the fork is inserted in the holes of the board, and a scraper is at once ready for use, figure 149. To store it, nail a cleat on the floor two inches from the wall, and secure the scraper behind this cleat; place one foot upon the board and withdraw the fork. Notches may be cut in the edge of the board opposite each hole, to assist in placing the tines.

Fig. 149. A "fork" barn scraper.

A METHOD OF CURING HAY

A method of curing hay which has been used for several years with entire satisfaction consists in taking four slender stakes six feet long, *a*, *a*, *a*, *a* (figure 150), fastened together at the upper ends with a loose joint similar to that of an ordinary tripod. One end of the fifth stake, *b*, rests on one of the four legs about a foot from the ground, the other end resting on the ground. The hay is stacked around this frame nearly to the top of the stakes, after which the stake, *b*, is withdrawn, and then the four upright stakes are removed. This is done by two men with hay forks, who raise them directly upwards. As soon as the legs are lifted from the ground the pressure of the hay brings them together, and they can be removed with ease, leaving a small stack of hay, as shown in figure 151, with an air passage running from the bottom upwards through the centre of the small stack, as indicated by the dotted lines.

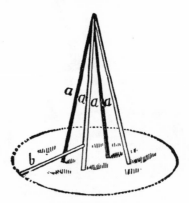

Fig. 150. The frame.

948

FIG. 151. The small stack.

GRANARY CONVENIENCES

The better plan for constructing grain bins is to have the upper front boards movable, that the contents may be more readily reached as they lessen. But as there are tens of thousands of granaries where the front bin

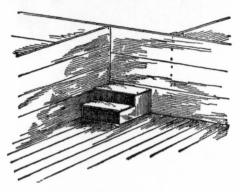

FIG. 152. Steps in a granary.

boards are firmly nailed, a portable step, like that shown in figure 152, is almost a necessity. It should have two steps of nine inches each, and be one foot wide, and two feet long on top. It is light and is easily moved about the granary.

Every owner of a farm needs a few extra sieves, which, when not in use, are usually thrown in some corner, or laid on a box or barrel to be knocked about and often injured by this rough handling, besides being frequently in the way. A little rack, which may be readily made above one of the bins in the granary, as shown in figure 153, is convenient to put sieves out of the way, and keep them from injury.

950

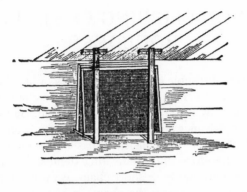

Fig. 153. A sieve rack.

Grain bags are too expensive and valuable to be scattered about the buildings. A simple mode of securing them is shown, which is at once cheap and safe. In the

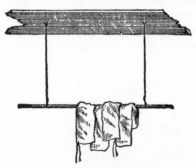

Fig. 154. A grain bag holder.

ceiling over the bins, staples are driven about four feet apart, to which are attached pieces of wire two feet in length. To these wires is fastened a pole five feet in length, over which the bags are thrown when not in use, and they are then out of reach of mice from the bins and wall, as shown in figure 154.

A NON-SLIPPING CHAIN FOR BOULDERS

One great trouble in hauling boulders or large stones with team and chain is the liability of the chain to slip off, especially if the stone is nearly round. By the use of the contrivance shown in figure 155, nearly all of this trouble is avoided. It consists in passing two log chains around the stone and connecting them a few inches above the ground by a short chain or even a piece of rope or wire. Connect the chains in a similar manner near the top of the stone. The ends of the draught chains are attached to the whipple-trees in any way desired. In hauling down an incline, or where the ground is very rough, it will be best to wrap each chain clear around the stone, connecting with whipple-trees by a single chain, thereby preventing a possibility of the chains becoming detached or misplaced in any way.

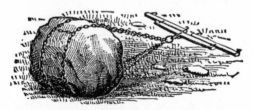

Fig. 155. Method of fastening chains on a boulder.

A PITCHFORK HOLDER

Having occasion to go into the barn one night, we received a very had wound from a pitchfork which had fallen from its standing position. This led us to construct

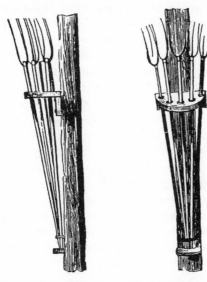

FIG. 156. Front view.   FIG. 157. Side view.

a holder, shown in the engravings. The fork-holder is made of an inch board, of a semicircular shape, with five holes large enough to admit a fork handle, bored near the curved side. This board is nailed to a standing post in the barn. A strap or curved bolt is placed some distance below to hold the handles in place, as they rest on a bottom board fixed for the purpose. Figure 156 shows the front view of the bolder; figure 157 gives the side view.

A CONVENIENT HOG LOADER

Figure 158 shows the "loader" attached to a wagon, with the rack. The bed-piece consists of two pine boards, six inches wide by nine feet long. These are fastened together by three cross-pieces of the same material, of proper length, so that the "bed" will just fit in between the sides of the wagon-box. A floor is laid on these cross-pieces, on which short strips of lath are nailed, to prevent hogs from slipping. At one end the sides are notched to fit on the bottom of the wagon-box. There are two staples on each side by which the sides are fastened on. The "rack" is made like an ordinary top-box, with the exception that each side is composed of three narrow boards about four inches apart, and nailed to three cleats (the two end cleats to be on the inside, and the middle one on the outside of the rack), and projecting down the side of the wagon-box. End-boards are made and fastened in like those of an ordinary wagon-box. For unloading the hogs nothing but the bed-piece need be used, which, being light, may be easily thrown on and taken with the wagon.

Fɪɢ. 158. Rack for loading hogs.

A HOME-MADE ROLLER

Take a log six or eight feet long, eighteen or twenty inches in diameter, and put pins in each end for journals, either of wood one and a half inch, or iron one inch. Make a frame of two by four scantlings, or flat rails three or four feet long to suit the size of the roller. Bore holes for journals a little back of centre, and also inchholes two inches from the back end of scantlings. Fasten these ends together with a chain or rope tight enough to keep the scantlings square with the ends of the log, figure 159. Fasten the front ends together with a stiff pole or rail, and put a heavy chain across the front, with one end around each front corner. Attach the double-tree at the middle of this chain. The draft chain and the pole will keep the front ends of the frame in position, and the chain behind will prevent the rear ends from spreading. When the roller goes faster than the team, the draft chain will slacken, and the front of the frame will drop and prevent the roller from striking the team. A roller is such a valuable implement that there should be one in use on every farm. Even a rough home-made roller is better than none, whether it is used to break up clods, or to compact the soil after sowing.

FIG. 159. A home-made holler.

A LAND SCRAPER

In districts where land needs draining, scrapers must be used. A very good one is shown in figure 160. It has one advantage over most scrapers: the team can star

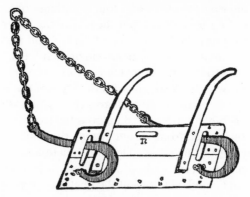

Fig. 160. A land scraper.

on the bank while the scraper is thrown into the ditch. When the ditch is a large one, fourteen feet or more wide at the top, it is only necessary to lengthen the chain. The

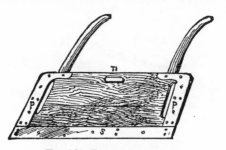

Fig. 161. Front of scraper.

956

scraper consists of two boards, twelve inches wide and three feet long, fastened firmly together by two strong iron plates, figure 161, *p*, *p*, bolts, and rod-iron nails. The scraper-edge is made of an old cross-cut saw, fastened on with rod-iron nails. Two notches are cut at *p*, *p*, for

FIG. 162. The hook-rod.

the hooks to pass through, also one at *n*, for a holder for lifting the scraper when necessary. To make the scraper work perfectly, the rod or hook should have the right bend as shown at *a*, figure 162. The hook is fastened to the scraper by two bolts, *b*, *b*, figure 162, and small pins, *c*, when the land scraper is complete.

A HOME-MADE BAG-HOLDER

This bag-holder is one of the most useful articles a man can have in his barn. It consists of a post, *a*, two by four inches, and five feet long, with six one-half inch holes near the upper end, as shown in figure 163. The bar, *b*, passes through a mortise and over the pin nearest the bag, and under the other pin. This bar can be moved

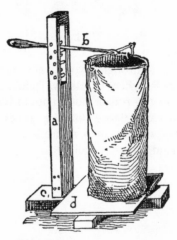

FIG. 163. A handy bag-holder.

up or down, to suit the length of the bag. The post, *a*, sets in a bed-piece, *c*, two by three inches and two feet long. A board, *d*, eighteen inches square, fastened upon the bed-piece, furnishes the necessary rest for the bag. The mouth of the bag is held open by means of hooks placed on the ends of the cross-bar, with another beneath the main bar.

A SAFETY EGG-CARRIER

In figure 164, *a* represents the bottom-board of the spring-box, near the edges of which are fastened six wire-coil springs or bed-springs. At *b* is represented a hole made in the board to receive the lower end of the spring, about half an inch of which is bent down for that purpose. Small staples are driven into the board to hold the springs in place. Scraps of leather or tin might be tacked or screwed down, instead of using staples; *d*,

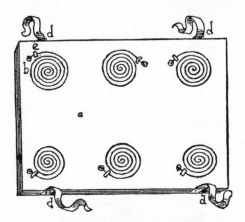

Fig. 164. Bottom-board of spring-box.

d, are leather straps, an inch or more wide, and long enough to reach from the bottom-board, where each one is fastened by two screws, to the egg-box, after being placed on the springs. Figure 105 represents the side and end boards, which, when placed over the bottom-board holding the springs, forms the spring-box; screws fasten the side and end-boards to the bottom-board of the spring-box, pieces of tin being nailed around the corners of the box, to give it proper strength, the nails being clinched on the inside.

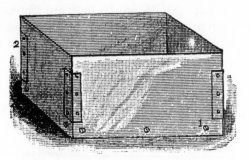

FIG. 165. Frame for holding egg-box.

After securing the springs and straps to the bottom board, the egg-box should be placed on the springs and the points of the springs placed in holes previously made in the bottom of the egg-box to receive them. Now put a sufficient weight in the egg-box to settle it down firmly on the springs, and fasten the upper ends of the straps to the box, being careful to have the box set level. Having

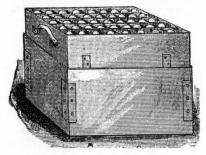

FIG. 166. Egg-box complete.

done this, take the part, figure 165, and put it down over the egg-box to its place, and make it fast to the bottom-board of the spring-box with heavy screws. The object of the bottom, figure 165, is to keep the springs from being strained to one side in going over rough ground. It should be made one-quarter inch or so larger than the egg-box, that the latter may have the benefit of the springs. Our former custom was to put a feeding of hay in the wagon-box, about midway from one end to the other, place the egg-box on the hay, and drive carefully over the rough places. But more or less eggs would be broken, the best we could do, whether they were packed in bran or put in paper "boxes" or cases. After setting the box on springs as described, place it on the bottom boards of the wagon-box, with one end directly over the forward axle of the wagon.

A BUSH-ROLLER

Figure 167 shows a device which has been made for clearing sage-bush land. It consists of a roller, eight feet long and two and a half feet in diameter, coupled by a short tongue—six feet is long enough—to the forward wheels of a wagon. A standard at each end of the roller-frame supports a cross-piece just clear of the roller. Upon this cross-piece, about four feet apart, and extending to the bolster of the wagon, are bolted two pieces of one and a quarter by six-inch spruce boards. A board is placed across the centre for a scat, thus making a complete and easy-inclining "buck-board." With a span of good horses and this machine, figure 167, one can roll from eight to ten acres of sage-bush in a day; and it is so easily killed, that in two or three weeks after such treatment, it will burn off like a prairie on fire.

Fɪɢ. 167. A home-made bush-roller.

BROOD-SOW PENS

Figures 168 and 169 represent a convenient arrangement for brood sows. The pens are not equal to the costly piggeries of wealthy breeders, but they answer a good purpose in a new country, where farmers are obliged to get along cheaply. Many who have built expensive houses say these pens answer a better purpose. First,

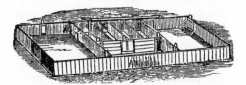

FIG. 168. Plan of cheap pig-pen.

there is a tightly-boarded pen (except in front;, sixteen feet by twelve feet. This is divided into four nests, twelve by four feet. A shed roof extends eight feet from

FIG. 169. Door to pen.

the rear. The tops of the nests are covered with boards, and the space between this room and the roof is filled with straw, making it wind-tight, except in front. When young pigs are expected during the cold weather of winter, hang a gunny sack in front of the nest.

The doors, figure 169, are the most convenient. The board door is slipped in from the top, between pairs of cross-boards in the pig-pen.

A RABBIT TRAP

Rabbits are a great nuisance both in the garden and orchard, and a trap of the following kind put in a blackberry patch, or some place where they like to hide, will thin them out wonderfully. A common salt barrel, with a notch sawed out at the top, is set in the ground level with the top. There is an entrance box, four feet long, with side pieces seven inches wide—top and bottom four and a half or five inches. The bottom board is cut in two at b, and is somewhat narrower than in front, that it may tilt easily on a pivot at c. A small washer should be placed on each side of the trap at c, that it may not bind in tilting. The distance from b to c should be somewhat longer than from c to d, that the board will fall back in place after being tipped. No bait is required, because a rabbit (hare) is always looking for a place of security. The bottom of the box should be even with the top of the ground at the entrance to the top of the barrel. The barrel should be covered closely with a board, as shown in figure 170. Remove the rabbits from the trap as fast as they are caught.

Fig. 170. A good rabbit trap.

964

WOODEN STABLE FLOOR

Elm makes an excellent and durable stable floor; the fibre of the wood is tough and yielding. The planks should be secured in position by wooden pins, as they are constantly liable to warp. Any of the soft oaks make a good floor, the hard, tough varieties are unyielding, and, until they have been in use several months, horses are liable to slip and injure themselves in getting up. Both pine and hemlock make good floors, being soft and yielding, but they are not as durable as many other woods. Planks for a stable floor should be two and a half inches in thickness, and not laid until quite thoroughly seasoned, and then always put down lengthwise of the stall, and upon another floor laid crosswise, as shown at *b*, *b*, *b*, figure 171. The planks of this floor, or cross

Fig. 171. Manner of laying a stable floor.

965

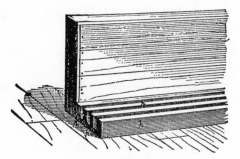

FIG. 172. A slatted stable floor.

floor, should be laid one inch apart, that they may the more readily dry off, and offer a better ventilation to the floor above. Unless the upper floor is of material liable to warp, it should not be nailed or pinned, but made as close-fitting as possible. It is not profitable or necessary to have the stall planks more than eleven and a half feet in length, or extend farther back than the stall partition, as shown at *e, e.* This plan leaves a wide smooth walk behind the stalls at *k,* so necessary for ease and rapidity in cleaning the manure from the stable.

FIG. 173. A cleaner for a slatted floor.

Some horse-keepers prefer a slatted floor, similar to that shown in figure 172. Material of the proper length, four inches wide and two inches thick, is set upon edge, as at *h,* with a strip three-quarters of an inch thick and one and a half inch wide placed between the slats, the whole made to fit the stall as closely as possible. By this method it is quite impossible for horses to become so dirty as when lying upon a common plank floor, as the space between the slats form a most admirable channel for carrying off the urine. A few days' constant use somewhat clogs the passages, but they are readily opened by using a home-made cleaner, like that shown in figure 173. Stable floors should have at least one inch descent in ten feet, and many make the descent three and even four inches in the same distance, but this is unnecessary. All stabled animals should stand upon floors as nearly level as is consistent with cleanliness.

A RAIL HOLDER OR "GRIP"

Drive two posts, *b, b,* figure 174, three feet long, firmly in the ground, four feet apart, between two parallel logs, *a, a.* A third post or "jaw," *c,* somewhat shorter, is mortised in a block placed between the logs, and out of line with, or to one side of the posts, *b, b,* so as to hold a rail, *d,* between the three. A lever, *e,* eight feet long, and heavy at the outer end, is mortised into another block, which is placed on the side of *a, b,* both blocks bearing against the posts. The lever and jaw are connected by a chain passing around the lever, over its block and through a hole in the jaw. An iron pin through a link couples them just enough apart to hold a rail firmly when the lever is on the ground. To remove the rail, raise the lever and rest it upon the small post, *f,* at the farther end, which slackens the chain.

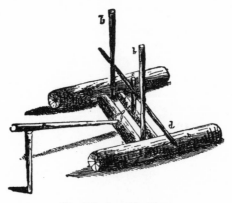

FIG. 174. A rail holder.

A CHEAP AND DURABLE GRINDSTONE-BOX AND HANGERS

A good grindstone, well hung, is one of the most valuable aids about the farm or workshop. Those who cannot afford to buy a very neat and handy grindstone frame of the hardware dealers, will find a frame and hangers shown in figure 175, that for convenience, cheapness, and durability is hard to excel. The frame consists of a well-seasoned "trough" of pine or other wood, fourteen inches square (or even one foot square), and from two and a half to three and a half feet in length, to which legs are nailed at *b*, *b*, four inches wide, an inch and a half thick, and bevelled at the top. Supports or hangers, *h*, *h*, are nailed firmly to the side, as indicated; they should be hard wood, and of a size to correspond with dimensions and weight of stone. The shaft may be of iron or wood; fit a piece of sheet lead, or piece of lead pipe, properly flattened out, in the top of each hanger; this will cause the shaft to turn easily, and prevent all squeaking for want of oil. The wooden plug at *r*, is for drawing off the water after each using of the stone, and should in no case be neglected. If one side of the stone is left standing in water, it softens, and the surface will soon wear quite uneven. After the box is completed, give it one heavy coat of boiled oil; then in a few days apply a coat of lead and oil, and with even common care, it will last a lifetime. When the stone becomes worn, it is kept down to the water by simply deepening the groove in the top of the hangers. Always buy a long shaft for a grindstone, for in this age of reapers and mowers, the cutting apparatus of which must be ground, a long shaft for a grindstone is almost a necessity, or truly a great convenience. If the grindstone is to stand out-doors, always cover it with a closely fitting wooden box when it is not in use.

Fig. 175. A box for a grindstone.

A "LADDER" FOR LOADING CORN

Take a plank two inches thick, ten inches wide, and eight feet long. Nail upon one side of it cleats, of one-inch by two-inch stuff, at easy stepping distances apart. At the upper end nail upon the underside of the plank a cleat projecting four inches upon either side, to which attach small ropes or chains, and suspend the ladder from the hind end of the rack, so that one end of the plank will rest upon the ground. This makes a very convenient step-ladder, up which a man can carry a large armful of fodder, and thus load his wagon to its full capacity with greater ease than two men could load it from the ground. I find it of great convenience to me when hauling corn fodder alone. The "Ladder" is shown in figure 176.

Fig. 176. A "ladder" for loading corn.

PROTECTING OUTLET OF DRAINS

One of the greatest annoyances in underdraining is the trouble arising from the outlet becoming choked or filled up by the trampling of animals, the action of frost, or even of water in times of freshets. This trouble is quite successfully overcome by the arrangement as shown in figure 177; it consists of a plank, ten or twelve inches in width, and five or six feet in length, with a notch cut in one side, near the centre. This plank is set upon edge at the outlet of the drain, with the notch directly over the end of the tile, and is held in position by several stakes

Fig. 177. End of tile drain.

on the outside, with earth or stone thrown against the opposite side This plan is best for all light soils, while for heavy clay land the one shown in figure 178 is just as good, and in most cases will prove more durable. It consists of two logs, eight or ten inches in diameter, and from three to ten feet in length, placed parallel with the drain, and about six inches apart; the whole is covered with plank twenty inches long, laid crosswise, Flat stones will answer and are more lasting than planks. The whole is covered with earth, at least eighteen inches in depth; two feet or more would be better, especially if the soil is to be plowed near the outlet.

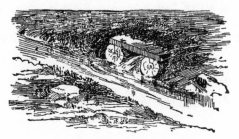

Fig. 178. Logs at end of drain.

A LOG BOAT

A convenient boat for dragging logs is shown in figure 179. The runners, *d, d,* are two by six inches and four and a half feet long; the plank is two by nine inches, and three and a half feet long. A mortise is made at *h* for the chain to pass through. The cross-piece, *c,* is four by seven inches, and three and a half feet long, and worked down to four and a half inches in the middle, Notches are cut into the cross-piece four inches wide and two inches deep, to receive the scantlings, *e, e,* two by four inches, and three feet long, which are fastened down by

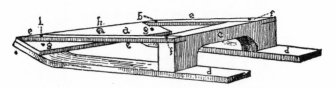

FIG. 179. A stout log boat.

strong bolts, as shown at the dotted lines, *f, f.* The two bolts in front, *b, b,* go through the scantling, plank and runner, while the bolts, *g, g,* pass only through the plank and runner.

It will be more convenient to load the logs by horses, as shown in the illustration, figure 180. The boat is raised with its upper side against the log. The chain is fastened to the cross-piece at *a,* with the large hook, and the other end is put around the log, under the runner and cross-piece at *b,* and pulled through between the runner and scantling at *c,* when the end of the chain, *d,*

974

is fastened to the whippletree. As the team is started, the boat tips over, with the log on top. Loosen the chain from the two-horse evener, and pull it back through the runner and scantling at *c*, and through the hole.

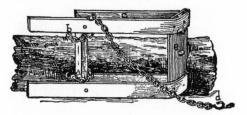

FIG. 180. Loading the logs.

CHEAP AND DURABLE
WAGON SEATS

It is tiresome to be jolted over rough roads, in a wagon without springs, with a simple board for a seat; but no farmer or cartman need adhere to this practice, when comfortable and portable seats can be so easily and cheaply made.

For a one-man seat, that shown in figure 181 is the simplest and most durable, and should be one foot

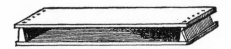

Fig. 181. Spring seat for one person.

longer than the wagon-box is wide; the connecting blocks should be four inches high, and placed near the ends. The one shown in figure 182 is arranged for two persons, the connecting block being placed in the centre, the ends being kept a uniform distance apart by

Fig. 182. A double spring seat.

bolts, with the nut upon the lower side, out of the way. The hole for the bolt through the lower board should be just large enough to allow the bolt to play freely.

In figure 183 is shown a seat a little more expensive, yet far more elastic. Both boards are eight inches longer than the width of the box upon which they rest. At each end of the top-board is mortised or nailed in a strip of hard wood, one inch thick, two inches wide, and about seven inches in length, which is made to pass freely up

FIG. 183. A coiled spring seat.

and down in a corresponding notch sawed in the end of the lower board. At or near each corner of the seat is placed a coiling spring. A pin, passed through the wooden strip near the bottom, keeps the seat-boards from separating.

A BAG-HOLDER ON PLATFORM SCALES

Figure 184 shows a contrivance which does away with the need of a second person in filling grain bags, and is both cheap and simple. It is attached to a platform scales for convenience in weighing, and consists of an iron hoop, nearly as large around as a bag. The hoop has four small hooks on it, at equal distances apart, to which the bag is fastened. Attached to the hoop is a piece of iron about six inches long, exclusive of the shank, which

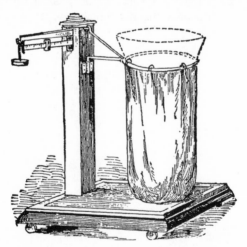

Fig. 184. A bag-holder on platform scales.

slips into a socket fastened to the front of the upright enclosing the rods, that run from the bottom of the scale to the weighing beam. This iron and hoop are fastened securely together. The shank should fit loosely in the socket, to let the hoop tilt down, so that the bag can be readily unhooked. There is an eye-bolt in the hoop where the iron rod joins it, and a rod with a hook on the upper end is fastened into it. This rod reaches to a staple fastened above the socket on the upright of the scales, as shown in figure 184. When the hook on the end of this rod is slipped into the staple, it lifts the hoop to a level position, and is of sufficient strength to hold a bag of grain. The hoop should be high enough to allow a bag to clear the platform of the scales. When filled, a sharp blow of the hand removes the hook of the sustaining rod, and lets the hoop tilt downward, when the bag rests on the platform. The hoop can be swung to one side, and entirely out of the way. We have a sort of hopper made out of an old dish pan with the bottom cut out. It is very convenient to keep grain from spilling while filling the bags.

MAKING BOARD DRAINS

On very many farms, wooden drains are used in place of tiles, but mostly in new districts where timber is cheap, and tiles cannot be purchased without much expense. They will answer the purpose well, without

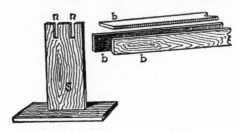

Fig. 185. Frame for folding boards.

much expense. Wooden drains, if laid deep enough, so that the frost will not affect them, will last many years. We know of an old drain that has been built twelve years, where the timber is still sound in some spots. To make wooden drains, two men are generally required—one to hold the boards, and another to nail them. This mode of constructing board drains can be improved upon by

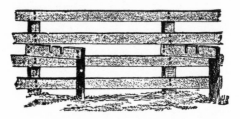

Fig. 186. Frame by a fence.

making a "standard," which consists of an upright board three feet high, having notches cut into it six inches apart, one inch wide, and several inches deep, to hold the boards firm. The boards b, b, figure 185, are laid into the notches, n, n, when the top board can be quickly and easily nailed on. Another method, shown in figure 18G, consists of two posts, driven into the ground about three feet from a fence, with a board nailed across from each post to the fence. Notches are then cut into each cross-board several inches deep, when it will be ready for use.

PUT THINGS IN THEIR PLACES

We have in mind an extensive and well-tilled farm, where a large space in the end of a wagon-shed is called a tool-room. The tools are deposited in the barn, wood-shed, crib, in the field, hung in trees, anywhere but in the right place. The tool-room floor is covered with heaps

FIG. 187. Section of a tool room.

of rusty iron, old leather, broken harness, fragments of tools, and other accumulations of forty years of farm life. The old iron should be sorted over, and any bolts, nuts, rings, hooks, etc., that are good, may be put in a box by themselves, and the rest should go to the junk dealer. There may be a few straps and buckles of the old harness worth saving. If so, oil the leather and lay it aside; throw the rest out of sight. Put a light scaffold near the roof-plates, and pile many small articles upon it; they will be out of the way and within easy reach. Make a drawer in a bench for holding small tools, and a row of pigeon-holes for nails, screws, etc. Across one end of the room, in front of the plate, fasten a long narrow board by pegs, so that a six-inch space will be between the plates and board. Let the pegs be a foot apart and stand out beyond the board some five or six inches, upon which to hang long-handled tools. About four feet from the floor make a similar rack for shovels, picks, chains, whippletrees, etc. Bring all the tools to this room, except those needed every day in the barn. There should be a paint-pot in the tool-house, to use on a rainy day for painting the tools. Figure 187 shows a section of a well-arranged tool-room.

Lay down this law to your man-servant and maidservant, to your son and daughter, to your borrowing neighbor and your good wife, to all that in your house abide, and to yourself: "That whoever uses a tool shall, when his or her work is done, return the tool to the tool-houseand place it where it was found."

WATER-SPOUT AND STOCK-TROUGH

The water-trough for the stock should not be immediately under the pump spout, but some ten or twelve feet distant, a spout being employed to convey the water. This spout (figure 188) is made of two good pieces of clean white pine, inch stuff. One piece is four inches and the other is three inches wide, nicely planed and jointed. If securely nailed, it will not leak for a long time, but when it does, let it dry, and then run hot pitch down the joint. The trough should be made of two-inch oak, or pine of the same. thickness may do, if kept well painted, inside and out. Instead of nailing on the sides to the ends have the ends fitted into grooves, and use rods, with burrs on them to bring the sides up tightly to their places. When the trough leaks, tighten up the burrs a little with a wrench, and the trouble generally ceases for the time. Even the best trough is by no mean3 very lasting, and its longevity is increased by keeping it thoroughly painted, inside and out, with good paint. Where there are horses that destroy the edges of the trough with their teeth, it is a good plan to rim it all around with thin iron. The spout, where it goes under the pump, can have a strap slipped over the nozzle of the pump.

Fig. 188. Water-trough.

984

A DESIRABLE MILKING SHED

(See Frontispiece.)

We recently observed a peculiarly constructed building used as a milking shed during the warmer portions of the year. It is a common frame structure, thirty-five feet in length and eighteen feet wide, with posts eight feet high. The sides and ends are boarded up and down with eight-inch stuff, leaving a space three inches wide between the boards for ventilation, light, etc. A row of common stanchions are placed along each side. A door is made at one end, through which the cows enter. If grain is fed, it is placed in position before the cows are admitted. A small quantity of salt is kept on the floor, immediately in front of the stanchions, thus allowing the cows to obtain a supply twice each day. This manner of salting is an inducement for the cows to enter the building and take their accustomed places; it also tends to keep them quiet while milking. This arrangement, for cleanliness, ventilation, etc., is far superior to the common basement stables, and is a great improvement over the usual plan of milking in the open yard, where broken stools, spilled milk, and irritable tempers are the rule rather than the exception. No matter how stormy it may be without, this shed always secures a dry place, with comparative quiet. A greater supply of milk is obtained with such a shed. The floor of the stable portion may be of earth, covered with coarse gravel.

WEAR PLATE FOR HARNESS TUGS AND COLLARS

In the manufacture of improved harness trimmings, devices are employed to prevent, as much as possible, the wear and breaking of the tugs where the buckle tongue enters them. This is quite an important point with those purchasing new harness. The simple contrivance, such as is shown in figure 189, consists of a thin iron plate a little narrower than the tug, and about two inches in length, with a hole for the reception of

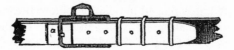

FIG. 189. Wear plate for tug.

the buckle-tongue when placed between the tug and the buckle. The strain from the buckle upon the tug is equally distributed over the entire surface against which the plate rests. A harness thus equipped will last many years longer than those not so provided. There is another part of the harness that is the cause of much trouble-mainly, the part where the tug comes in contact with the collar. The tug and its fastenings to the hame soon wear through

FIG. 190. Wear plate for hames.

the collar, and compress the latter so much that during heavy pulling the horse's shoulder is often pinched, chafed, and lacerated. This is worse than carelessness on the part of the teamster, as the collar should be kept plump at this point, by re-filling when needed; yet, very much of this trouble may be avoided by tacking to the underside of the hame a piece of leather, as shown in figure 190. It will be found not only to save the collar, but prevent chafing of the shoulder.

POTABLE WATER FENCE

The water fence, shown in figure 191, is one of the best we have ever used, and those who live near or on tide-water will find such an one very useful. This fence is made usually of pine; the larger pieces, those which lie on the ground and parallel with the "run" of the fence, are three by four-inch pieces, hemlock or pine, and connected by three cross-bars, of three by four-inch pieces, mortised in, three feet apart. Into the middle of these three cross-pieces (the upright or posts), are securely mortised, while two common boards are nailed underneath the long pieces to afford a better rest for the structure when floating on the water, or resting on the ground. Stout wires are stretched along the posts, which are four feet high.

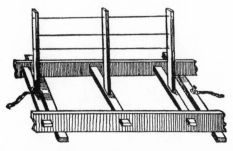

Fig. 191. Section of a water fence.

DITCH CLEANER AND DEEPENER

Open ditches require constant attention to prevent their being choked with weeds and accumulations of silt. Keeping them cleaned out with a hoe is a difficult and laborious task, while drawing a log down them is unsatisfactory and ineffective. To run a plow along the bottom is not only a disagreeable task, but frequently does more harm than good. In view of these facts we devised the simple and effective implement shown in figure 192.

Fig. 192. A cleaner for ditches.

The centre-piece is six by eight-inch oak, eight feet long, and shaped as shown in the cut. The wings, or scrapers, are made of oak, or other tough wood; boards ten inches wide. They are attached to the centre-piece at the forward end by an inch bolt that passes through all three pieces. They are connected at the rear end by a strong cross-bar of hard-wood. Twelve or fifteen inches back of this bar the end of the lever is attached to the centre-piece by an eye and staple. A short chain is fastened underneath the centre of the cross-bar, with an

989

eye-bolt passing through it. The chain is attached to the lever with a hook, and may be lengthened or shortened as required.

The implement is drawn by two horses, one on each side of the ditch. A man stands on the centre-piece, and handles the lever. If the ditch is narrow and deep, the rear ends of the wings or scrapers will naturally be forced upward to a considerable hight, and the lever chain should be lengthened accordingly. In wide, shallow ditches, the cross-bar will nearly rest on the centre-piece, and the chain must be short. The scrapers are forced down hard by bearing on the lever. If the bottom of the ditch is hard, two men may ride on the implement. Long weeds catching on the forward end must be removed with a fork. A strap of iron is fastened across the forward ends of the scrapers where the bolt passes through to prevent them from splitting. The horses may be kept the proper distance apart by means of a light pole fastened to the halter rings.

HOW TO BUILD A DAM

A form of crib, shown in figure 193, is built of logs, about eight feet square for ordinary streams. The bottom should have cross-pieces pinned on the lowest logs. The stones that till the crib rest on these cross-pieces and hold everything secure. The crib can be partly built on shore, then launched, and finished in its place in the dam. All the logs should be firmly pinned together.

FIG. 193. A crib for a dam.

The velocity of the stream will determine the distance between the cribs. The intervening spaces are occupied with logs, firmly fastened in their places. Stone is filled in between the logs, and the bottom is made water-tight with brush and clay.

A dam without cribs, built of timbers spliced together, and reaching quite across the stream, is shown in figure 194. The frame is bound together with tiers of

FIG. 194. Log frame for a dam.

cross-timbers about ten feet apart. The sides of this framework of spliced logs are slanting and nearly meet at the top. The interior is filled with stone and clay, and planked over tightly, both front and rear. For a small stream with an ordinary current, this is perhaps the cheapest and most durable dam made. The engravings fully illustrate the construction of the two forms.

DRIVING HOP AND OTHER POLES

The usual method of driving stakes, etc., is to strike them on the upper end with a sledge or other heavy article; but in the case of hop or other long poles this mode is impracticable. Hop poles are usually set by making a

Fig. 195. Driving block.

hole with an iron bar and forcing into it the lower end of the pole. Poles and other long stakes often need to be driven deeply in the ground, and this may be done quickly, and without a high step or platform, by using a device shown in figure 195. This consists of a block of tough wood, one foot in length, four or five inches square at the top, made tapering, as shown, with the part next the pole slightly hollowed out. Take a common trace chain, wind closely about the block and pole, and hook it in position. With an axe, sledge, or beetle, strike heavy blows upon the block. Each blow serves only to tighten the grip of the chain upon the pole. In this way, quite large poles or stakes may be quickly driven firmly in the ground. To keep the chain from falling to the ground when unfastened from the pole, it should pass through a hole bored through the block.

A CONVENIENT GRAIN BOX

The box here represented, figure 196, is at the foot, and just outside of the bin. It serves as a step when emptying grain into the bin. The front side of it is formed by two pieces of boards, hung on hinges at the outside corners, and fastened at the middle with a hook and staple. The contrivance opens into the bin at the back, thus allowing the grain to flow into it. When a quantity of grain is to be taken from the bin, the cover is fastened up, the front pieces swung round, giving a chance to use the scoop-shovel to fill bags or measures. The box is a

Fig. 196.   Grain boxes.   Fig. 197.

foot deep and sixteen inches wide. Its length is the same as the width of the bin. The first four boards, forming the front of the bin, may be made stationary by this arrangement, as, at that convenient hight, bags may be emptied over by using the box as a step. The cost of this is about seventy-five cents. An improvement has the front piece and ends nailed together, and the whole fastened to the bin-posts by hooks and staples from the end-pieces, as shown in figure 197. Then the whole could be removed by unhooking the fastenings, and the cover could be let down, to form the lower board on the front of the bin, if desired.

A ROAD-SCRAPER

A road-scraper is shown in figure 198, which consists of a heavy plank or hewn log, of oak or any other hard timber, six feet long, six inches in thickness, and ten inches wide. A scantling, 5, two by four inches thick and six feet long, and the brace, *c*, are secured to the log, *a*, by a strong bolt. The edge of the scraper is made of an old drag-saw, and secured by rod-iron nails. The scantling serves as a reach, and is attached to the front part of a heavy wagon, when in use. When the road is very hard, it becomes necessary sometimes for the driver to stand on the scraper, to make it take better hold. The scraper should be shaped about as shown at *d*, in the engraving, so as to make it run steady, and cause the loose dirt to slide to one side, and leave it in the middle of the road.

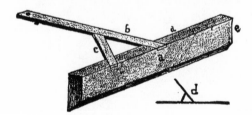

Fig. 198. A road-scraper.

AIDS IN DIGGING ROOT CROPS

Figure 199 shows a carrot and sugar beet lifter, made in the following manner: Take a piece of hard wood, two and a half by three inches, and six feet long, for the main piece, *a*, into which make a mortise two feet from the wheel end, to receive the lifting foot (figure 200); attach two handles, *b*, *b*, at one end, and a wheel, *c*, at the other.

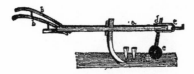

Fɪɢ. 199. A root lifter.

This wheel can be set high or low as desired, by the set screw, *d*, in the clevis, *e*. Figure 201 shows the lifting "foot" separate from the machine. This is made of flat iron or steel, five-eighths inch thick and three inches wide, with a steel point and a small wing at the bottom. It is in the curved form seen in the engraving. The roots are first topped with a sharp hoe or sickle, two rows of tops being thrown into one, which leaves one side of the rows clear for the lifter. The horse walks between the

Fɪɢ. 200.

Fɪɢ. 201.

997

rows and the foot of the implement enters the ground at the side of the roots in a slanting direction, as shown in figure 201, lifting the roots so they may be rapidly picked up. The implement is very easily made to run deep or shallow, by simply changing the wheel and lifting, or pressing down upon the handles. A "foot," made in the form of figure 200, may be placed in the centre arm of a common horse hoe with sides closed, and used as above.

THE WOOD-LOT IN WINTER

A few acres in trees is one of the most valuable of a farmer's possessions; yet no part of the farm is so mistreated, if not utterly neglected. Aside from the fuel the wood-lot affords, it is both a great saving and a great convenience to have a stick of ash, oak, or hickory on hand, to repair a break-down, or to build some kind of rack or other appliance. As a general thing, such timber as one needs is cut off, without any reference to what is left. By a proper selection in cutting, and the encouragement of the young growth, the wood-lot will not only continue to give a supply indefinitely, but even increase in value. A beginnings and often the whole, of the improvement of the wood-lot, is usually to send a man or two to "brush it," or clean away the underbrush. This is a great mistake. The average laborer will cut down everything; fine young trees, five or six years old, go into the heap with young poplars and the soft underbrush. The first point in the management of the wood-lot is, to provide for its continuance, and generally there are young trees in abundance, ready to grow on as soon as given a chance. In the bracing winter mornings one can find no more genial and profitable exercise than in the wood-lot. Hard-wooded and useful young trees should not have to struggle with a mass of useless brush, and a judicious clearing up may well be the first step. In timber, we need a clean, straight, gradually tapering and thoroughly sound trunk. In the dense forest, nature provides this. The trees are so crowded that they grow only at the upper branches. The lower branches, while young, are starved out and soon perish, the wounds soon healing over are out of sight. In our open wood-lots, the trees have often large heads, and the growth that should be forming the trunk is scattered over a great number of useless branches. Only general rules can be given in pruning neglected timber trees; the naked trunk, according to age, should be from one-third to one-half the whole hight of the tree; hence some of the lower branches may need to be cut away. All the branches are to be so shortened in or cut back as to give the head an oval or egg-shaped outline. This may sometimes remove half of the head, but its good effects will be seen in a few years. In removing branches, leave no projecting stub on the timber, and cover all large wounds with coal-tar. Whosoever works in this manner thoughtfully cannot go far astray.

SWINGING-STALL FRONTS

The value of swinging-stall fronts is appreciated by those who have used them. They prevent the animals from putting their heads out into the alleys, and endangering themselves thereby. The "cribber," or "wind-sucker," has been made such by want of a contrivance like the one shown in figure 202. Anyone with a moderate knowledge of the use of tools can put it up, as the

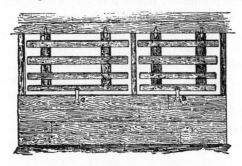

Fig. 202. Front of stalls.

engraving shows how it is made; *a, a,* being straps to fasten the "fronts" down into place when they are not raised to feed the stock. Inch stuff constitutes the material. The cleats to which the strips are attached should be four inches wide, with the sharp, exposed edges taken off with a plane. The strips should be from two to two and a half inches wide, and attached with screws or wrought nails. The hinges can either be of wrought iron or of heavy leather. If more durable fronts are desired, oak, or yellow pine can be used, though it is much more expensive. Unplaned lumber will answer, but to make a neat, workmanlike job had better use planed lumber.

SAVE ALL CORN FODDER EVERYWHERE

The profits of farming, as in other business, is the margin between receipts and expenditures. The receipts are largely augmented by saving wastes; these wastes in farming are enormous in the aggregate. The losses in this direction, that might be saved, would make the business very profitable, where it is now barely paying, or not doing that. Take corn stalks, for example. The leaves and a portion of the stems that produce each bushel of corn have a certain amount of nutriment that would support and increase the weight and growth of animals. Yet of our great corn crop, seventeen hundred to two thousand million bushels annually, only a very small part of the fodder is turned to much account. At the very lowest estimate, the stalks yielding one bushel of corn are on the average worth ten cents for feed, even including the great corn regions—a total of two hundred million dollars. At the South, generally, little value is attached to corn stalks as fodder. At the West, many farmers let their cattle roam in the fields, pick off some leaves, eat a little of the stalk, and trample the rest down; they pack the earth so much in trampling on it, that the damage thus done to many fields surpasses the value of the food obtained.

Nearly the whole of a corn stalk, except a very little of the thin, hard outside coating, affords nutritious fodder, if it is cut at the proper time, is well cured and judiciously fed. It needs to be cut when not so green as to mould in the shock, but not so ripe as to lose all its succulence and become woody. Experience and observation will generally indicate to every one the proper time of cutting it.

In shocking corn, the stalks should be kept straight and parallel. The shocks should be large enough to not have too many stalks exposed to the weather, yet small enough to dry and cure through. For somewhat heavy corn, twelve hills square (one hundred and forty-four hills), is abundant for one shock. A good mode of shocking is this: When the shocks are set nearly perpendicular, draw the tops together very firmly with a rope, and tie temporarily—two men working together. Bind with straw or with stalks. For the latter choose tough, nearly ripe, long, slender stalks. "Bend-break" the top with the thumb and finger every two or three inches. Thrust the butt end into the shock and towards the centre nearly two feet, and carefully bend-break it at the surface to a right angle. Insert a similar top-broken stalk two feet distant; bring the top of the first one firmly around the shock, bend it around the second stalk close to the shock, and then bend the second stalk around and over a third one; and so on, using as many stalks as required by size of shock and length of binders. Bring the end of the last one over the bend in the first, and tuck it under the binder into a loop, into which insert a stalk stub, pushing it into the shock to hold the loop. All this is more quickly done than described.

IMPROVED BRUSH RAKE

One of the most disagreeable tasks connected with a hedge fence is gathering and burning the annual or semi-annual trimmings. It is generally done with pitchforks, and often causes pain. To have a long shoot, covered with thorns an inch long, spring out from a roll of brush and hit one square across the countenance, is exasperating in the extreme. To avoid this danger, many expedients are resorted to. Among the best of these is a long, strong rail, with a horse hitched to each end by means of ropes or chains eight or ten feet long. A boy is placed on each horse, and two men with heavy sticks, eight or ten feet long, follow. The horses walk on each side of the row of brush, and the men place one end of their

Fig. 203. A brush rake.

sticks just in front of the rail, and hold them at an angle of about forty-five degrees, to pre rent the brush from sliding over it. When a load is gathered, the horses are turned about, and the rail withdrawn from the brush.

The device shown in figure 203 is an improvement on this method. A good, heavy pole, eight to twelve feet long, has four or five two-inch hard-wood teeth set in it, as seen in the cut. These teeth may be twelve to twenty inches long, and slide on the ground in front of the pole similar to those of a revolving hay rake. The handles are six to eight feet long, of ash or other tough wood, and fit loosely into the holes in the pole. Two horses are employed, one at each end of the rake. One man holds the handles, and raises or lowers the teeth as necessary. When a load is gathered, the handles are withdrawn, the ends of the teeth strike the ground, throw the pole up, and it passes over the heap. After a little practice, a man can handle this rake so as to gather up either large or small brush perfectly clean, and do it rapidly.

DIGGING MUCK AND PEAT

A dry fall often furnishes the best time in the whole circle of the year for procuring the needed supply of muck or peat for absorbents in the sty and stable. The use of this article is on the increase among those farmers who have faithfully tried it, and are seeking to make the most of home resources of fertilizers. Some who have used muck only in the raw state have probably abandoned it, but this does not impeach its value. All that is claimed for it has been proved substantially correct, by the practice of thousands of our most intelligent cultivators, in all parts of the land. There is considerable difference in its value, depending somewhat upon the vegetable growth of which it is mainly composed, but almost any of it, if exposed to the atmosphere a year before use, will pay abundantly for digging. This dried article, kept under cover, should be constantly in the stables, in the sties and sinks, and in the compost heap. So long as there is the smell of ammonia from the stable or manure heap, you need more of this absorbent. Hundreds of dollars are wasted on many a farm, every year, for want of some absorbent to catch this volatile and most valuable constituent of manure. In some sections it is abundant within a short distance of the barn. The most difficult part of supplying this absorbent is the digging. In a dry fall the water has evaporated from the swamps, so that the peat ped can be excavated to a depth of four or five feet at a single digging. Oftentimes ditching, for the sake of surface draining, will give the needed supply of absorbents. It will prove a safe investment to hire extra labor for the enlargement of the muck bank. It helps right where our farming is weakest—in the manufacture of fertilizers. It is a good article not only for compost with stable manure, but to mix with other fertilizerd, as butcher's offal, night soil, kainite, ashes, bone dust, fish, rock weed, kelp, and other marine products. Dig the muck when most convenient and have it ready.

A CLEANER FOR HORSES' HOOFS

The engraving herewith given shows a simple and convenient implement for removing stones and other substances from between the frog and the ends of a horse's shoe. Its value for this and other purposes will be quickly appreciated by every driver and horse owner.

FIG. 204. A hoof-cleaner.

When not in use, the hook is turned within the loop of the nandle, and the whole is easily carried in the pocket. The engraving shows the implement open, two and one-half times reduced in size. If horsemen keep this cleaner within easy reach, it will often serve a good turn, and be of greater value than a pocket corkscrew.

COLD WEATHER SHELTER FOR STOCK PROFITABLE

Not one farmer in a hundred understands the importance of shelter for stock. This has much to do with success or failure of tens of thousands of farmers. Animals fairly sheltered consume from ten to forty per cent. less food, increase more in weight, come out in spring far healthier; and working and milk-producing animals are much better able to render effective service. The loss of one or more working horses or oxen, or of cows, or other farm stock, is often a staggering blow to those scarcely able to make the ends of the year meet, and the large majority of such losses of animals are traceable to diseases due, directly or indirectly, to improper protection in autumn, winter, or spring. Of the food eaten, all the animals use up a large percentage in producing the natural heat of the body at all seasons, and heat enough to keep up ninety-eight degrees all through the body is absolutely essential. Only what food remains after this heat is provided in the system can go to increase growth and strength, and to the manufacture of milk in cows and of eggs in fowls. When heat escapes rapidly from the surface, as in cold weather, more heat must be produced within, and more food be thus consumed. In nature this is partly guarded against by thicker hair or fur in winter.

Any thinking man will see that an animal either requires less food, or has more left for other uses, if it is protected artificially against winds that carry off heat rapidly, and against storms that promote the loss of heat by evaporation of moisture from the surface of the body. A dozen cows, for example, will consume from two to six tons more of hay if left exposed from October to April, than if warmly sheltered, and in the latter case they will be in much better health and vigor, and give much more milk. Other cattle, horses, sheep and swine will be equally benefited by careful protection.

GOOD STONE TROUGHS OR TANKS

Figure 205 shows an unpatented stone water tank, or trough, neat, effective, and readily constructed by almost any one. These troughs may be of any length, width and depth desired, according to their position, use, and the size of stones available. Here are the figures of the one shown: The two side-pieces are flagging stones, six feet long and twenty-seven inches wide. The bottom-piece is four feet ten inches long, two feet wide; and the two

FIG. 205. A stone trough.

end-pieces, two feet long, twenty inches wide, or high. These stones were all a little under two inches thick. Five rods, of three-eighths inch round iron, have a flat head on one end, and screw and nut on the other; or there may be simply a screw and nut on each end; they must not extend out to be in the way. Five holes are bored or drilled through each side-piece, which is easily done with brace and bit in ordinary stone. The middle hole is four to five inches above the bottom edge, so that the rod through it will fit under and partially support the bottom stone. The end rods are about four inches from the ends of the side-pieces, and stand clear of the end stones in this case so that the dipper handles hang upon them; but they may run against the end stones. When setting up, the stones being placed nearly in position, newly-mixed hydraulic cement is placed in all the joints, and the rods screwed up firmly. The mortar squeezed out in tightening the rods is smoothed off neatly, so that when hardened the whole is almost compact solid stonework—if good water-lime be used. Almost any fiat stones will answer, if the edges of the bottom and end-pieces be dressed and a somewhat smooth groove be cut in the side-pieces for them to fit into or against. The mortar will fill up any irregularities. A little grooving will give a better support to the bottom-piece and the ends than the simple cement and small rods. It will be noted that the side-pieces extend down, like sleigh runners, leaving an open space below. A hole can be drilled in a lower edge to let out the water in hard freezing weather, and be stopped with a wooden plug. Such tanks will keep water purer than wood, and last a century or longer, if not allowed to be broken by freezing. Any leakage can be quickly stopped by draining off the water and applying a little cement mortar where needed. When flagging or other flat stones are plentiful, the work and cost would be little, if any, more than for wooden tanks. They can be set in the ground if desired. The iron rods need painting, or covering with asphalt, to prevent rusting.

ARTIFICIAL FEEDING OF LAMBS

It frequently happens that artificial feeding of lambs is necessary, and to do it successfully good judgment is required. The point is to promote a healthy and rapid growth, and not allow the lambs to scour. The milk of some cows, especially Jerseys, is too rich, and should be diluted with a little warm water. Farrow cows' milk, alone, is not a good feed, since it frequently causes constipation. It may be given by adding a little cane molasses. Milk, when fed, should be at about its natural temperature, and not scalded. Lambs, and especially "pet" lambs, are often "killed with kindness." Feed only about a gill to a half pint at first. After the lamb has become accustomed to the milk, it may be fed to the extent of its appetite. When old enough, feed a little flax seed and oats, or oil-meal if early fattening is desired. There are various methods of feeding young lambs artificially. A satisfactory way is to use a one-quart kerosene oil can with the spout fixed so as to attach a nipple; the milk flows more freely from this than from a bottle, on account of the vent. Let ewes and lambs have clean, well-ventilated apartments. When the weather is mild and warm turn them out into the yard. If it is not convenient to let the ewes out, arrange partitions and pens, so that the lambs may enjoy the outside air and sunlight

A CONVENIENT BAILED BOX

The common box with a bail, or handle, is a useful farm appliance; it answers the purpose of a basket, is much more durable, and a great deal cheaper. Instead of a flat bail, we would suggest, for heavy work, a green hickory or other tough stick, to be chamfered off where it is nailed to the sides of the box, the portion for the hand being, of course, left round. It will be found useful

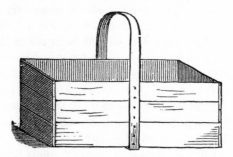

FIG. 206. A bailed box.

to have these boxes of a definite size, to hold a half-bushel or a bushel. A legal bushel is two thousand one hundred and fifty (and a fraction) cubic inches. A box may be made of this capacity of any desired shape. Ends a foot square, and side-pieces and the bottom eighteen and a half-inches long, will make a bushel box. If desired narrower, make the ends eight inches high and fourteen inches wide, with the sides and bottom two feet long. Such a box, shown in figure 206, holds a very little more than an even bushel. It is inexpensive.

SAWDUST FOR BEDDING

We have tried for two years dry sawdust in the cows' stable, and on the whole like it better than any bedding we have ever tried. It makes a more comfortable bed, completely absorbs the urine, and the cow is kept clean with less labor than when any other is used. The objection to salt-marsh sods, dried, or to headlands, and dry muck, is that they soil the cow, and make it necessary to wash the bag before milking. Straw, of all sorts, soon becomes foul, and, without more care than the ordinary hired man is likely to bestow, soils the cow's bag also. Dry sawdust is clean, and makes a soft, spongy bed, and is an excellent absorbent. The bag is kept clean with the aid of a coarse brush without washing A charge of fifteen bushels in a common box-stall, or cow stable, will last a month, if the manure, dropped upon the surface, is removed daily. The porous nature of the material admits of perfect drainage, and of rapid evaporation, of the liquid part of the manure. The sawdust is not so perfect an absorbent of ammonia as muck, but it is a much better one than straw, that needs to be dried daily, in the sun and wind, to keep it in comfortable condition for the animals. In the vicinity of saw and shingle mills, and of ship-yards, the sawdust accumulates rapidly, and is a troublesome waste that mill-owners are glad to be rid of it can be had for the carting. But even where it is sold at one or two cents a bushel, a common price, it makes a very cheap and substantial bedding. The saturated sawdust makes an excellent manure, and is so fine that it can be used to advantage in drills. It is valuable to loosen compact clay soils, and will help to retain moisture on thin, sandy and gravelly soils. There is a choice in the varieties of sawdust for manure, but not much for bedding. The hard woods make a much better fertilizer than the resinous timber. To keep a milch cow in clean, comfortable condition, we have not found its equal.

A CHEAP ENSILAGE CART

The adoption by many farmers of the silo method of preserving fodder, has made it necessary to change the manner of feeding live stock. When the ensilage is removed twenty feet or more from the silo to the feeding rack, it is best to have some means of conveying it in quantities of from one to two hundred pounds at a time.

FIG. 207. An ensilage cart.

This can be done cheaply and quickly by a small handcart, one of which any farmer having the tools can make in half a day. A good form of ensilage cart is shown in figure 207, and is simply a box eighteen inches wide, three feet long, and two and a half feet in hight. A wooden axle, of some tough fibre, is nailed to the bottom, ten inches from the end, and wheels from one to two feet in diameter are placed upon the axle. Suitable wheels can be made from planks, with cleats nailed on to keep them from splitting. Handles and legs are attached as shown in the engraving. The axle being near the centre, throws nearly the whole weight of the load upon it while being moved. It will be found easier to handle than a barrow, and not so liable to upset when unequally loaded. It is a cheap arrangement, and may be used for various other purposes as well as for moving ensilage.

MILKING AND MILKING TIME

Any one who has had to do with dairy farming knows that there are a great many poor milkers, against a few who understand and practice the proper method of removing the milk from a cow. It is a well-known fact that some persons can obtain more milk from a cow with greater ease and in quicker time than others. In the first place, there must be an air and spirit of gentleness about the milker, which the cow is quick to comprehend and appreciate. It is not to be expected that a cow, and especially a nervous one, will have that easy, quiet condition so necessary to insure an unrestrained flow of milk, when she is approached in a rough way, and has a person at her teats that she justly dislikes. There must be a kindness of treatment which begets a confidence before the cow will do her best at the pail. She should know that the milker comes not as a thief to rob her, but simply to believe her of her burden, and do it in the quickest, quietest, and kindest way possible. The next point in proper milking is cleanliness: and it is of the greatest importance if first-class milk and butter are the ends to be gained in keeping cows. No substance is so easily tainted and spoiled as milk; it is particularly sensitive to bad odors or dirt of any kind, and unless the proper neatness is observed in the milking, the products of the dairy will be faulty and second-class. Those persons who can and will practice cleanliness at the cow, are the only ones who should do the milking. It matters not how much care is taken to be neat in all the operations of the dairy, if the milk is made filthy at the start; no strainer will take out the bad flavor. Three all-essential points are to be strictly observed in milking: kindness, quickness, and neatness. Aside from these three is the matter of the time of milking. It should be done at the same hour each and every day, Sundays not excepted. It is both cruel and unprofitable to keep the cows with their udders distended and aching an hour over their time. We will add another *ness* to the essentials already given, namely: promptness.

A REVOLVING SHEEP HURDLE

An easily moved feeding hurdle is shown in figure 208. It consists of a stout pole or scantling of any convenient length, bored with two series of holes, alternating in nearly opposite directions, and twelve inches apart. Small poles five or six feet long are so placed in the holes that each adjoining pair makes the form of the letter X. These hurdles are arranged in a row across the field, and the sheep feed through the spaces between the slanting poles. The hurdles are moved forward by revolving them, as shown in the engraving. By using two rows of these hurdles, sheep may be kept on a narrow strip of land, and given a fresh pasture daily by advancing the lines of hurdles. This method of feeding off a forage crop is one of the most effective and inexpensive for enriching worn-out land, especially if a daily ration of grain or oil-cake is given to the sheep.

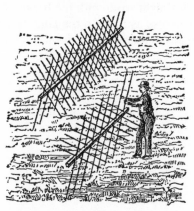

Fig. 208. A revolving hurdle fence.

LIGHTS IN THE BARN

It is estimated that nine-tenths of all fires are caused by carelessness. Winter is the season when the lantern is frequently used in the barn, and we give a word of caution. Never light a lamp or lantern of any kind in the barn. Smokers may include their pipes and cigars in the above. The lantern should be lighted in the house or some out-building where no combustibles are stored. A lantern which does not burn well should never be put in order in the hay-mow. There is a great temptation to strike a match and re-light an extinguished lantern, wherever it may be. It is best to even feel one's way out to a safe place, than to run any risks. If the light is not kept in the hand, it should be hung up. Provide hooks in the various rooms where the lights are used. A wire running the whole length of the horse stable, at the rear of the stalls, and furnished with a sliding hook, is very convenient for night work with the horses. Some farmers are so careless as to keep the lamp oil in the barn, and fill the lantern there while the wick is burning. Such risks are too great, even if the buildings are insured.

A NEST FOR SITTING HENS

The nest box shown in figure 209 can be made to contain as many nests as desired, and be placed in the poultry house or any other convenient place. When a hen is set in one of the nests, the end of the lever is slid from under the catch on top of the box, and the door falls over the entrance to keep out other hens. They rarely

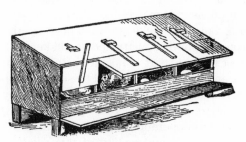

Fig. 209. Box of hens' nest.

molest the sitting hen after she has held exclusive possession three or four days, and the drop may be raised again. The box legs should not be over six inches long. The step in front of the nests, four to six inches wide, is a continuation of the bottom of the box. It is a vast improvement on old barrels, broken boxes, and other makeshift hens' nests so generally employed.

BARN-YARD ECONOMY

A dark stream, often of golden color, always of golden value, flows to waste from many an American barn-yard. This liquid fertility often enters the side ditch of the farm lane, sometimes of the highway, and empties into a brook, which removes it beyond the reach of plants that would greatly profit by it. Mice may gnaw a hole into the granary and daily abstract a small quantity of grain, or the skunks may reduce the profits of the poultry yards, but these leaks are small in comparison with that from the poorly-constructed and ill-kept barn-yard. The most valuable part of manure is that which is very soluble, and unless it is retained by some absorbent, or kept from the drenching rains, it will be quickly out of reach. Manure is a manufactured product, and the success of all farm operations in the older States depends upon the quantity and quality of this product. Other things being equal, the farmer who comes out in the spring with the largest amount of the best quality of manure will be the one who finds farming pays the best. A barn-yard, whether on a side-hill or on a level, with all the rains free to fall upon the manure heap, should be so arranged as to lose none of the drainage. Side-hill barn-yards are common, because the barns thus located furnish a convenient cellar. A barrier of earth on the lower side of the yard can be quickly thrown up with a team and road-scraper, which will catch and hold the drenchings of the yard above, and the coarse, newly-made manure will absorb the liquid and be benefited by it. It would be better to have the manure made and kept under cover, always well protected from rains and melting snows. Only enough moisture should be present to keep it from fermenting too rapidly. An old farmer who let his manure take care of itself, once kept some of his sheep under cover, and was greatly surprised at the increased value of the manure thus made. In fact, it was so "strong" that when scattered as thickly as the leached dung of the yard, it made a distinct belt of better grain in the field. The testimony was so much in favor of the stall-made manure that this farmer is now keeping all his live stock under cover, and the farm is yielding larger crops and growing richer year by year. If it pays to stop any leak in the granary, it is all the more important to look well to the manure that furnishes the food, that feeds the plants, that grow the grain, that fills the grain bin. At this season the living mills are all grinding the hay and grain, and yielding the by-products of the manure heap. Much may be saved in spring work by letting this heap be as small as out-door yard feeding and the winds and rains can make it, but such saving is like that of the economic sportsman who went out with the idea of using as little powder and lead as possible. In farming, grow the largest possible crops, even though it takes a week or more of steady hard work to get the rich, heavy, well-prepared manure upon the fields. More than this, enrich the land by throwing every stream of fertility back upon the acres which have yielded it. Watch the manure heap as you would a mine of gold.

A CHEAP MANURE SHED

Many farmers waste much of their stable manure by throwing it out of doors to be acted upon by sun and rain. We recently saw a very cheap, sensible method of almost wholly preventing such loss. A board roof, ten feet square, is supported by posts eight feet long above ground, which are connected inside by a wall of planks (or of poles, as the one examined was). Near the post at each end, stakes *a, a* (figure 210), are set, against which one end of the end-planks rest. This allows the front planks, *d, d,* to be removed in filling or loading. It is placed near the stable, preferably, so that the manure from the stable can be thrown directly into one corner, whence it is forked to the opposite corner in a few days, to prevent too violent fermentation. A frequent addition of sods, leaves, and other materials that will decompose, will increase the heap, and improve its value, supplying a manure superior to many of the commercial fertilizers, at less cost.

Fig. 210. A shed for manure.

A SHEEP RACK

The dimensions of the rack (fig. 211) are: length twelve feet, width two feet nine inches, and hight three feet. The materials are: ten boards twelve feet long, eight of them ten inches wide, one seven inches wide, and one eight inches wide; four boards, two feet nine inches long and twelve inches wide; six posts three by four inches, three feet long; sixty-four slats, sixteen inches long and one inch square; and two strips, twelve feet long and two and a half inches wide. Nail the two narrower boards in the shape of a trough, turn it bottom up, and draw a line through the middle of each side. Set the dividers to four and a half inches, and mark along the lines for holes with a three-quarter-inch bit, and bore the narrow strips to match. Set the slats into the trough, and fasten the strips on their upper ends. Nail two of the boards to the posts on each side, as seen in the sketch, and also the short boards on the ends. Lay in a floor one foot from the ground, and set in the trough as shown in the engraving. Fit a board from the slats up to the top of the outside of the frame. The floor need not cover the middle under the trough.

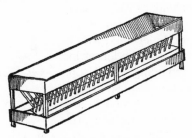

Fig. 211. Fodder rack for sheep.

A GOOD PICKET POINTER

On many farms a picket pointer might never be of use, but anyone wishing to put up a picket fence a hundred or more feet long would save time by making one for the occasion. Pickets may be purchased ready pointed, but true economy consists in doing as much of the work as

possible at home. This arrangement does not concern the fancy-topped pickets sometimes seen, but simply the popular square picket with pyramidal point, which makes, after all, one of the neatest fences that can be found for the yard. The waste material from building or fence-making, and an hour's time, will suffice for its making.

A bit of studding material, 30 inches long, has a hardwood strip

Fig. 212. Front view.

three inches wide nailed on each side so as to project half of its width forward, thus forming a groove in which the picket is held, as will be seen later. They extend lower down than the central piece and with it form the front leg. The left strip, instead of extending to the top, however, is there replaced by a broader bit of hardwood board five or six inches long and pro-

Fig. 213. Side view.

jecting forward three inches, after which the projecting

edges on both sides are sawed off at the proper angle for the picket points, say a little lower than 45 degrees.

The two rear legs are strips of lath five feet long, fastened near the top of the front leg and braced so that the forward part is not quite vertical. A block or seat 18 inches long is fastened across them 32 inches from the lower end, and so adjusted as to hold them one foot apart at the ground. The clamp by which the pickets are held in place consists of a half cylindrical block suspended by short lengths of strap iron and connected by a wire on each side to a foot lever, the action of which need be but slight.

Measure from the bevel at the top, down just the length the pickets are to be made, and place a block transversely in the groove at that point, for the stick to rest on. The groove should be at least one-fourth inch wider than the pickets, but a small wedge is inserted at the bottom on the left, so that as they fall into position they are crowded over to the right side.

To do the pointing, first cut all the pickets in a miter box to the right length, and at the proper angle to fit the water ledge over the baseboard, then place one in the groove of the pointer, thrust it down past the clamp, which it will push out, till it reaches the block at the bottom. Apply a little pressure on the foot lever to hold it in place, and then, with a sharp drawing knife, bevel the top, keeping the blade flat on the guides of hardwood; lift the picket, turn one quarter to the right, thrust down and cut again, and so on until it is finished. With poplar pickets one and one-fourth inches square, I have seen them pointed at a little more than one per minute, which is certainly much better than to lay off each one and cut with a chisel, as I have known a carpenter to do.

STERILIZING OVEN AND BOTTLE TRUCK

Both oven and truck for milk can be made by any carpenter and tinner. Fig 214 represents the sterilizing oven. It is made on a light frame, of matched lumber; the inside is lined with zinc soldered at the joints. The door should be double, with beveled edges fitting loosely and

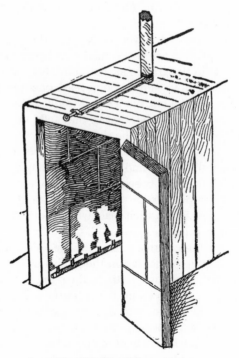

FIG. 214. Sterilizing oven.

having felt, rubber or asbestos packing all around the outside. No threshold or extra floor is required. Drainage must be supplied, preferably through the floor.

Steam is introduced by a row of jets eight to 12 inches apart in a steam pipe laid on or near the floor on the two sides and back and connected with steam supply. A valve just outside regulates the amount to be used. The pipes at the end just inside the door are capped so that no steam escapes except at the short nipples, or simply

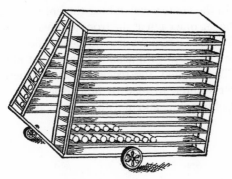

Fig. 215. Bottle truck.

holes drilled in pipe, which will answer very well. A flue opens out of the top of the oven, made of tin, three or four inches in diameter and long enough to go out at the roof. This flue is closed by a damper just above the oven; except after sterilizing it is opened to hasten the cooling and assist in drying off the bottles which are inside. Such an oven is never to be used for the heating of milk, but in it may profitably be placed not only bottles, but tinware, stirrers, faucets, dishcloths, in fact, anything movable that comes in contact with the milk.

A convenient method of handling a large number of bottles is illustrated by Fig. 215. This consists of shelves so arranged that when the bottles are placed on them, necks inside, they are inclined sufficiently for the water to drain out of them readily, and the dust does not as readily enter them as it would if they were in an upright position.

The truck is of such a size that when loaded it will readily enter the oven and admit of the door being closed. A good way to mount such a truck is to place it on two wheels in the center, which bear the entire weight. The little wheels, one each at the front and rear, do not quite touch the floor when the truck is level; these latter are also fixed so as to turn around in a socket like a table caster. Thus rigged, the truck may be pushed around wherever wanted to load or unload and saves a vast amount of handling and inevitable breakage.

INEXPENSIVE BUILDING CONSTRUCTION

Many farmers would like to put up a small building for some purpose or other but are deterred by the expense, the shingling or clapboarding of walls and the shingling of the roof being a large item in the expense account, both for labor and materials. The cut shows a simple and inexpensive plan that will give good satisfaction. The frame of the building is put up and covered, roof

Fig. 216. Battened building.

and sides, with red resin-sized building paper stretched tightly and lapping so as to shed water if any should ever reach the paper. This costs only $1 per 500 square feet. The boarding is then put on, "up-and-down," and the cracks battened, as shown. Cover the boards and battens with a cheap stain or paint, and they will last for many years. Such a building will not only be inexpensive but it will be very warm, and in later years can, if desired, be clapboarded and shingled by simply removing the old battens.

COVER FOR SAP BUCKETS

A good cover for sap buckets may be made at a cost of less than one cent by taking a wide shingle *(a)*, sawing off four inches of the tip end and fastening to it a small spring wire, as shown in the illustration. The wire can be made fast to the shingle by little staples, or by using a narrow cleat like a piece of lath. The wire should be about 30 inches long and will cost less than half a cent. When done, spring the ends of wire apart and it will hug the tree firmly.

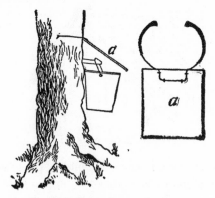

Fig. 217. Cover for sap buckets.

A HANDY TROUGH

For watering or feeding cattle in the barn a handy trough is illustrated, gotten up by a practical farmer. It may be of any desired dimensions, but is usually about four feet long and one and one-half feet wide. If

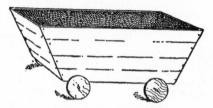

Fig. 218. A handy trough.

built slanting, stock can eat up clean any feed in it, or the trough can be readily cleaned. It is very handy for watering cattle in winter, as the trough full of water can be rolled down in front of the cattle, and from one to another as soon as they are through drinking. Where running water is handy, it can be let into this tub and quickly rolled in front of the cattle. With wheels made of hard wood this device will last for years, and can also be used for a variety of other purposes about the barn. It is one of those handy contrivances that save labor and add to the pleasure and profit of farming.

SUBSTITUTE FOR FLOOD GATE

When a flood gate cannot be used, the device shown in the illustration is very desirable; *a* represents the posts or trees to which the device is attached; *b* is a piece of iron in the shape of a capital L, the lower end of which is driven into the post. Further up is a small iron with

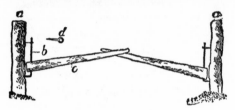

FIG. 219. Flood gate substitute.

an eye which fits over the upper end of this iron. This is driven in or turned in after the poles, *c*, have been placed in position. It is best to make the poles or rails, *c*, of some good timber. Use enough of these to make the fence or gate sufficiently high. These swing around on the rods as the water forces them apart. When the water recedes these can be again placed in position, and there is no loss of fence material. The ends are laid on each other, as in building up a rail fence.

HOOKS FOR SHOP OR STORE HOUSE

A handy arrangement for hanging up articles, as for instance, tools in the shop, or meats and other eatables in the storeroom, is shown in the accompanying sketch. This plan is particularly to be commended where it is desired to get the articles up out of the reach of mice, rats or cats. Suspend a worn-out buggy wheel to the ceiling by an iron bolt, with a screw thread on one end and a nut or head upon the other. The wheel can be hung as high or as low as desired. Hooks can be placed all about the rim and upon the spokes, in the manner shown, giving room in a small space for the hanging up of a great many articles. This arrangement is convenient, also, from the fact that one can swing the wheel about and bring all articles within reach without moving.

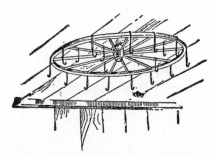

Fig. 220. Cheap support for hooks.

IMPROVING A PASTURE SPRING

The average pasture spring is apt to be a mud hole because not protected from the cattle's feet. Where a spring is to furnish the sole supply of water for a pasture year after year, it is worth while to make the most of it. If there is an old iron kettle with a break in the bottom, it can be utilized after the fashion shown in the cut, provided the source of the spring is a little higher than the point where it issues from the ground. With rough stones and cement, build a water-tight wall about the spring, setting the rocks well down into the ground. Set the kettle with the opening in the bottom, so that the water will rise to its top. A pure supply will thus always be at hand for the stock and a permanent improvement made to the pasture.

Fig. 221. A spring walled up.

A GENERAL FARM BARN

The ground plan shown in the illustration, fig. 222, provides sufficient stable room for ten cows, three horses, and a box stall, besides a corn crib and a tool house. These are all on the first floor. The building is 40x30, with a feed way running through the middle four feet wide. The building can be made any desired hight, but 20-foot posts are usually most desirable. On the second floor is space for hay, sheaf oats, corn fodder or other coarse food. There should also be on the second floor a bin for oats or ground feed. This is spouted down to the feed way, where it can be easily given out. The corn crib, of course, can be divided, if it is thought necessary, so that ground feed can be kept in a portion of it. There are plenty of windows in front and back, so that the building is well lighted. This barn can be built cheaply, and is large enough for a small dairy farm.

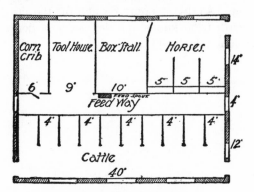

Fig. 222. Ground plan of barn.

HANDY CLOD CRUSHER AND LEVELER

One who has not tried it would be surprised to find how much execution the device shown in the cut will accomplish. Insert a narrow plank in front of the rear

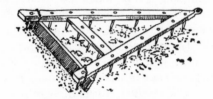

Fig. 223. Clod crusher.

teeth of an A harrow, and the land will be harrowed, the lumps crushed and the surface leveled, at one operation. One can also, by stepping on and off the crosspiece, drag earth from knolls and deposit it in depressions, thus grading the land very nicely.

GIVING SEEDS AN EARLY
START IN THE GARDEN

The ground is often cold when the seed is put into the garden plot. To get the earliest vegetables, have a few boxes without bottoms and with a sliding pane of glass for a top, as shown in the cut. Let the top slope toward the sun. Shut the slide entirely until the plant breaks ground, then ventilate as one would in a hotbed, as suggested in the right-hand sketch. A few such boxes will make some of the garden products ten days earlier— worth trying for.

FIG. 224. Forcing boxes.

A POST ANCHOR

Where temporary wire fences are used to any considerable extent, the corner or end posts may be anchored, as shown in the illustration. The large rock, *a*, is sunk into the ground as deep as the post is placed and the earth is solidly trampled above it. Place the wire around the stone before it is put into the ground, then pass it around the top of the post. By using a stick, *b*, the wire can be tightened if there is any tendency to become

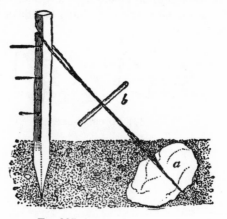

Fig. 225. Anchor for end post.

loose. To move the fence, loosen the lower strand from the posts. Begin at one end and make a coil about two feet across. Roll this on the ground, crossing and recrossing the strand of wire with the roll, about every foot of length on the strand. The barbs will hold it and keep the roll together. When the roll is as large as is convenient to handle, cut the wire and begin again. When replacing, fasten one end to the post where the top wire is to stay and roll along the ground close to the posts. Follow with the second one a little further off, and then the third. Experience has proved to me that this is the easiest, quickest and best plan to remove wire fence, as after some practice it can be done quickly.

STONEBOAT FROM TWO BOARDS

Most of the stoneboats in use are made with runners. I prefer to secure two boards the length desired for the boat, about 15 inches wide and three inches thick. I then measure 12 inches on top of the board and 18 inches on the opposite side, as shown in Fig. 1. Saw through on the dotted line, turn the end of the board over and with four bolts fasten it as shown in Fig. 2. Do this with both boards, place them side by side and fasten with strong crosspieces. This makes a good boat, and in my experience is more desirable than any other kind. They can not only be used for hauling about the place, but are excellent for breaking roads during the winter.

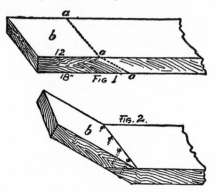

FIG. 226. Easily-made drag.

A HANDY GARDEN BARROW

A great improvement on the ordinary garden wheel-barrow is shown in the cut. The wheels have broad tires, are light and run beneath the body—just in the position to balance the load when the handles are raised. This barrow can be dumped from the side, as in the case of the ordinary barrow. It is thus possible to make over one of the old-fashioned wheelbarrows into the style shown, and that, too, at but small trouble and expense.

FIG. 227. Improved barrow.

HOMEMADE TRUCKS AND WHEELS

Low trucks are constantly of service on the farm. Now it is a feed car for the barn, or a two-wheel barrow for the garden, or it may be that low wheels are needed for one end of a crate for moving sheep or hogs. The cut shows how to make any of them. With a "key-hole" saw cut circles from inch boards and screw them together with the grain at right angles, as shown. Two-inch hoop iron binds the edges and keeps them from splitting. Large iron washers help to hold such wheels firmly in place on the axles.

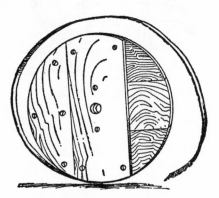

Fig. 228. Homemade wheel.

A ROLLER FROM MOWING MACHINE WHEELS

Cast-off mowing machine wheels may be utilized very readily for making a land roller. Use narrow strips of plank with slightly beveled edges, putting them around the wheels in the manner shown in the cut, making slots in the planks to fit the cogs on the rims of the wheels. These strips are held firmly in place by "shrinking on" two iron hoops at the ends, as shown. The frame is attached in the usual manner.

Fig. 229. Side view.

MAKING A PICKET FENCE HEN-TIGHT

On many farms the hens could be given free range if the garden fence were a sufficient barrier to the fowls. The cut shows a picket fence with a picket extending upward for fifteen inches every twelve feet. To these extended ends of the pickets is stretched a twelve-inch strip of wire netting, as shown in the sketch. In the prominence of the pickets the fowls do not clearly notice the netting until they fly against it. After a few trials they will give up the attempt to fly over. Poultry yard fence can be constructed in this way, using ordinary pickets, and above them any needed width of netting, according as the fowls are Brahmas, Plymouth Rocks or Leghorns.

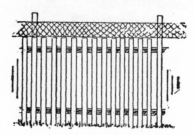

FIG. 230. Picket fence.

BARREL STRAWBERRY CULTURE

Probably many readers have heard of the plan of raising strawberries on the outside of a barrel. If one has only a small city or village lot, or "back yard," the experiment is well worth trying. The accompanying illustration shows one or two wrinkles that may help make the experiment a success. First bore the holes all about the barrel, then put inside a drain pipe made of four strips of board, reaching from the top to the bottom. The joints should not be tight. Now fill in earth about the pipe and set out the strawberry plants in all the holes and over the top. Put the barrel on a bit of plank, on the bottom of which wide casters have been screwed. The barrel can then turned about every few days to bring the sun to all the plants. An ordinary flour barrel will answer very well for trying this interesting experiment.

Fig. 231. View of barrel.